ABSTRACT ALGEBRA

ABSTRACT ALGEBRA: A First Course

LARRY JOEL GOLDSTEIN
Department of Mathematics
University of Maryland

PRENTICE-HALL, INC., *Englewood Cliffs, New Jersey*

Library of Congress Cataloging in Publication Data

GOLDSTEIN, LARRY JOEL.
Abstract algebra; a first course.

1. Algebra, Abstract. I. Title.
QA266.G55 512'.02 72-12790
ISBN 0-13-000851-6

10 9 8

Printed in the United States of America

PRENTICE-HALL INTERNATIONAL, INC., *London*
PRENTICE-HALL OF AUSTRALIA, PTY, LTD., *Sydney*
PRENTICE-HALL OF CANADA, LTD., *Toronto*
PRENTICE-HALL OF INDIA PRIVATE LIMITED, *New Delhi*
PRENTICE-HALL OF JAPAN, INC., *Tokyo*

For Melissa
and Jonathan

CONTENTS

Logical Interdependence of Chapters

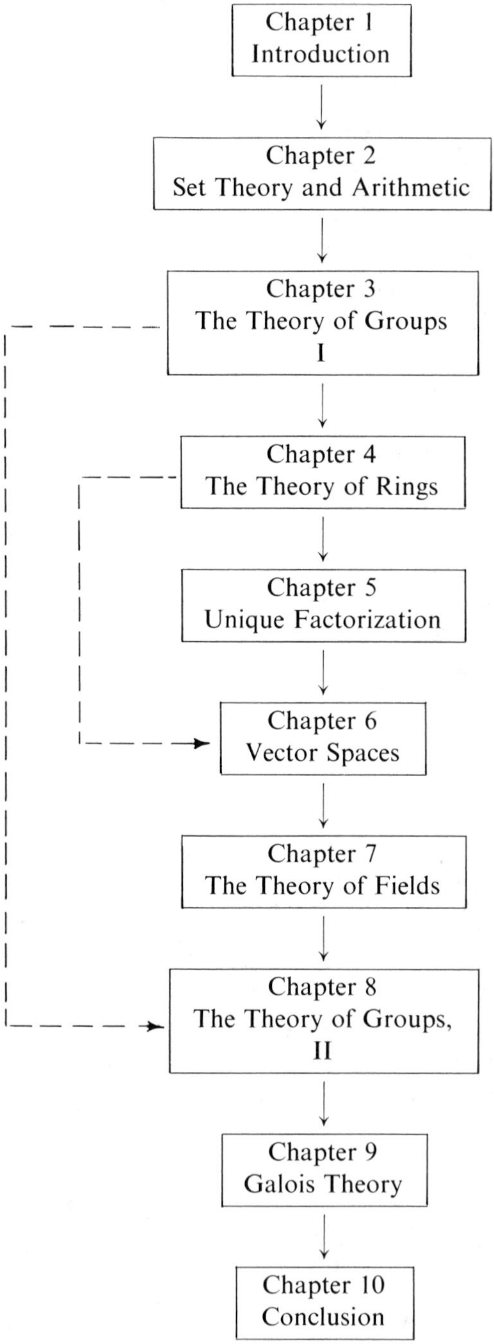

PREFACE

The present book grew out of lectures which I gave at the University of Maryland to junior level mathematics students. My goal was to present the subject of algebra from a historical viewpoint, concentrating on some of the important questions that contributed to the historical development of the subject. Among the problems that I considered were: the solution of equations of low degree in radicals, insolubility of the quintic, constructions with straight edge and compass, Fermat's last theorem, arithmetic in the Gauss and Eisenstein fields, the fundamental theorem of algebra, the theory of partitions. At all times, I attempted to emphasize concrete computations as the ultimate goals of general theories. It seems that this approach is far superior to the pure axiomatic approach that, although undeniably more efficient, leads students to the erroneous impression that any syllogism deduced from any set of hypotheses constitutes interesting mathematics.

The present book is designed for a one year course in algebra and presumes no prior background in set theory or algebra. Linear algebra is treated superficially in Chap. 6, but I have decided not to dwell on the structure of linear transformations in this book since it seems that this more advanced linear algebra had best be treated in some other context (e.g., in a basic course in functional analysis).

Most sections are followed by exercises that range from simple computations to more difficult, starred problems. In some instances, I have provided whole theories for the student to work out in small steps.

During the preparation of this book, the author has received invaluable assistance from a large number of people. The manuscript was expertly typed by Patricia Berg and Julie Smith. The text has benefitted greatly from the suggestions of many people who have read it and/or taught from it. I would like to particularly thank Drs. William Adams, Gertrude Ehrlich and Kenneth Hoffman for their suggestions, and Ms. Eileen Thompson for her editorial assistance. Finally, I would like to thank Robert Martin and Arthur Wester, both of Prentice-Hall, the former for helping to initiate this project and the latter for guiding it through to completion.

LARRY JOEL GOLDSTEIN
College Park, Maryland

ABSTRACT ALGEBRA

1 *INTRODUCTION*

1.1 *Some Historical Perspectives*

The subject of algebra began thousands of years ago at the hands of the ancient Egyptians and Babylonians. Although their methods of solving algebraic problems have been lost to history, it seems clear that the accomplishments of their highly developed civilizations necessitated being facile with a wide spectrum of algebraic tools. For example, their vast architectural programs, astronomical calculations, and calendar computations must have required a practical knowledge of the solutions of linear and quadratic equations. The mathematics of the Babylonians and Egyptians is characterized by its empirical nature: They were more interested in the solutions to specific problems than in general techniques for solving them.

The great contribution of the Greeks to mathematics was, of course, the axiomatic method, which must rank as one of the most original creations in the history of human thought. According to the Greek conception, geometry consisted of theorems which are deduced, via an accepted system of logic, from a set of axioms. Of course, for the Greeks, there was a preferred set of axioms—the axioms of Euclidean geometry. They felt that this set of axioms had a certain metaphysical significance, reflecting the "true" geometrical state of the universe. And if the Greek's failure to recognize other axiom systems was a failure of Greek mathematics, then it is but a slight shortcoming when weighed against the successful development of the axiomatic method.

The geometric viewpoint of the Greeks led to certain shortcomings in their development of algebra. For example, the Greeks solved quadratic equations by "completing the square." However, negative roots were ignored since they did not have any geometrical significance. The introduction of negative numbers was due to the Hindus in the first centuries after Christ. Algebra prospered at the hands of the Hindus, who were much more oriented toward arithmetic than the Greek geometers. The algebra of the Hindus was brought westward by the Moslems, who embellished it considerably and gave the subject its current name. One of the most significant algebraic results which dates from the Moslem period is the *quadratic formula*, which is learned by every student of high school algebra. The formula asserts that the quadratic equation

$$aX^2 + bX + c = 0$$

has the roots

$$X = \frac{-b \pm \sqrt{D}}{2a},$$

where $D = b^2 - 4ac$ is the *discriminant* of the equation.

Mathematicians searched for several hundred years to find a "cubic formula" which could be used to solve cubic equations in much the same way that the quadratic formula is used to solve quadratic equations. Such a formula was finally discovered by the Italian mathematician Tartaglia (1506?–1557). However, a contemporary and fellow-countryman of Tartaglia, Cardan (1501–1576), first published Tartaglia's solution of the cubic equation (without Tartaglia's permission), and history cruelly gives credit for the solution of the general cubic equation to Cardan. The general cubic equation is of the form

$$X^3 + bX^2 + cX + d = 0. \tag{1}$$

One of Tartaglia's discoveries was the observation that if one performs the change of variables

$$Y = X + \frac{b}{3},$$

Equation (1) can be put into the form

$$Y^3 + pY + q = 0. \tag{2}$$

Therefore, if suffices to consider the cubic equations of the form (2). Let†

$$\omega = \frac{-1 + \sqrt{-3}}{2}.$$

Then ω is a cube root of 1; that is, $\omega^3 = 1$. In fact, the three cube roots of 1 are given by

$$1, \omega, \omega^2.$$

† We mean by $\sqrt{-3}$ the complex number $i\sqrt{3}$, where $i = \sqrt{-1}$.

Let X_1, X_2, and X_3 be the roots of the cubic (2). Then Tartaglia, by using very clever arguments, found that

$$X_1 = \sqrt[3]{-\frac{q}{2} + \sqrt{D}} + \sqrt[3]{-\frac{q}{2} - \sqrt{D}},$$

$$X_2 = \omega\sqrt[3]{-\frac{q}{2} + \sqrt{D}} + \omega^2\sqrt[3]{-\frac{q}{2} - \sqrt{D}},$$

$$X_3 = \omega^2\sqrt[3]{-\frac{q}{2} + \sqrt{D}} + \omega\sqrt[3]{-\frac{q}{2} - \sqrt{D}},$$

where

$$D = \frac{q^2}{4} + \frac{p^3}{27}$$

is the discriminant of the cubic.† The reader is strongly urged to verify, by direct calculation, that Tartaglia's solution of the cubic really works. On examining Tartaglia's formulas, it is hard not to admire their beauty and to envy the cleverness of the man who discovered them. The above formulas for X_1, X_2, X_3, appear very strange, especially because of the presence of the complex number ω. Nothing analogous occurs in the case of quadratic equations. Nevertheless, it is not possible to find a solution of the general cubic which does not involve complex numbers!

In 1545, Cardan's pupil, Ferrari, obtained the solution of the general fourth-degree (or *biquadratic*) equation. Ferrari's formulas are along the same lines as Tartaglia's, only much more complicated. We will delay stating Ferrari's results until later in this book, where they will fit more naturally.

After Tartaglia's and Ferrari's successes in treating equations of the third and fourth degree, there was a feeling of optimism in the mathematical community that within a short time, a "general solution" of equations of the fifth and higher degree would be discovered. We should ponder for a moment what we mean by a "general solution" of an equation of degree n. Based on our experience with equations of degree 2 and 3, let us understand by a "general solution" of the nth-degree equation a set of formulas which allow calculation of the n roots of the equation of degree n as functions of the coefficients. Moreover, the formulas should involve only arithmetic operations (addition, subtraction, multiplication, and division) and extraction of roots. Much effort was expended in search of such a general solution. However, two centuries elapsed without any real progress.

The first real breakthrough was due to Joseph-Louis Lagrange, who, in the late eighteenth century, gave a uniform technique for solving equations of degree at most 4, thereby incorporating in a single method all that was known

† The cube roots in the formula must always be chosen so that $\sqrt[3]{-q/2 + \sqrt{D}} \cdot \sqrt[3]{-q/2 - \sqrt{D}}$ is a real number.

about general solutions to equations. His idea was to reduce the solution of a given equation to the solution of certain auxiliary equations, called *resolvents*. In case the degree of the given equation is at most 4, the resolvents are of lower degree than the given equation, so that by starting with quadratic equations, the method of Lagrange can be used to give, successively, the solution to quadratic, cubic, and biquadratic equations. However, Lagrange's method breaks down in the case of fifth-degree (quintic) equations, because the resolvent which his method associated with a quintic is of sixth degree.

The failure of Lagrange's method to solve the quintic suggested the startling possibility that no general solution of the quintic exists. This is, indeed, the case, and was first proved† by Neils Heinrik Abel in 1828. What Abel's theorem states is that it is impossible to find a general formula for the roots of a quintic equation if the formula is only allowed to use arithmetic operations and extraction of roots. However, what is even more surprising is the fact that there exist particular quintic equations, with ordinary integers as coefficients, whose roots cannot be expressed in terms of arithmetic operations and the extraction of roots. It is important to recognize that Abel's theorem required a very novel proof. It is fairly easy to determine whether or not a given formula provides a solution of a given equation. But to prove that *no* formula can be written down is quite a different matter!

A necessary and sufficient condition that a given equation be *solvable in radicals* (i.e., in terms of arithmetic operations and extraction of roots) was given by Evariste Galois in 1830. A child prodigy, who was killed in a duel before he was 21, Galois laid the foundations for the modern theory of equations. Galois' brilliant ideas were incredibly sophisticated for his day and were not acknowledged by his contemporaries. It was not until several decades after the death of Galois that his great contributions were first appreciated. At this point, we cannot state Galois's results with any kind of precision; nevertheless, let us try to get the flavor of Galois' ideas by considering some examples.

Let us consider a polynomial

$$X^n + a_1 X^{n-1} + \cdots + a_n$$

having roots

$$X_1, \ldots, X_n,$$

where we have labeled the roots in some fixed (but otherwise arbitrary) order. A *permutation* of the roots is a rearrangement of $X_1, \ldots, X_n$ of the form

$$X_{i_1}, X_{i_2}, \ldots, X_{i_n},$$

where the set $i_1, \ldots, i_n$ consists precisely of the integers $1, 2, \ldots, n$, each

† Actually Ruffini had given what he thought was a proof in 1813. This proof had a gap, which was filled in 1876.

appearing once. It is often convenient to think of a permutation as replacing X_1 by X_{i_1}, X_2 by X_{i_2}, and so forth. For this reason, it is often convenient to denote a permutation by the symbol

$$\begin{pmatrix} 1 & 2 & \cdots & n \\ i_1 & i_2 & \cdots & i_n \end{pmatrix},$$

where the notation indicates that X_j is to be replaced by X_{i_j} $(1 \leq j \leq n)$. The collection consisting of all permutations of $X_1, \ldots, X_n$ is denoted S_n.

The basic idea of Galois was to associate to the polynomial

$$X^n + a_1 X^{n-1} + \cdots + a_n$$

a certain collection of permutations belonging to S_n and depending on the coefficients. This set of permutations forms a mathematical structure known as a *group* and is called the *Galois group* of the polynomial. At this point we do not wish to digress to precisely define the notion of a group, nor do we wish to describe how we arrive at the Galois group. However, to get the flavor of this concept, let us consider a few examples.

EXAMPLE 1: $X^2 - 2$. Here the Galois group is S_2 and consists of the permutations

$$\left\{\begin{array}{rcr} \sqrt{2} & \longrightarrow & \sqrt{2} \\ -\sqrt{2} & \longrightarrow & -\sqrt{2} \end{array}\right\} \quad \text{and} \quad \left\{\begin{array}{rcr} \sqrt{2} & \longrightarrow & -\sqrt{2} \\ -\sqrt{2} & \longrightarrow & \sqrt{2} \end{array}\right\}.$$

EXAMPLE 2: $X^3 - 2$. Here the Galois group is S_3. We leave it to the reader to list the elements of the Galois group. (There are six of them.)

EXAMPLE 3: $(X^2 - 2)(X^2 + 1)$. Here the Galois group is a subgroup of S_4 consisting of the permutations

$$\left\{\begin{array}{rcr} \sqrt{2} & \longrightarrow & \sqrt{2} \\ -\sqrt{2} & \longrightarrow & -\sqrt{2} \\ \sqrt{-1} & \longrightarrow & \sqrt{-1} \\ -\sqrt{-1} & \longrightarrow & -\sqrt{-1} \end{array}\right. \qquad \left\{\begin{array}{rcr} \sqrt{2} & \longrightarrow & -\sqrt{2} \\ -\sqrt{2} & \longrightarrow & \sqrt{2} \\ \sqrt{-1} & \longrightarrow & \sqrt{-1} \\ -\sqrt{-1} & \longrightarrow & -\sqrt{-1} \end{array}\right.$$

$$\left\{\begin{array}{rcr} \sqrt{2} & \longrightarrow & -\sqrt{2} \\ -\sqrt{2} & \longrightarrow & \sqrt{2} \\ \sqrt{-1} & \longrightarrow & -\sqrt{-1} \\ -\sqrt{-1} & \longrightarrow & \sqrt{-1} \end{array}\right. \qquad \left\{\begin{array}{rcr} \sqrt{2} & \longrightarrow & \sqrt{2} \\ -\sqrt{2} & \longrightarrow & -\sqrt{2} \\ \sqrt{-1} & \longrightarrow & -\sqrt{-1} \\ -\sqrt{-1} & \longrightarrow & \sqrt{-1} \end{array}\right.$$

The algebraic properties of a polynomial are mirrored by the Galois group. For example, solvability by radicals of the equation determined by a polynomial is translated into a very simple property of the Galois group. And when a given equation is solvable in radicals, the properties of the Galois group enable one to write down the solution in radicals. In fact, if a given equation is solvable in radicals, the solution can always be obtained from the Galois group using a fixed procedure. And when this fixed procedure cannot be carried out, the equation cannot be solved in radicals. In this way, we can establish both Abel's theorem on the insolubility of the quintic and the formulas for the quadratic, cubic, and biquadratic equations. If all this sounds somewhat obscure, we hope to clarify matters in later chapters.

The problem of solving polynomial equations led mathematicians to study groups and is thus one of the historical cornerstones on which contemporary algebra is based. The introduction of groups into mathematics and the great triumphs in the theory of equations had a profound influence on many fields of mathematics which were undergoing development in the early nineteenth century. One of these fields was geometry. Although the influences of algebra on geometry were many, let us be content to cite three examples.

The Greek geometers were interested in the problem of constructing geometrical figures using a straightedge and compass, and by the time of Euclid, many such constructions were known. For example, the Greeks knew how to bisect a line segment, bisect an angle, construct a line perpendicular to a given line, and even construct a regular pentagon. However, there were three seemingly elementary constructions which the Greeks were unable to solve. The first was to trisect a given angle; that is, to construct an angle equal to one third of a given angle. The second was to construct a cube whose volume was twice that of a given cube. The third was to construct a square whose area was the same as that of a given circle. It was finally shown in the nineteenth century that these three classical problems have no solution. That is, the required constructions cannot be carried out in general, at least if the only tools allowed are straightedge and compass. It was rather amazing that the proof of the impossibility of the three classical problems used the ideas of algebra and not those of geometry. Perhaps even more surprising was the fact that the relevant algebraic ideas came from the theory of equations and the work of Galois. We will give a detailed description of these classical geometry problems in Chapter 7.

The nineteenth century was a time of great ferment in the field of geometry. The chief feature of this period was the emergence of many new systems of geometry, which satisfied all the axioms of Euclid's geometry, except for the axiom which asserts that if we are given a line and a point off the line, there is one and only one line passing through the given point and parallel to the given line. Once the first such geometry was created, many more followed, and in the middle of the nineteenth century there was a fundamental confusion about the axioms and goals of geometry. In the latter part of the nineteenth century, Felix Klein proposed the idea of unifying all these different geometries around the notion of a group, a notion which had been created, as we have seen, for an entirely different purpose. Klein's conception of geometry was called the *Erlangen program.*

A third point of contact between algebra and geometry in the nineteenth century was the *theory of algebraic curves*, which received great impetus through the brilliant ideas of the German mathematician Bernhard Riemann. Roughly speaking, an algebraic curve is the set of all ordered pairs of complex numbers (x, y) which satisfy an equation of the form

$$y^n + a_1(x)y^{n-1} + a_2(x)y^{n-2} + \cdots + a_n(x) = 0, \tag{3}$$

where $a_1(x)$, $a_2(x)$, . . . , $a_n(x)$ are polynomials with complex coefficients. Examples of algebraic curves are given by the solutions (in complex numbers) of the equations

$$X^2 + Y^2 = 1,$$
$$XY = 1,$$
$$X^3 = Y^2 + Y^3 + XY.$$

The reader is probably accustomed to think of an algebraic curve as the set of solutions in real numbers (x, y) of an equation of the form (3). The reason for considering complex solutions is to prevent embarassing examples of curves with no points on them! For example, the equation $X^2 + Y^2 = -1$ has no solution in real numbers, but it has many in complex numbers—for example $(i, 0)$, $(0, i)$. Riemann was able to formulate many of the geometric properties of an algebraic curve in purely algebraic terms, so that the machinery of algebra could be applied to solve the geometric problems. Out of Riemann's work has grown the contemporary field of mathematics called *algebraic geometry*, which is a field receiving much current research interest. Throughout the last century, the development of algebra and the development of algebraic geometry have proceeded side by side, with developments in algebra suggesting geometric insights and conversely.

Nor was geometry the only field which experienced a fruitful interchange with the field of algebra in the nineteenth century. Another such field was the *theory of numbers*. Roughly speaking, the theory of numbers is a branch of mathematics which studies the properties of the integers:

$$\ldots, -3, -2, -1, 0, 1, 2, 3, \ldots .$$

It is one of the most difficult fields of mathematics, which for centuries has been the source of countless ideas for other branches of mathematics. Although it is impossible to describe completely all the algebraic ideas which arose in the consideration of problems in the theory of numbers, let us at least give one example.

In the early part of the seventeenth century, Pierre Fermat, an amateur mathematician, claimed to have proved that if n is an integer greater than or equal to 3, it is impossible to find nonzero integers x, y, and z such that

$$x^n + y^n = z^n.$$

Fermat never wrote down his proof, and it seems very unlikely that he actually had a proof. Indeed, three and one-half centuries later, no one can currently prove or disprove Fermat's assertion! A particularly good try at proving Fermat's statement was made by the German mathematician Ernst Kummer in 1835. It was subsequently pointed out to Kummer by Lejeunne Dirichlet that Kummer's alleged proof was wrong, but for a rather intriguing reason. Kummer then made a very detailed study of his proof in order to better understand why it was wrong. As a result of this study, Kummer was able to make his proof correct for certain values of n by using a new algebraic idea, the concept of an *ideal*. Today, the notion of an ideal is one of the basic ideas in the theory of rings. We will study about rings and ideals in Chapter 4, and we will retrace Kummer's reasoning (at least in the case $n = 3$) in Chapter 5.

In this introduction we have tried to give some indication of the origins of contemporary algebra. Let us now begin a discussion of the algebra itself.

1.1 *Exercises*

1. Solve the cubic equation $X^3 + X + 1 = 0$.
2. Solve the cubic equation $X^3 - 2X^2 + X + 5 = 0$.
3. Verify that Tartaglia's formulas provide a solution of the cubic.

2 SET THEORY AND ARITHMETIC

Toward the end of the nineteenth century, a branch of mathematics called "set theory" was developed by the German mathematician George Cantor. Although Cantor's original investigations were undertaken in order to clarify the notion of infinity in mathematics, it was soon discovered that the theory of sets provided a convenient framework in which to discuss all of mathematics. More precisely, it was discovered that one can deduce all known mathematics from a list of axioms about sets. Thus, it is not surprising to find that set theory has become an indispensible tool in discussing all branches of mathematics, algebra included. In this chapter we will describe some of the most basic facts about set theory and their application to the study of the most fundamental algebraic system—the integers. However, we will not embark on a formal study of set theory and its relations to the foundations of mathematics. Rather, for us, set theory will be merely a convenient language in which to state algebraic results.

2.1 Sets

In mathematics, one naturally meets collections of various kinds of objects. For example, the collection of all points on a line, the collection of all positive integers less than 100, the collection of all lines passing through a given point of the plane, are all collections which arise naturally in elementary mathematics. These collections are typical examples of *sets*. Roughly speaking, a set is any well-defined collection of objects. When we say "well-defined," we mean that it is possible to tell, without ambiguity, whether or not a given object belongs to the set. The reason we have defined the notion of a set only "roughly" is that in more formal developments of set theory, "set" is an undefined term, whose properties are described by a system of axioms. Thus, the notion of "set" in set theory is analogous to the notions of "point" and "line" in Euclidean geometry.

Cantor was certainly not the first mathematician to use sets. Sets had come up in connection with particular mathematical problems centuries before Cantor. But it was Cantor who first considered the properties of general sets in a systematic way. And to understand the impact of the theory of sets on mathematics, it is necessary to consider the state of mathematics in the early nineteenth century. The Greek concept of mathematics as a deductive science was firmly entrenched. But the axioms from which mathematicians deduced their results were never quite spelled out. For example, mathematicians worked incessantly with real numbers without ever knowing precisely what a real number was! One of the great triumphs of Cantor was to give a definition of a real number in terms of sets, which could be described by a rather simple list of properties. It is precisely the simplicity and generality of the notion of a set which makes it so natural a concept on which to lay the foundations of mathematics.

An object belonging to a set A is called an *element* of A. If x is an element of A, then we write $x \in A$; if x is not an element of A, we write $x \notin A$. If every element of the set A is contained in the set B, then we say that A is a *subset* of B, denoted $A \subseteq B$. The two sets A and B are the same if they contain the same elements; in this case, we write $A = B$. There is a simple relationship between equality of sets and the notion of subsets, which is expressed in the following result.

PROPOSITION 1: Let A and B be sets. If $A = B$, then $A \subseteq B$ and $B \subseteq A$. Conversely, if $A \subseteq B$ and $B \subseteq A$, then $A = B$.

Proof: If $A = B$, then A and B contain the same elements, so every element of A is contained in B and every element of B is contained in A. Thus, $A \subseteq B$ and $B \subseteq A$. Conversely, if $A \subseteq B$ and $B \subseteq A$, then every element of A is contained in B and every element of B is contained in A. Therefore, A and B contain the same elements and $A = B$.

Proposition 1 is extremely useful, because it gives a convenient way of proving that two sets A and B are equal, by proving that $A \subseteq B$ and $B \subseteq A$. This simple principle will be used over and over throughout this book.

If A is a set, we will often display the elements of a set by inserting them in braces. Thus, if A is the set consisting of the first five letters of the English alphabet, then we will write $A = \{a, b, c, d, e\}$, and if B is the set of natural numbers (that is, the set of all counting numbers, including zero), we will write $B = \{0, 1, 2, \ldots\}$. If A is a set, then we can form subsets of A as follows: Suppose that for every $x \in A$, there is given a proposition $P(x)$, which may be true or false. Then we can form the subset of A consisting of all those elements of A for which $P(x)$ is true. We will denote this set by $\{x \in A \mid P(x)\}$. For example, let B be as above and let $P(x)$ be the proposition "x is even." Then

$$\{x \in B \mid x \text{ is even}\} = \{0, 2, 4, \ldots\}.$$

The set which contains no elements will be called the *null set* (or *empty set*) and will be denoted $\varnothing$. Note that $\varnothing$ is a subset of every set A, because every element of $\varnothing$ is contained in A.

Let $\mathcal{C}$ be a collection of sets—that is, let $\mathcal{C}$ be a set (always assumed to be nonempty) whose elements are sets $A, B, C, \ldots$. It does not matter for the present whether or not the number of sets in $\mathcal{C}$ is finite. The *union* of the collection $\mathcal{C}$ is the set whose elements are those elements which belong to at least one set of the collection $\mathcal{C}$. The *intersection* of the collection is the set whose elements are those elements which belong to every set of the collection $\mathcal{C}$. The union and intersection of the collection $\mathcal{C}$ are denoted

$$\bigcup_{A \in \mathcal{C}} A$$

and

$$\bigcap_{A \in \mathcal{C}} A,$$

respectively. If $\mathcal{C}$ consists only of a finite number of sets, say

$$\mathcal{C} = \{A_1, \ldots, A_n\},$$

then we will denote the union and intersection of $\mathcal{C}$ by

$$A_1 \cup \cdots \cup A_n$$

and

$$A_1 \cap \cdots \cap A_n,$$

respectively.

Let us consider a few examples of the above definitions. Let $\mathbf{R}$ denote the set of all real numbers and define C_1, C_2, and C_3 by

$$C_1 = \{x \in \mathbf{R} \mid 0 \leq x < 1\}$$
$$C_2 = \{x \in \mathbf{R} \mid 0 < x < 3\},$$
$$C_3 = \{x \in \mathbf{R} \mid -1 \leq x \leq \tfrac{1}{2}\}.$$

Then

$$C_1 \cup C_2 \cup C_3 = \{x \in \mathbf{R} \mid -1 \leq x < 3\},$$
$$C_1 \cup C_2 = \{x \in \mathbf{R} \mid 0 \leq x < 3\},$$
$$C_1 \cup C_3 = \{x \in \mathbf{R} \mid -1 \leq x < 1\}.$$
$$C_1 \cap C_2 \cap C_3 = \{x \in \mathbf{R} \mid 0 < x \leq \tfrac{1}{2}\},$$
$$C_1 \cap C_2 = \{x \in \mathbf{R} \mid 0 < x < 1\}.$$

Let A and B be sets. Then the set $\{x \in A \mid x \notin B\}$ is called the difference of A and B, denoted $A - B$. The difference of A and B consists of precisely those elements of A which are not contained in B.

Thus, using the same sets C_1, C_2, and C_3 as above, we can immediately see that

$$C_1 - C_3 = \{x \in \mathbf{R} \mid \tfrac{1}{2} < x < 1\},$$
$$C_1 - C_2 = \{0\},$$
$$C_2 - C_3 = \{x \in \mathbf{R} \mid \tfrac{1}{2} < x < 3\}.$$

There are a number of elementary properties of unions, intersections, and differences which are very useful in manipulating with sets. Let A, B, and C be sets. Then

$$A \cup (B \cup C) = (A \cup B) \cup C, \tag{1}$$
$$A \cup B = B \cup A, \tag{2}$$
$$A \cap (B \cap C) = (A \cap B) \cap C, \tag{3}$$
$$A \cap B = B \cap A, \tag{4}$$
$$A - (B \cap C) = (A - B) \cup (A - C), \tag{5}$$
$$A - (B \cup C) = (A - B) \cap (A - C). \tag{6}$$

Formulas (5) and (6) are called *de Morgan's formulas*. All the above formulas are proved using Proposition 1. Let us illustrate the technique of proof by giving a proof of (5). Suppose that $x \in A - (B \cap C)$. Then $x \in A$ and $x \notin B \cap C$. Therefore, $x \in A$ and either $x \notin B$ or $x \notin C$. If $x \notin B$, then $x \in A - B$; if $x \notin C$, then $x \in A - C$. Thus, in any case, $x \in (A - B) \cup (A - C)$, and $A - (B \cap C) \subseteq (A - B) \cup (A - C)$. Suppose that $x \in (A - B) \cup (A - C)$. Then either $x \in A - B$ or $x \in A - C$, so that $x \in A$ and x does not belong to both B and C. In other words, $x \in A$ and $x \notin B \cap C$, so that $x \in A - (B \cap C)$. Thus, $(A - B) \cup (A - C) \subseteq A - (B \cap C)$. Thus, by Proposition 1, we see that (5) holds. We leave the proof of the remainder of the formulas to the exercises.

Let A and B be sets. The *Cartesian product* of A and B is the set whose elements are pairs (a, b), where $a \in A$ and $b \in B$. The Cartesian product of A and B is denoted $A \times B$. Since we must make a distinction between the first element a and the second element b of the pair (a, b), the elements of $A \times B$ are called *ordered pairs*. For example, if $A = \{1, 2\}$, $B = \{1, 3\}$, then

$$A \times B = \{(1, 1), (1, 3), (2, 1), (2, 3)\}.$$

The most familiar example of a Cartesian product is the plane of analytic geometry, which consists of all ordered pairs (a, b) of real numbers a and b, and is thus the Cartesian product $\mathbf{R} \times \mathbf{R}$.

Let us end this section with a few comments about logic. If P and Q are logical propositions, we will denote the implication "if P, then Q" by $P \Longrightarrow Q$ or $Q \Longleftarrow P$. We will denote the logical equivalence "P if and only if Q" by $P \Longleftrightarrow Q$. In order to prove an equivalence $P \Longleftrightarrow Q$, it is necessary to prove the two implications $P \Longrightarrow Q$ and $P \Longleftarrow Q$.

2.1 Exercises

1. Let $A = \{1, 2, 3, 4, 5, 6\}$, $B = \{1, 2, 17\}$. Describe the following sets (that is, determine their elements):
   (a) $A \cup B$.
   (b) $A \cap B$.
   (c) $A - B$.
2. Let A, B, C denote the points on or inside the circles of the diagram.

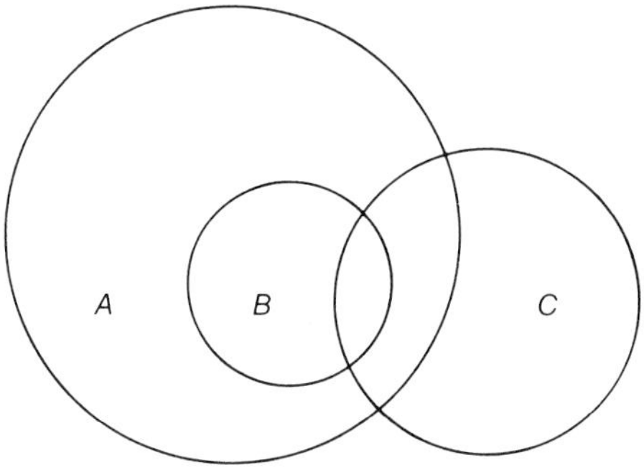

   Determine which points belong to $A \cap B \cap C$, $(A \cup B) \cap C$, $A \cap (B \cup C)$, $(A \cap B) \cup C$.
3. Let $A = \{0, 1, 2, 3\}$.
   (a) Find all subsets of A.
   (b) Give an example of two subsets B and C of A such that B is not contained in C and C is not contained in B.
4. Let $A = \{1, 2, 3, \ldots, 10\}$, $B = \{1, 2\}$.
   (a) Determine the elements of $A \times B$ and $B \times A$.
   (b) Is $A \times B = B \times A$? Why or why not?
   (c) Let $C \subseteq A$, $D \subseteq B$. Show that $C \times D \subseteq A \times B$.
   (d) Give an example of a subset of $A \times B$ which is not of the form $C \times D$, with C and D as in (c).
5. Prove (1)–(4).
6. Prove (6).
7. Let k be a positive integer and let $A_k = \{0, k, 2k, 3k, 4k, \ldots\}$. Describe the sets $A_1 \cap A_2 \cap \cdots \cap A_k \cap \cdots$; $A_1 \cup A_2 \cup \cdots \cup A_k \cup \cdots$.

   In Exercises 8–10, let A, B, and C be arbitrary sets.
8. Prove that $(A \cup B) - A = B - A$.
9. Prove that $B - (A \cap B) = \varnothing \Leftrightarrow A \supseteq B$.
10. Prove that $(A \cap B \cap C) = (A \cap B) - [(A - C) \cup (B - C)]$.
11. Let $A = \{\text{all real numbers} \mid -1 \leq x \leq 1\}$, $B = \{\text{all real numbers} \mid -5 \leq x < 7\}$. Describe the following sets:
   (a) $A \times B$.

(b) $B \times A$.
(c) $(A \times B) \cap (B \times A)$.
(d) $(A \times B) - (B \times A)$.
(e) $(A \times B) \cup (B \times A)$.

12. Prove that $A \subseteq B \Leftrightarrow A \cup B = B$.

2.2 Functions

Almost as basic to modern mathematics as the concept of a set is the idea of a function. If A and B are sets, a *function f from A to B* is any correspondence (or rule) which associates to each element x of A an element $f(x)$ of B. If f is a function from A to B, we write $f\colon A \longrightarrow B$. If $x \in A$, then $f(x)$ is called the *image of x under f*. As was the case with sets, functions appeared in mathematics long before the general concept was formulated. Prior to the eighteenth century, a function was thought of as a "formula" which allowed computation of one numerical quantity in terms of others. Thus, typical functions were $f_1(x) = x^2$, $f_2(x) = \pi x^2$, $f_3(x) = 1/x$, and $f_4(x) = \sin(e^x)$. With the development of calculus of functions of a complex variable or of several real variables, the concept of a function was somewhat enlarged. Mathematicians began to consider functions which associated to every point (x, y) of the plane a point $(f(x, y), g(x, y))$ of the plane, where $f(x, y)$ and $g(x, y)$ were given by "formulas." As (x, y) traces out a curve C in the plane, the point $(f(x, y), g(x, y))$ traces out a curve C'. Thus, mathematicians began to think of such functions as "transforming" C into C'. These functions were often called *transformations* or *mappings*. This terminology has persisted and today the terms "transformation" and "mapping" are synonomous with "function."

If $f\colon A \longrightarrow B$, and $x \in A$, then we often say that *f maps x into $f(x)$* and we often suggestively write $x \longrightarrow f(x)$ [read: x is mapped into $f(x)$].

Mathematicians also thought of a function f as "operating" on x in order to produce $f(x)$. This terminology arose in the situation where A and B are sets of functions and f "operates" on a function x in A to yield a function $f(x)$ in B. For example, if $A = B =$ the set of all functions which are differentiable arbitrarily often at every point of the real line, then the function f which maps the function x in A onto its derivative x' is a typical example of such an operator. Even today, functions are sometimes referred to as "operators." During the eighteenth and nineteenth centuries it was realized that not every function which mathematicians wished to study could be defined by a formula. For example, if P denotes the set of positive integers, let the function $f\colon P \longrightarrow P$ be defined by $f(n) =$ the number of positive divisors of n. Then there is no simple formula expressing $f(n)$ in terms of n. Thus, mathematicians were led to extend the notion of a function to include any rule or correspondence, not just those given by formulas.

Let us give a few examples of functions which occur rather frequently.

EXAMPLE 1: Let A and B be any nonempty sets and let $b_0 \in B$. The function $f: A \longrightarrow B$ defined by $f(a) = b_0$ for all $a \in A$ is called a *constant function*.

EXAMPLE 2: Let A be any nonempty set. The function $f: A \longrightarrow A$ defined by $f(a) = a$ for all $a \in A$ is called the *identity function* and will be denoted i_A.

If $f: A \longrightarrow B$ is a function, the set A is called the *domain* of f, and the set B is called the *range* of f. A function is defined by specifying its value for each element belonging to the domain. Two functions f, g from A to B are said to be *equal* if $f(x) = g(x)$ for all $x \in A$.

Suppose that $g: A \longrightarrow B$, $f: B \longrightarrow C$ are functions. If $x \in A$, then we may define a function from A into C by first mapping x into $g(x)$ and then mapping $g(x)$ into $f(g(x))$. This function is called the *composite of f and g* and will be denoted fg. According to our definition,

$$(fg)(x) = f(g(x)) \qquad (x \in A).$$

The reader is probably well acquainted with the notion of the composite of two functions from his study of the chain rule in calculus.

Suppose that $h: A \longrightarrow B$, $g: B \longrightarrow C$, and $f: C \longrightarrow D$ are functions. Then we have the following *associative law:*

$$(fg)h = f(gh). \tag{1}$$

In order to prove (1), we must show that the functions on each side of (1) have the same value for all $x \in A$. But $f(gh)(x) = f(gh(x)) = f(g(h(x)))$, and $(fg)h(x) = (fg)(h(x)) = f(g(h(x)))$. Therefore, (1) holds.

Let $f: A \longrightarrow B$ be a function. Then f is said to be *injective* (or *one to one*) if, whenever $f(x) = f(y)$, we have $x = y$. Thus, for example, if $f: \mathbf{R} \longrightarrow \mathbf{R}$ is the function $f(x) = 3x$, then f is injective, since $3x = 3y$ implies that $x = y$. The function $g(x) = x^2$ is not injective, since $g(4) = g(-4)$.

A function $f: A \longrightarrow B$ is said to be *surjective* (or *onto*) if for every $y \in B$, there exists $x \in A$ such that $f(x) = y$. Thus, for example, $f(x) = 3x$ is surjective, since for every $y \in \mathbf{R}$, we have $f(y/3) = y$. However, $f(x) = \sin(x)$ is not surjective, since there does not exist $x \in \mathbf{R}$ such that $f(x) = 17$ [since $-1 \leq \sin(x) \leq 1$ for all $x \in \mathbf{R}$].

A function $f: A \longrightarrow B$ which is both injective and surjective is said to be *bijective*. If $f: A \longrightarrow B$ is a bijection, then for any $b \in B$, there is exactly one $a \in A$ such that $f(a) = b$. Therefore, we may define the function $f^{-1}: B \longrightarrow A$ by $f^{-1}(b) = a$. The function f^{-1} is called the *inverse* of f. We clearly have

$$ff^{-1} = i_B, \qquad f^{-1}f = i_A.$$

A few words of caution about inverse functions. First, in order for us to be able to define an inverse function, f must be a bijection. Second, if $f^{-1}(x)$ is a real number, then it is not usually true that $f^{-1}(x) = 1/f(x)$. For example, let $f: \mathbf{R} \longrightarrow \mathbf{R}$ be defined by $f(x) = 3x$. Then f is a bijection and $f^{-1}(x) = x/3$. Thus, $f^{-1}(2) \neq 1/f(2)$.

Let $f: A \longrightarrow B$ be a function and let $C \subseteq A$. Let us define the *image* $f(C)$ *of* C *under* f by

$$f(C) = \{y \in B \mid y = f(x) \text{ for some } x \in C\}.$$

If $D \subseteq B$, let us define the *inverse image* $f^{-1}(D)$ *of* D *under* f by

$$f^{-1}(D) = \{x \in A \mid f(x) \in D\}.$$

Let A be a set. Then a function $f: A \times A \longrightarrow A$ is called a *binary operation* on A. We may think of a binary operation as defining an operation of multiplication, denoted $\cdot$, among the elements of A. Namely, if $a, b \in A$ then we define the "product" $a \cdot b$ to be $f((a, b))$. Examples of binary operations come from elementary arithmetic. For example, addition and multiplication of real numbers are examples of binary operations on the set of real numbers. We will meet many examples of binary operations in the later sections of this book.

Let $\cdot$ be a binary operation on a set A. Then we say that $\cdot$ is *associative* if $a \cdot (b \cdot c) = (a \cdot b) \cdot c$ for all $a, b, c \in A$. We say that $\cdot$ is *commutative* if $a \cdot b = b \cdot a$ for all $a, b \in A$. Finally, we say that $i \in A$ is an *identity* with respect to $\cdot$ if $i \cdot a = a \cdot i = a$ for all $a \in A$.

EXAMPLE 3: Let $\mathbf{Z} = \{\ldots, -3, -2, -1, 0, 1, 2, 3, \ldots\}$ be the set of integers, and let us consider the binary operation $+$ of addition on $\mathbf{Z}$. Then $+$ is associative since $(a + b) + c = a + (b + c)$ for all integers a, b, c. Moreover, $+$ is commutative since $a + b = b + a$ for all integers a and b. Also, 0 is an identity for $+$ since $a + 0 = 0 + a = a$ for all $a \in \mathbf{Z}$.

EXAMPLE 4: Again, let $\mathbf{Z}$ denote the set of integers, but now let us consider the binary operation $\cdot$ of multiplication on $\mathbf{Z}$. Then $\cdot$ is associative since $a \cdot (b \cdot c) = (a \cdot b) \cdot c$ for all integers a, b, c. Moreover, $\cdot$ is commutative since $a \cdot b = b \cdot a$ for all integers a, b. Finally, 1 is an identity for $\cdot$ since $1 \cdot a = a \cdot 1 = a$ for all $a \in \mathbf{Z}$.

EXAMPLE 5: Again let $\mathbf{Z}$ denote the set of integers. This time, however, let us consider the binary operation $*$ on $\mathbf{Z}$ defined by $a * b = ab^2$. Then $*$ is not associative, since, for example, $(1 * 2) * 3 = 36$, while $1 * (2 * 3) = 324$. Moreover, $*$ is not commutative, since, for example, $2 * 3 = 18$, whereas $3 * 2 = 12$. Does $*$ have an identity?

2.2 Exercises

1. Let $A = \{0, 1, 2\}$, $B = \{a, b\}$.
   (a) Find all functions $f: A \longrightarrow B$.
   (b) Determine which are surjective.
   (c) Determine which are injective.
   (d) Determine which are bijective.

2. Let **R** denote the set of all real numbers, $f: \mathbf{R} \longrightarrow \mathbf{R}$ the function $f(x) = x^2 + 1$, $g: \mathbf{R} \longrightarrow \mathbf{R}$ the function $g(x) = x + 2$.
   (a) What is the image of **R** under f? Under g?
   (b) What is $f^{-1}(\{3, 2\})$?
   (c) What is $f(g^{-1}(C))$, where $C = \{x \in \mathbf{R} \mid -1 \leq x \leq 1\}$?
   (d) What is $(fg)(x)$?
   (e) Is f injective?
   (f) Is g injective?
3. Let $*$ denote the binary operation on **R** defined by $a * b = ab + 3b$.
   (a) Is $*$ associative?
   (b) Is $*$ commutative?
4. Let $*$ denote the binary operation on **R** defined by $a * b = ab + b^2$.
   (a) Does **R** contain an identity with respect to $*$?
   (b) Is $*$ associative?
   (c) Is $*$ commutative?
5. Let $f: A \longrightarrow B$, $g: B \longrightarrow C$. If f and g are surjective, then show that gf is surjective. Is the converse true?
6. Let $f: A \longrightarrow B$, $g: B \longrightarrow C$. If f and g are injective, then show that gf is injective. Is the converse true?
7. Let $f: A \longrightarrow B$ be a function, and let $\mathfrak{C}$ be a collection of subsets of A. Show that $f(\bigcup_{D \in \mathfrak{C}} D) = \bigcup_{D \in \mathfrak{C}} f(D)$.
8. Let the notation be the same as in Exercise 7, and let A_1 and A_2 be subsets of A. Then $f(A_1 \cap A_2) \subseteq f(A_1) \cap f(A_2)$. [It may be that $A_1 \cap A_2 = \varnothing$. Therefore, we make the convention $f(\varnothing) = \varnothing$.] Is it true that $f(A_1 \cap A_2) = f(A_1) \cap f(A_2)$?
9. Let $\mathfrak{D}$ be a collection of subsets of B. (The notation of Exercise 7 is still in effect.) Then
   (a) $f^{-1}(\bigcup_{D \in \mathfrak{D}} D) = \bigcup_{D \in \mathfrak{D}} f^{-1}(D)$.
   (b) $f^{-1}(\bigcap_{D \in \mathfrak{D}} D) \subseteq \bigcap_{D \in \mathfrak{D}} f^{-1}(D)$.
   (c) Does equality necessarily hold in (b)?
10. Let $f: A \longrightarrow B$ be a function, and let $C \subseteq D \subseteq B$. Then $f^{-1}(C) \subseteq f^{-1}(D)$.

2.3 *The Integers*

One of the most important sets in all of mathematics is the set **Z** of *integers:*

$$\mathbf{Z} = \{\ldots, -3, -2, -1, 0, 1, 2, 3, \ldots\}.$$

In spite of centuries of effort by the greatest mathematical minds, there are many simple problems concerning **Z** which cannot be settled at the present time. For example, consider the problem of Fermat which was mentioned in Chapter 1, concerning the nonexistence of nonzero solutions of the equation $x^n + y^n = z^n$, x, y, z integers, $n \geq 3$. The reader is undoubtedly familiar with the integers, and some of the properties of integers will seem like obvious consequences of his everyday experience with them. It is our intention to make use of this experience by assuming a certain number of basic properties of the integers. We will then make a number of deductions from our basic properties and will thereby prove a number of facts concerning the integers which are, in some sense, less obvious than our basic properties.

It is possible to construct the integers in terms of sets and to verify all the basic properties from the properties of sets. However, this is a rather drawn-out process and would take us rather far afield from our central area of interest, so we have resisted the temptation to provide a complete development of the integers.

The Basic Properties of $\mathbf{Z}$

In the following discussion, we will present five basic sets of properties of the integers. Mathematicians have distilled these properties from the myriad properties of the integers during centuries of investigations. The properties which we will assume have two convenient features: They agree with our preconceived notions about the integers and they form a natural point of departure for the development of the less-elementary properties of the integers, which will be taken up in Section 2.4.

On the set of integers, there are defined two binary operations, one called *addition* (denoted $+$) and one called *multiplication* (denoted $\cdot$). Our first three sets of properties of the integers are just some familiar properties of $+$ and $\cdot$.

I. Properties of Addition

A1. *Associativity:* If $a, b, c \in \mathbf{Z}$, then $(a + b) + c = a + (b + c)$.

A2. *Commutativity:* If $a, b \in \mathbf{Z}$, then $a + b = b + a$.

A3. *Identity:* If $a \in \mathbf{Z}$, then $a + 0 = 0 + a = a$.

A4. *Inverses:* If $a \in \mathbf{Z}$, then there exists a unique integer b such that $a + b = b + a = 0$. b is called the *inverse* of a with respect to addition and is denoted $-a$ (read: minus a). Note that our notation for integers was chosen in such a way that the inverse of 2, say, is the integer -2.

II. Properties of Multiplication

M1. *Associativity:* If $a, b, c \in \mathbf{Z}$, then $(a \cdot b) \cdot c = a \cdot (b \cdot c)$.

M2. *Commutativity:* If $a, b \in \mathbf{Z}$, then $a \cdot b = b \cdot a$.

M3. *Identity:* If $a \in \mathbf{Z}$, then $a \cdot 1 = 1 \cdot a = a$.

III. Distributive Law

Addition and multiplication in **Z** are related by the *distributive law* which asserts that if $a, b, c \in \mathbf{Z}$, then

$$a \cdot (b + c) = a \cdot b + a \cdot c.$$

At this point, let us cite some rules for calculating with integers. These rules are familiar from high school algebra and can easily be deduced from I, II, and III. Let $x, y, z \in \mathbf{Z}$. Then we have

$$-0 = 0, \qquad 0 \cdot x = 0, \tag{1}$$

$$-(x + y) = -x + (-y), \tag{2}$$

$$-(-x) = x, \tag{3}$$

$$(-1) \cdot x = -x, \tag{4}$$

$$(-x) \cdot (-y) = x \cdot y, \tag{5}$$

$$x \cdot (-y) = (-x) \cdot y = -(x \cdot y). \tag{6}$$

By way of illustration, let us prove (3). We will leave (1)–(6) as exercises for the reader. By A4, $-y$ is the unique integer which has the property $y + (-y) = 0$, In particular, if we set $y = -x$, then $-(-x)$ is the unique integer having the property $-x + [-(-x)] = 0$. However, by A4, $-x + x = 0$. Therefore, by the uniqueness, we have $x = -(-x)$. Hereafter, we will use rules (1)–(6) without explicit reference to them.

The elements of the set

$$\mathbf{P} = \{1, 2, 3, \ldots\}$$

are called *positive integers*. The set **P** of positive integers is a subset of **Z**. Our next set of properties concerns the positive integers.

IV. Properties of Positive Integers

P0. Let $a \in \mathbf{Z}$. Then exactly one of the following is true: $a \in \mathbf{P}$, $a = 0$, $-a \in \mathbf{P}$.

P1. If $x, y \in \mathbf{P}$, then $x + y \in \mathbf{P}$.

P2. If $x, y \in \mathbf{P}$, then $x \cdot y \in \mathbf{P}$.

The set

$$\mathbf{N} = \mathbf{P} \cup \{0\} = \{0, 1, 2, 3, \ldots\}$$

is called the set of *natural numbers*.

Our last basic property of **Z** is the *principle of mathematical induction*.

V. Principle of Mathematical Induction

Let S be a nonempty subset of the set of positive integers **P**. Assume that (a) $1 \in S$ and (b) whenever $n \in S$ we have $n + 1 \in S$. Then S contains all positive integers.

In more formal developments of the integers, the principle of mathematical induction is the basic tool available to prove theorems and, in fact, all the preceding basic properties may be deduced by using the principle of mathematical induction and a set of axioms for **N**, called Peano's axioms.

The principle of mathematical induction can be used to prove theorems as follows: Suppose that for each positive integer n, we are given a proposition P_n. Suppose further that we are able to show that P_1 is true and that whenever P_n is true, P_{n+1} is true. Then P_n is true for every positive integer n. Indeed, if $S = \{n \in \mathbf{P} \mid P_n \text{ is true}\}$, then $1 \in S$, and if $n \in S$, then $n + 1 \in S$, so that $S = \mathbf{P}$ by V. A useful variant of the principle of mathematical induction is given by the following:

V′. Variant of Mathematical Induction Principle
Suppose that for each positive integer n there is given a proposition P_n. Suppose that (a) P_1 is true and that (b) whenever P_m is true for all $m \leq n$, m a positive integer, we have P_{n+1} is true. Then P_n is true for every positive integer n.

We leave it to the reader to prove V' from V.

Let us illustrate the use of the principle of mathematical induction by means of a number of examples.

EXAMPLE 1: Let us prove that if n is a positive integer, then

$$1 + 2 + \cdots + n = \frac{n(n+1)}{2}.$$

Let P_n denote this proposition. It is clear that P_1 is true, since $1 = [1 \cdot (1 + 1)]/2$. If P_n is true, then

$$1 + 2 + \cdots + n = \frac{n(n+1)}{2}.$$

But then by adding $n + 1$ to both sides of the equation, we derive that

$$1 + 2 + \cdots + n + (n+1) = \frac{n(n+1)}{2} + (n+1) = \frac{(n+1)(n+2)}{2},$$

and thus P_{n+1} is true. Thus, by the principle of mathematical induction, P_n is true for all $n \in \mathbf{P}$.

EXAMPLE 2: Let S be a set containing n elements. Then S contains 2^n subsets.

Denote this proposition by P_n. It is clear that P_1 is true since if $S = \{a\}$, then the subsets of S are $\{a\}$ and $\varnothing$. Assume that P_n is true, and let $S = \{a_1, \ldots, a_n, a_{n+1}\}$. By the principle of mathematical induction, it suffices to show that S has 2^{n+1} subsets, assuming P_n. Let $S_0 = \{a_1, \ldots, a_n\}$. By our assumption S_0 has 2^n subsets $T_1, \ldots, T_{2^n}$. Let T be a subset of S. There are two cases:

If T is contained in S_0, then T is one of $T_1, \ldots, T_{2^n}$. If T is not contained in S_0, then $a_{n+1} \in T$ and $T - \{a_{n+1}\}$ is contained in S_0. Thus, in the second case, $T - \{a_{n+1}\}$ is one of $T_1, \ldots, T_{2^n}$ and T is one of

$$T_1 \cup \{a_{n+1}\}, \ldots, T_{2^n} \cup \{a_{n+1}\}.$$

We have exhibited $2 \cdot 2^n = 2^{n+1}$ subsets of S. Thus, P_{n+1} holds and we are done.

EXAMPLE 3: Definition by induction. It is possible to use mathematical induction to define various mathematical concepts. The basic idea of this process of "definition by induction" is as follows: Suppose that it is required to define a function $f: \mathbf{P} \longrightarrow A$, where A is some given set. This may be done by (a) specifying $f(1)$ and (b) specifying $f(n+1)$ in terms of $f(1)$, $f(2)$, $\ldots, f(n)$. If $S = \{n \in \mathbf{P} \mid f(n) \text{ is defined}\}$, then by V', $S = \mathbf{P}$ and f is defined for all positive integers n. For example, let us define the nth power a^n of an integer a as follows:

$$a^0 = 1,$$
$$a^1 = a,$$
$$a^{n+1} = a^n \cdot a.$$

By the principle of mathematical induction, a^n is defined for n a positive integer or 0.

EXAMPLE 4: Let $x, y \in \mathbf{Z}$ and let m and n be positive integers or 0. Then the following *laws of exponents* hold:

$$(x \cdot y)^m = x^m \cdot y^m, \tag{7}$$
$$x^m \cdot x^n = x^{m+n}, \tag{8}$$
$$(x^m)^n = x^{m \cdot n}, \tag{9}$$

Properties (7)–(9) can be easily verified by induction and are left as exercises for the reader.

Order

Let a and b be integers. We say that a is *greater than* b, denoted $a > b$, if $a - b$ is a positive integer. Thus, in particular, if $a \in P$, then $a > 0$, and conversely, if $a > 0$, then $a \in P$. In other words, the positive integers are precisely those which are greater than 0. If a is greater than b, then we say that b is *less than* a, and we write $b < a$. We will use the notation $a \leq b$ to mean that either $a < b$ or $a = b$. Similarly, $a \geq b$ will mean that either $a > b$ or $a = b$. Let us verify some of the elementary properties of the relations "greater than" and "less than."

O1. *Exactly one of the following holds:* $x < y$, $x = y$, $x > y$.

For, by PO, exactly one of the following holds:

$$y - x \in \mathbf{P}, \qquad y - x = 0, \qquad -(y - x) \in \mathbf{P}.$$

In the first case, $x < y$, while in the second case, $x = y$. Noting that $-(y - x) = x - y$, we see that in the third case, $x > y$. Thus, at least one of the conditions $x < y$, $x = y$, $x > y$ holds. We leave it to the reader to show that no two of the conditions hold simultaneously. Therefore, O1 is proved.

O2. *If* $x < y$, $y < z$, *then* $x < z$.

O3. *If* $x < y$ *and* $z < w$, *then* $x + z < y + w$.

We leave the proofs of O2 and O3 as exercises for the reader. A simple application of O2 and O3 shows that there are no integers between 0 and 1. More precisely, we have

PROPOSITION 1: Suppose that b is a positive integer. Then b is greater than or equal to 1.

Proof: Let us proceed by induction. The assertion is clearly true for $b = 1$. Let us assume that the assertion is true for b—that is, $b \geq 1$. By O3, we see that $b + 1 \geq 1 + 1$. However, 1 is positive, so that $1 > 0$. [For if 1 were not positive, -1 would be positive by P0, and therefore, by P2, $1 = (-1)\cdot(-1)$ is positive, which is a contradiction.] Therefore, by O3, $1 + 1 > 1 + 0 = 1$. Thus, we see that

$$b + 1 \geq 1 + 1, \qquad 1 + 1 > 1.$$

Therefore, by O2

$$b + 1 > 1,$$

which implies our assertion for $b + 1$. Therefore, the proposition is proved.

COROLLARY 2: Let x and y be integers. If $x > y$, then $x \geq y + 1$.

Proof: If $x > y$, then $x - y$ is positive, so that by Proposition 1, $x - y \geq 1$. Thus, by O3, $x \geq y + 1$.

We will require one additional basic property of inequalities:

O4a. *If* $x < y$ *and* $z > 0$, *then* $xz < yz$.

O4b. *If* $x < y$ *and* $z < 0$, *then* $xz > yz$.

For if $x < y$ and $z > 0$, then z and $y - x$ belong to $\mathbf{P}$, so that by *P2*, $(y - x)\cdot z = yz - xz$ belongs to $\mathbf{P}$. Therefore, $xz < yz$, whence O4a. If $x < y$ and $z < 0$, then $-z > 0$ by PO, so that by O4a, we see that $x(-z) <$

$y(-z)$. In other words, $xz - yz = y(-z) - x(-z) \in \mathbf{P}$, which implies that $xz > yz$.

Another easy consequence of Proposition 1 is given by

COROLLARY 3: Let a and b be positive integers. Then there exists a positive integer n such that $a \leq nb$.

Proof. Set $n = a + 1$. For then $n \cdot b = ab + b$. However, by Proposition 1 and O4, $ab \geq a \cdot 1 = a$, $b \geq 1 > 0$. Therefore, by O3,

$$nb \geq a + 0 = a.$$

A very important property of $\mathbf{Z}$ is the fact that if a and b are nonzero integers, then $a \cdot b$ is nonzero. The simplest way to see this is to divide the reasoning up into four cases: $a > 0, b > 0$; $a > 0, b < 0$; $a < 0, b > 0$; $a < 0, b < 0$. In each case, we may apply O4 to conclude that either $a \cdot b > 0$ or $a \cdot b < 0$. In any case, $a \cdot b \neq 0$. Let us record this important fact for later reference.

THEOREM 4: Let a and b be nonzero integers. Then $a \cdot b \neq 0$.

If S is a subset of $\mathbf{Z}$, then an integer s is said to be the *least* (or *smallest*) *element of* S if $s \in S$ and $s \leq t$ for all t belonging to S. Similarly s is said to be the *greatest* (or *largest*) *element of* S is $s \in S$ and $s \geq t$ for all t belonging to S. Not every subset of $\mathbf{Z}$ has a greatest or least element. For example, if $S = \mathbf{Z}$, then S has neither a largest or smallest element. However, it is reasonably clear from property O1 that we have the following result.

PROPOSITION 5: Let S be a finite, nonempty subset of $\mathbf{Z}$. Then S has a largest and a smallest element.

Proof: Exercise. (*Hint:* Use induction on the number of elements in S.)

A somewhat more difficult result to prove is the *well-ordering principle:*

THEOREM 6: Let S be a nonempty (finite or infinite) subset of $\mathbf{N}$. Then S contains a smallest element.

Proof: The following argument makes rather interesting use of the principle of mathematical induction. Let $A = \{y \in \mathbf{N} \mid y \leq x \text{ for all } x \in S\}$. Since $S \subseteq \mathbf{N}$, $0 \leq s$ for all $s \in S$, so that $0 \in A$ and $A \neq \varnothing$. Moreover, $A \neq \mathbf{N}$. For if $x \in S$, then $x + 1 > x$ and thus $x + 1 \notin A$. Since $0 \in A$ and $A \neq \mathbf{N}$, the principle of mathematical induction implies that there exists $y \in A$ such that $y + 1 \notin A$. (If, whenever $y \in A$, we have $y + 1 \in A$, then, since $0 \in A$, we would have $A = \mathbf{N}$ by induction.) Let us prove that y is the smallest element of S. Since $y \in A$, $y \leq x$ for all $x \in S$. Therefore, it suffices to show that $y \in S$. Since $y + 1 \notin A$, there exists $x_0 \in S$ such that $y + 1 > x_0$. But then, by Corollary 2, we see that $y + 1 \geq x_0 + 1$, so that $y \geq x_0$ by O3. However, since $y \leq x$ for all $x \in S$, we see that $y \leq x_0$. Combining this latter inequality with the inequality $y \geq x_0$, we see that $y = x_0 \in S$.

The Division Algorithm

Let a be a nonnegative integer and let b be a positive integer. In elementary arithmetic, we learned how to "divide" a by b, thereby deriving a quotient q plus a remainder of the form r/b, where

$$0 \leq r/b < 1. \tag{10}$$

This process of division can be expressed by the equation

$$a/b = q + r/b. \tag{11}$$

It is possible to formulate the process of division, as described by (10) and (11), solely in terms of integers.

THEOREM 7 (Division Algorithm): Let a be a nonnegative integer and let b be a positive integer. Then there exist natural numbers q and r such that $0 \leq r < b$ and

$$a = bq + r.$$

Proof: Of all the multiples

$$0 \cdot b, 1 \cdot b, 2 \cdot b, 3 \cdot b, \cdots$$

of b, there are only finitely many less than or equal to a. Indeed, Corollary 3 asserts that there exists a positive integer n such that $a < nb$. Then, the only multiples of b which are less than or equal to a are among $0 \cdot b$, $1 \cdot b$, $\ldots$, $(n - 1) \cdot b$. Among the multiples of b which are less than or equal to a, let $q \cdot b$ be the largest. Then, by the choice of $q \cdot b$, we have $q \cdot b \leq a < (q + 1) \cdot b$. Let $r = a - qb$. Then $r \geq 0$ since $q \cdot b \leq a$; and

$$r = a - qb < (q + 1) \cdot b - qb = b$$

since $(q + 1) \cdot b > a$. Therefore, we have shown that $a = qb + r$, where $0 \leq r < b$.

The division algorithm, although it appears to be a rather superficial feature of the integers, is actually one of the most important results about $\mathbf{Z}$. In Section 2.5 we will see that the division algorithm can be applied in rather ingenious ways to yield arithmetic facts about the integers.

2.3 Exercises

1. Prove formulas (1)–(6).
2. Prove the laws of exponents (7)–(9).
3. Let n be a positive integer. Prove that
   (a) $1^2 + 2^2 + \cdots + n^2 = n(n+1)(2n+1)/6$.
   (b) $1^3 + 2^3 + \cdots + n^3 = (1 + 2 + 3 + \cdots + n)^2$.
4. Give an example of a set of propositions P_n ($n = 1, 2, 3, \ldots$) such that $P_n \Longrightarrow P_{n+1}$ ($n = 1, 2, 3, \ldots$), but nevertheless all are false. (Thus, in order to apply induction, it is crucial that P_1 be true.)
5. Prove that if $x \in \mathbf{Z}$, then $x^2 \geq 0$.
6. Prove that if $x, y \in \mathbf{Z}$, then $2xy \leq x^2 + y^2$.
7. Prove the following *cancelation law*: Let $a,b,c \in \mathbf{Z}$, $c \neq 0$. Show that if $ac = bc$, then $a = b$.

2.4 Number Theory in Z

In this section we will study some of the arithmetic properties of integers. The properties which we establish will motivate many of the abstract notions which we will consider later in this book. *Throughout this section, lower case roman letters will always denote integers.*

We say that $a \mid b$ (read "a divides b") if $b = k \cdot a$ for some k. If $a \mid b$, we say that *b is a multiple of a and that a is a divisor of b.* The following properties of divisibility are more or less obvious:

$$a \mid b,\ b \mid c \Longrightarrow a \mid c, \tag{1}$$

$$1 \mid a \qquad \text{for all } a, \tag{2}$$

$$a \mid b,\ a \mid c \Longrightarrow a \mid bx + cy \qquad \text{for all integers } x, y, \tag{3}$$

$$a \mid b,\ b \mid a \Longrightarrow b = a \quad \text{or} \quad b = -a. \tag{4}$$

We will leave (1) and (2) as exercises. To prove (3), let us prove the two auxiliary assertions

$$a \mid b,\ a \mid c \Longrightarrow a \mid b + c, \tag{3a}$$

$$a \mid b \Longrightarrow a \mid bx \qquad \text{for all } x. \tag{3b}$$

Assertions (3a) and (3b) imply (3) since if $a \mid b$, $a \mid c$, then $a \mid bx$, $a \mid cy$ by (3b), which implies that $a \mid bx + cy$ by (3a). To prove (3a), notice that if $a \mid b$, $a \mid c$, there exist k and m such that $b = k \cdot a$, $c = m \cdot a$. Therefore, $b + c = (k + m) \cdot a$, so that $a \mid b + c$. The proof of (3b) is left as an exercise.

Let us now prove (4). If $a \mid b$, then $\pm a \mid \pm b$, so that it suffices to consider the case where $a > 0$, $b > 0$. In this case, $a \mid b$ implies that $b = k \cdot a$ with

$k \geq 0$. Therefore, by the order properties of $\mathbf{Z}$, $b \geq a$. But, since $b \mid a$, we deduce similarly that $a \geq b$. Therefore, $a = b$.

We say that c is a *greatest common divisor* of a and b, provided that

1. $c > 0$.
2. $c \mid a$, $c \mid b$.
3. Whenever d is an integer such that $d \mid a$, $d \mid b$, then $d \mid c$.

Property (2) asserts that c is a divisor of both a and b, that is, a common divisor of a and b. Property (3) asserts that among all common divisors of a and b, c is the "greatest" in the sense that every common divisor of a and b is also a divisor of c. We denote c by (a, b).

EXERCISE 1: Suppose that a and b have a greatest common divisor c. Show that c is the largest common divisor of a and b.

It is not obvious that two integers have a greatest common divisor. Nor is it obvious that two integers have only one greatest common divisor. Let us clarify the second point: If a greatest common divisor exists, it must be unique. For suppose that c_1 and c_2 both satisfy (1)–(3). Then, by (2), $c_2 \mid a$, $c_2 \mid b$. Thus, by (3) for c_1, we see that $c_2 \mid c_1$. Similarly, $c_1 \mid c_2$. Therefore, by (4), $c_1 = \pm c_2$. However, since both c_1 and c_2 are positive, $c_1 = c_2$. Let us now prove that a greatest common divisor exists.

PROPOSITION 1: Suppose that a and b are not both zero. Then a and b have a greatest common divisor.

Proof: We noted that the proposition is not at all obvious. In order to prove it, a tremendously ingenious idea is required. Let

$$S = \{ax + by \mid x, y \in \mathbf{Z}\}.$$

We will prove that S contains a greatest common divisor of a and b! If $f = ax + by \in S$, then $-f = a(-x) + b(-y) \in S$. Moreover, since $a \in S$ and $b \in S$, we see that S contains elements distinct from zero, and therefore S contains positive elements. Let $S_+ = \{z \in S \mid z > 0\}$. By the well ordering of $\mathbf{N}$, S_+ contains a smallest element c. Then $c > 0$ and we assert that c is a greatest common divisor of a and b. Since $c \in S$, there exist x and y such that

$$c = ax + by. \tag{5}$$

Therefore, if $d \mid a$, $d \mid b$, we see that $d \mid c$ by (3). Thus, properties (1) and (3) hold. In order to prove (2), let us prove the following

CLAIM: Each element of S is a multiple of c.

It clearly suffices to show that every positive element of S is a multiple of c. Therefore, let $f = ax' + by' \in S_+$. By the division algorithm, there exist q and r such that $0 \leq r < c$ and such that

$$f = qc + r.$$

By equation (5) we see that

$$r = a\cdot(x' - qx) + b\cdot(y' - qy),$$

which implies that $r \in S$. If $r > 0$, then $r \in S_+$, which is a contradiction to the assumption that c is the smallest element in S_+. Therefore, $r = 0$ and $f = qc$; that is, f is a multiple of c. This establishes the claim.

In order to complete the proof of the proposition, we note that since $a \in S, b \in S$, we have $c \mid a$, $c \mid b$ by the claim.

COROLLARY 2: Let a and b be integers, at least one of which is nonzero, and let c be the greatest common divisor of a and b. Then there exist integers x and y such that

$$c = ax + by.$$

We say that two nonzero integers are *relatively prime* if their greatest common divisor is 1. By the preceding corollary, we immediately derive

COROLLARY 3: If a and b are relatively prime, then there exist integers x and y such that

$$ax + by = 1.$$

Henceforth, we will denote by (a, b) the greatest common divisor of a and b.

DEFINITION 4: A *prime* is a natural number $p > 1$ whose only divisors are ± 1 and $\pm p$.

For example, 2 and 3 are primes, but $6 = 2\cdot 3$ is not a prime. The primes are the basic building blocks from which all integers can be built. We have

LEMMA 5: Let n be an integer greater than 1. Then n can be expressed as a product of primes.

Proof. Let us reason by induction on n. The assertion is clearly true for $n = 2$. Thus, assume that the assertion is true for all integers $< n$, and let us prove from this assumption that n can be expressed as a product of primes. If n is prime, then n is clearly a product of primes. Thus, assume that n is not prime. Then n can be written in the form $n = n'n''$, $1 < n', n'' < n$. By the induction assumption, both n' and n'' can be written as products of primes, so n can be written as a product of primes and the induction is complete.

As a simple application of the lemma, we can prove

THEOREM 6 (Euclid): There exist infinitely many primes.

Proof. Let us reason by contradiction. Suppose that $\{p_1, \ldots, p_N\}$ were a complete list of the primes. The integer

$$M = p_1 \cdots p_N + 1$$

is greater than 1, and is therefore divisible by some prime, by Lemma 5. But M is not divisible by any one of $p_1, \ldots, p_N$. Thus, there must exist a prime different from $p_1, \ldots, p_N$, and a contradiction is reached.

There is a simple way of constructing a table of the primes, first discovered by the Greek mathematician Eratosthenes. Write down all the natural numbers, beginning with 1:

1, 2, 3, 4, 5, 6, 7, 8, 9, ...

Cross out all numbers divisible by 2, except for the first one, 2, to get

1, 2, 3, ~~4~~, 5, ~~6~~, 7, ~~8~~, 9, ~~10~~, ...

Next, cross out all multiples of 3, except for the first one, to get

1, 2, 3, ~~4~~, 5, ~~6~~, 7, ~~8~~, ~~9~~, ~~10~~, 11, ~~12~~, ...

Four and its multiples have already been crossed out, so next cross out all multiples of 5, except for the first, and continue in this way. The natural numbers remaining will be precisely the primes and 1.

Lemma 5 asserts that every nonzero integer can be expressed as a product of primes. Our last major result in this section will be to show that this expression is essentially unique, that is, unique except for rearrangements of the factors. This fact is known as the *fundamental theorem of arithmetic.*

The notion of a prime was already known to the Greeks. In fact, the fundamental theorem of arithmetic had its origins in Greek mathematics. The proof that there are infinitely many primes was known to Euclid, who included a proof in his *Elements*. If one looks at a table of primes, one of the interesting features which one notices is the rather irregular way in which the primes are distributed. For example, the primes below 100 are

2, 3, 5, 7, 11, 13, 17, 19, 23, 29, 31, 37, 41,

43, 47, 53, 59, 61, 67, 71, 73, 79, 89, 93, 97.

In 1798 Legendre made some highly significant empirical investigations concerning the distribution of the primes. For $x \geq 1$, let $\pi(x)$ denote the number of primes $\leq x$. By experimentation, Legendre advanced the conjecture that $\pi(x)$ is "approximately"

$$\frac{x}{(A \log x + B)},$$

where A and B are numerical constants, not depending on x. By "approximately" Legendre meant that as x tends to infinity, the ratio

$$\frac{\pi(x)}{x/(A \log x + B)}$$

tends to 1. Actually, Legendre did not know that Gauss had done the same empirical experiments in 1792 and 1793 without publishing them. Legendre was quite specific. He conjectured that $A = 1$ and $B = 1.08366\cdots$. In particular, Legendre's conjecture implies that

$$\lim_{x\to\infty} \frac{\pi(x)}{x/\log x} = 1. \tag{6}$$

Statement (6) is called the *prime-number theorem.*

The first substantial progress toward a proof of (6) was made by the Russian mathematician Tchebycheff in 1851–1852. He showed, among other things, that for all sufficiently large x,

$$0.92 \leq \frac{\pi(x)}{x/\log x} \leq 1.106.$$

However, Tchebycheff's ideas could not be pushed far enough to prove the prime-number theorem.

The year 1860 was the turning point in the efforts to give a proof of the prime-number theorem. In that year Riemann published a paper in which he connected the function $\pi(x)$ with a certain function of a complex variable, subsequently called the *Riemann zeta function.* The analytic properties of the zeta function reflect themselves in the behavior of $\pi(x)$. Riemann made many conjectures about the zeta function, the most famous of which is the celebrated *Riemann hypothesis*, which predicts the points at which the zeta function is zero. This conjecture has been the object of an incredible effort by many of the finest minds of the last 100 years, but it remains unproved. If the Riemann hypothesis holds, it is possible to state the prime-number theorem much more precisely than (6) by estimating how good an approximation to $\pi(x)$ is given by the function $x/\log x$.

Although Riemann was not able to prove the prime-number theorem, his ideas figured prominently in the first proof, which was discovered independently by Hadamard and de la Vallee Poussin, both in 1896.

Let us now turn to the fundamental theorem of arithmetic. Before we will be able to give a proof of this basic result, we must first establish two preliminary facts which are important in their own right. Again, the ideas are Euclid's.

LEMMA 7: Let a, b, and c be nonzero. If a divides bc and a and b are relatively prime, then a divides c.

Proof: By Corollary 3 there exist x and y such that $ax + by = 1$. Since $a \mid bc$, there exists k such that $bc = ka$. Then

$$\begin{aligned} acx + bcy &= c, \\ \Longrightarrow acx + kay &= c, \\ \Longrightarrow a(cx + ky) &= c, \\ \Longrightarrow a &\mid c. \end{aligned}$$

THEOREM 8 (Euclid): Let p be a prime. If p divides a product ab, then either p divides a or p divides b.

Proof: The result is clear if a or b is zero. Therefore, assume both a and b are nonzero. Assume that p does not divide a (denoted $p \nmid a$). Then, since p is prime, a and p are relatively prime. Therefore, by Lemma 7, $p \mid b$.

The reader should have no difficulty showing that if $p > 1$ is not prime, there exist nonzero integers a and b such that $p \nmid a, p \nmid b$ but nevertheless $p \mid ab$. Thus, Theorem 8 gives an equivalent definition of a prime. A prime is an integer $p > 1$ such that for any integers a and b having the property that $p \mid ab$, we have either $p \mid a$ or $p \mid b$. In more abstract contexts it turns out that this is the "proper" definition of a prime.

A useful consequence of Theorem 8 is

COROLLARY 9: Let p be a prime. If p divides $a_1 a_2 \cdots a_r$, then p divides a_i for some i $(1 \leq i \leq r)$.

Proof: Follows by induction from Theorem 8.

Let us now address ourselves to the fundamental theorem.

THEOREM 10 (Fundamental Theorem of Arithmetic): Let n be an integer greater than 1. Then n can be represented as a product of primes

$$n = p_1 \cdots p_r.$$

Moreover, the representation of n is unique up to rearrangement of the factors. That is, if $n = q_1 \cdots q_s$ is another representation of n as a product of primes, then $r = s$ and we may renumber $q_1, \ldots, q_s$ so that $p_i = q_i$ for $1 \leq i \leq r$.

Proof: The first assertion is true by Lemma 5. Let us proceed by induction on r. If $r = 1$, then n is prime and the assertion is clear. Thus, assume that $r > 1$ and that the theorem holds for n which are products of $r - 1$ primes. Since $p_1 \mid n$, we see that $p_1 \mid q_1 \cdots q_s$. Therefore, by Corollary 9, $p_1 \mid q_i$ for some i. Without loss of generality, suppose that $p_1 \mid q_1$. Then, since p_1 and q_1

are both primes, $p_1 = q_1$. Therefore,

$$p_1 p_2 \cdots p_r = p_1 q_2 \cdots q_s,$$

which implies that

$$p_2 \cdots p_r = q_2 \cdots q_s \tag{7}$$

by Exercise 7, Section 2.3. By our induction assumption and Equation (7), we see that $r - 1 = s - 1$and that $q_2, \ldots, q_s$ may be renumbered so that $p_i = q_i$ for $i = 2, \ldots, r$. Thus, since $p_1 = q_1$, the theorem has been proved for r and the induction is complete.

2.4 *Exercises*

1. Use the sieve of Eratosthenes to determine all primes less than 200.
2. Prove properties (1) and (2) of the divisibility relation.
3. Let I be a subset of $\mathbf{Z}$ having the properties: $x, y \in I \Rightarrow x - y \in I$, $I \neq \varnothing$. Show that there exists $c \in \mathbf{Z}$ such that $I = c\mathbf{Z} = \{cx \mid x \in \mathbf{Z}\}$
4. If p is not prime, there exist integers a and b, $a \neq 0$, $b \neq 0$, such that $p \mid ab$, but $p \nmid a$ and $p \nmid b$. (Assume $p > 1$.)
5. Here is a procedure for determining the greatest common divisor of a, b, where $a > 0$, $b > 0$. Use the division algorithm to find q_1 and r_1 such that

   $$b = aq_1 + r_1 \qquad (0 \leq r_1 < a),\ a \leq b.$$

   If $r_1 = 0$, then $(a, b) = a$. Otherwise, find q_2 and r_2 such that

   $$a = r_1 q_2 + r_2 \qquad (0 \leq r_2 < r_1).$$

   If $r_2 = 0$, then $(a, b) = r_1$. Otherwise, find q_3 and r_3 such that

   $$r_1 = r_2 q_3 + r_3 \qquad (0 \leq r_3 < r_2).$$

   Assume that we go on in this way to find $(q_4, r_4), \ldots, (q_n, r_n)$ such that

   $$r_i = r_{i+1} q_{i+2} + r_{i+2} \qquad (0 \leq r_n < r_{n-1} < \cdots < r_1).$$

   Eventually, this process terminates. (Why?) Suppose that $r_n = 0$. Then $r_{n-1} = (a, b)$. This process is called the *Euclidean algorithm*. Prove that it works. Use the Euclidean algorithm to write (a, b) in the form $ax + by$.
6. Use Exercise 5 to find the greatest common divisor of 120 and 336. Write (120, 336) in the form $120x + 336y$.
7. Let $\mathbf{Z}[\sqrt{-5}] = \{a + b\sqrt{-5} \mid a, b \in \mathbf{Z}\}$. Define addition and multiplication in $\mathbf{Z}[\sqrt{-5}]$ as follows: $(a + b\sqrt{-5}) + (c + d\sqrt{-5}) = (a + c) + (b + d)\sqrt{-5}$; $(a + b\sqrt{-5}) \cdot (c + d\sqrt{-5}) = (ac - 5bd) + (bc + ad)\sqrt{-5}$. Show that addition and multiplication are commutative and associative, that there exist identities under addition and multiplication. Show that the distributive law holds. Show that there exist inverses under addition.
8. Let us keep the same notation as in Exercise 7. Define the norm mapping

   $$N\colon \mathbf{Z}[\sqrt{-5}] \longrightarrow \mathbf{Z}$$

by

$$N(a + b\sqrt{-5}) = a^2 + 5b^2.$$

Show that

(a) $N(xy) = N(x)N(y) \qquad (x, y \in \mathbf{Z}[\sqrt{-5}])$.
(b) $N(x) = 0 \Leftrightarrow x = 0 + 0\sqrt{-5}$.
(c) $N(x) = 1 \Leftrightarrow x = \pm 1 + 0\sqrt{-5}$.

9. Show that the elements $3, 2, 1 + \sqrt{-5}$, and $1 - \sqrt{-5}$ are primes in $\mathbf{Z}[\sqrt{-5}]$. [*Hint:* $N(3) = 9$. Therefore, if x is a divisor of 3, $N(x)$ is a factor of 9, so that $N(x) = 9$, 3, or 1. In the first case, $x = \pm 3$. In the second case, x cannot exist, since $a^2 + 5b^2 = 3$ has no solutions in integers a and b. In the third case, $x = \pm 1$. Fill in the details.]

10. Note that $3 \cdot 2 = (1 + \sqrt{-5})(1 - \sqrt{-5})$. In the light of Exercise 9, what can you say about the uniqueness of factorization in $\mathbf{Z}[\sqrt{-5}]$?

2.5 Congruences

In this section we will introduce two algebraic systems which can be constructed from the integers. These systems will provide us with enlightening examples of many algebraic phenomena and will prove to be important in their own right. Throughout this section let n be a positive integer.

DEFINITION 1: Let $a, b \in \mathbf{Z}$. We say that a is *congruent to b modulo n* [denoted $a \equiv b \pmod{n}$] if $n \mid a - b$.

EXAMPLE 1: $5 \equiv 7 \pmod{2}$; $17 \equiv 11 \pmod{3}$.

If n is clear from context, we will sometimes write $a \equiv b$ instead of $a \equiv b \pmod{n}$. If a is not congruent to b modulo n, we write $a \not\equiv b \pmod{n}$. Let us establish some of the elementary properties of congruence.

PROPOSITION 2: (1) For all $a \in \mathbf{Z}$, $a \equiv a \pmod{n}$.
(2) If $a \equiv b \pmod{n}$, then $b \equiv a \pmod{n}$.
(3) If $a \equiv b \pmod{n}$ and $b \equiv c \pmod{n}$, then $a \equiv c \pmod{n}$.
(4) If $a \equiv b \pmod{n}$ and $c \equiv d \pmod{n}$, then $a + c \equiv b + d \pmod{n}$ and $ac \equiv bd \pmod{n}$.

Proof: (1), (2)—Clear.

(3) If $a \equiv b \pmod{n}$, $b \equiv c \pmod{n}$, then $n \mid a - b$, $n \mid b - c$. Therefore, by Equation (3) of Section 2.4, $n \mid (a - b) + (b - c) \Rightarrow n \mid a - c \Rightarrow a \equiv c \pmod{n}$.

(4) If $a \equiv b \pmod{n}$, $c \equiv d \pmod{n}$, then $n \mid a - b$, $n \mid c - d$. Therefore, by Equation (3) of Section 2.4, we have

$$n \mid (a + c) - (b + d) \Longrightarrow a + c \equiv b + d \pmod{n}.$$

Also, we have $n \mid c(a-b)$, $n \mid b(c-d)$. Therefore, by the definition of congruence, we see that

$$ac \equiv bc \pmod{n}, \qquad bc \equiv bd \pmod{n}.$$

Therefore, by part (3), we have

$$ac \equiv bd \pmod{n}.$$

Many facts concerning congruences are quite old. For example, in the beginning of the seventeenth century, the jurist and amateur mathematician Fermat proved that if a is an integer and p is a prime such that $p \nmid a$, then $a^{p-1} \equiv 1 \pmod{p}$. However, the first organized study of congruences was undertaken by Gauss in his book *Disquisitiones Arithmeticae*, written in 1801. In broad terms, one of Gauss's principal goals in the *Disquisitiones* was to study the solutions (if any exist) of the congruence

$$a_n x^n + a_{n-1} x^{n-1} + \cdots + a_0 \equiv 0 \pmod{m},$$

where $m, a_0, \ldots, a_n$ are given integers and x is to be determined. The problem as stated is incredibly difficult and much of nineteenth-century research in number theory was aimed at its solution, which is not complete even today.

Just to get a feel for Gauss's work, let us consider the particular congruence

$$x^2 \equiv a \pmod{p},$$

where a is a given integer and p is a prime. This congruence may or may not have a solution x. For example, if $p = 3$ and x is any integer, then $x \equiv 0, 1, 2 \pmod{3}$, so that $x^2 \equiv 0, 1, 2^2 \pmod{3}$. Thus, if x is any integer, $x^2 \equiv 0$ or $1 \pmod{3}$, so that the congruence $x^2 \equiv 2 \pmod{3}$ has no solutions. Legendre defined the symbol (a/p) as follows:

$$\left(\frac{a}{p}\right) = \begin{cases} 0 & \text{if } p \mid a, \\ 1 & \text{if } p \nmid a \text{ and } x^2 \equiv a \pmod{p} \text{ has a solution,} \\ -1 & \text{otherwise.} \end{cases}$$

One of Gauss's greatest achievements is the law of *quadratic reciprocity*, which asserts that if p and q are distinct, odd primes, then

$$\left(\frac{p}{q}\right)\left(\frac{q}{p}\right) = (-1)^{(p-1)(q-1)/4},$$

$$\left(\frac{2}{p}\right) = (-1)^{(p^2-1)/8},$$

$$\left(\frac{-1}{p}\right) = (-1)^{(p-1)/2}.$$

Using the quadratic reciprocity law, it is possible to determine precisely when the congruence $x^2 \equiv a \pmod{p}$ has a solution. Given the complexity of the above formulas, the reader should have no difficulty accepting the assertion that the problem of solving congruences of higher degree is indeed very complicated.

Much work on congruences of higher degree was undertaken during the nineteenth century and these investigations have led to some of the most significant developments in twentieth-century mathematics, which are still objects of research.

In this section we will concentrate on the theory of linear congruences. If k is an arbitrary integer, let $\bar{k}$ denote the set of all integers congruent to k modulo n. Then

$$\bar{k} = \{k + ns \mid s \in \mathbf{Z}\}.$$

The sets $\bar{k}$ are called *residue classes modulo n*. Suppose that $\bar{k}$ is a given residue class. The division algorithm says that there exist $s, k_0 \in \mathbf{Z}$ such that $k = ns + k_0, 0 \leq k_0 \leq n - 1$. Therefore, since

$$\{k + ns \mid s \in \mathbf{Z}\} = \{k_0 + nt \mid t \in \mathbf{Z}\},$$

we see that $\bar{k}$ is the same as one of the residue classes

$$\bar{0}, \bar{1}, \ldots, \overline{n-1}.$$

Note that none of these latter residue classes are equal. For if $\bar{i} = \bar{j}$ with $0 \leq i < j < n$, then

$$\{i + ns \mid s \in \mathbf{Z}\} = \{j + nt \mid t \in \mathbf{Z}\},$$

so that $j = i + ns$ for some $s \in \mathbf{Z}$. But then $j - i$ is divisible by n, which contradicts the fact that $0 < j - i < n$. Thus, we have shown that there are precisely n different residue classes modulo n and that these are given by

$$\bar{0}, \bar{1}, \bar{2}, \ldots, \overline{n-1}.$$

Let us denote the set of all residue classes modulo n by $\mathbf{Z}_n$.

From our above discussion, it is clear that every integer is contained in some residue class modulo n, and therefore

$$\mathbf{Z} = \bar{0} \cup \bar{1} \cup \cdots \cup \overline{n-1}.$$

Moreover, no integer is contained in two of the residue classes $\bar{0}, \bar{1}, \ldots, \overline{n-1}$. Indeed, if $x \in \bar{i} \cap \bar{j}, 0 \leq i < j < n$, then x can be written in the form

$$x = i + ns = j + nt$$

for suitable $s, t \in \mathbf{Z}$. But then $j - i$ is a positive integer less than n and $j - i = n(t - s)$, so that $j - i$ is divisible by n, which is a contradiction. Thus, we have shown that if $0 \leq i < j < n$, then

$$\bar{i} \cap \bar{j} = \varnothing.$$

Let us define addition and multiplication of elements of $\mathbf{Z}_n$ as follows:

$$\bar{a} + \bar{b} = \overline{a + b},$$
$$\bar{a} \cdot \bar{b} = \overline{a \cdot b}.$$

That is, the sum (respectively, product) of two residue classes $\bar{a}$ and $\bar{b}$ is the residue class containing $a + b$ (respectively, $a \cdot b$). We must, of course, check that these definitions depend only on the residue classes $\bar{a}$ and $\bar{b}$ and not on the choice of the elements a and b. But this is easily accomplished using Proposition 2, part (4). For if $\bar{a} = \bar{c}$ and $\bar{b} = \bar{d}$, then

$$\begin{aligned} &a \equiv c \pmod{n}, \qquad b \equiv d \pmod{n}, \\ &\Longrightarrow a + b \equiv c + d \pmod{n}, \\ &\Longrightarrow \overline{a + b} = \overline{c + d}. \end{aligned}$$

Therefore, the operation of addition of residue classes is consistent. A similar argument works for multiplication.

Addition and multiplication are commutative and associative and the two operations obey the distributive law

$$\bar{a} \cdot (\bar{b} + \bar{c}) = \bar{a} \cdot \bar{b} + \bar{a} \cdot \bar{c}.$$

The residue class $\bar{0}$ is the identity element with respect to addition:

$$\bar{a} + \bar{0} = \bar{0} + \bar{a} = \bar{a} \qquad (\bar{a} \in \mathbf{Z}_n).$$

The residue class $\bar{1}$ is the identity element with respect to multiplication:

$$\bar{a} \cdot \bar{1} = \bar{1} \cdot \bar{a} = \bar{a} \qquad (\bar{a} \in \mathbf{Z}_n).$$

Every residue class $\bar{a}$ has an inverse $-\bar{a}$ with respect to addition. In fact, we may set $-\bar{a} = \overline{-a}$. Then

$$\bar{a} + \overline{-a} = \overline{-a} + \bar{a} = \bar{0}.$$

Thus, in most respects, the algebraic system $\mathbf{Z}_n$ possesses the same properties as $\mathbf{Z}$. However, there is at least one important difference. Recall that we proved that the product of two nonzero integers is nonzero. This property of $\mathbf{Z}$ does not carry over to $\mathbf{Z}_n$. For example, let us consider $\mathbf{Z}_6$. We clearly have $\bar{3} \neq \bar{0}$, $\bar{2} \neq \bar{0}$; nevertheless,

$$\bar{3} \cdot \bar{2} = \bar{6} = \bar{0}.$$

In this circumstance we say that $\bar{3}$ and $\bar{2}$ are *divisors of zero*. We leave to the exercises the verification of the fact that if n is prime, then $\mathbf{Z}_n$ has no divisors of zero, while if n is > 1 and not prime, then $\mathbf{Z}_n$ always has divisors of zero.

Note that if $(a, n) = 1$, then $(a + kn, n) = 1$ for all $k \in \mathbf{Z}$. Therefore, if one element of a residue class modulo n is relatively prime to n, every element of the residue class is relatively prime to n. When this occurs, we say that the residue class is *reduced*. Let $\mathbf{Z}_n^\times$ denote the set of all reduced residue classes modulo n.

EXAMPLE 2: $\mathbf{Z}_6^\times = \{\bar{1}, \bar{5}\}$; $\mathbf{Z}_7^\times = \{\bar{1}, \bar{2}, \bar{3}, \bar{4}, \bar{5}, \bar{6}\}$; $\mathbf{Z}_{10}^\times = \{\bar{1}, \bar{3}, \bar{7}, \bar{9}\}$.

Since $\mathbf{Z}_n = \{\bar{0}, \bar{1}, \bar{2}, \ldots, \overline{n-1}\}$, and since $\bar{a} \in \mathbf{Z}_n^\times$ if and only if $(a, n) = 1$, we see that the number of elements in $\mathbf{Z}_n^\times$ is equal to the number of positive integers a such that $a < n$ and $(a, n) = 1$. This number is usually denoted

$\phi(n)$. From the above examples, we see that

$$\phi(6) = 2, \qquad \phi(7) = 6, \qquad \phi(10) = 4.$$

The function $\phi(n)$ is defined for positive integers n and is called Euler's *phi function*.

If

$$n = p_1^{a_1} \cdots p_r^{a_r}$$

is the decomposition of n into a product of powers of distinct primes, then

$$\phi(n) = \prod_{i=1}^{r} p_i^{a_i - 1}(p_i - 1). \tag{1}$$

The reader should verify that this formula works in the special cases of the examples given above. A complete proof of (1) will be outlined in the exercises.

PROPOSITION 3: Suppose that $\bar{a}, \bar{b} \in \mathbf{Z}_n^\times$. Then $\bar{a}\cdot\bar{b} \in \mathbf{Z}_n^\times$.

Proof: Since $\bar{a}\cdot\bar{b} = \overline{a\cdot b}$, it suffices to show that if $(a, n) = 1$, $(b, n) = 1$, then $(a\cdot b, n) = 1$. Suppose that $(a\cdot b, n) > 1$. Then $(a\cdot b, n)$ is divisible by some prime p. Therefore, $p \mid a\cdot b$ and $p \mid n$. However, by Theorem 8 of Section 2.4, either $p \mid a$ or $p \mid b$. Suppose that $p \mid a$. Then, since $p \mid n$, $p \mid (a, n)$, which is a contradiction to the assumption $(a, n) = 1$.

By Proposition 3, $\cdot$ defines a law of multiplication among reduced residue classes. Since $\bar{1} \in \mathbf{Z}_n^\times$, we see that $\mathbf{Z}_n^\times$ contains an identity element with respect to $\cdot$.

PROPOSITION 4: Let $\bar{a} \in \mathbf{Z}_n^\times$. Then there exists $\bar{b} \in \mathbf{Z}_n^\times$ such that

$$\bar{a}\cdot\bar{b} = \bar{b}\cdot\bar{a} = \bar{1}.$$

That is, $\bar{a}$ has an inverse with respect to multiplication.

Proof: It suffices to find $b \in \mathbf{Z}$ such that $(b, n) = 1$ and

$$a\cdot b \equiv 1 \pmod{n}.$$

Since $(a, n) = 1$, there exist $x, y \in \mathbf{Z}$ such that

$$a\cdot x + n\cdot y = 1.$$

We clearly have $(x, n) = 1$, since any common divisor of x and n divides $a\cdot x + n\cdot y = 1$. Moreover,

$$a\cdot x \equiv 1 \pmod{n}.$$

Therefore, we may set $x = b$.

Proposition 4 has an interesting application. Let $a, b, c \in \mathbf{Z}$ and suppose that it is required to solve the equation

$$ax + by = c \tag{2}$$

in integers. Such an equation is called a *Diophantine equation*, after the Greek mathematician Diophantus, who first studied its solutions. If $d = (a, b)$, then $d \mid ax + by \Rightarrow d \mid c$. Therefore, in order for (2) to have any solutions, we must have $c = dc'$. If we set $a = a'd$, $b = b'd$, we see that (2) is equivalent to the equation

$$a'x + b'y = c'.$$

But $(a', b') = 1$ since $d = (a, b)$ (Exercise 8). Therefore, without loss of generality, we may restrict ourselves to equations of the form (2) for which $(a, b) = 1$. Equation (2) is equivalent to the congruence

$$ax \equiv c \pmod{b}. \tag{3}$$

In turn, the congruence (3) is equivalent to the equation

$$\bar{a} \cdot \bar{x} = \bar{c} \tag{4}$$

among residue classes of $\mathbf{Z}_b$. By Proposition 4, since $(a, b) = 1$, there exists $\overline{a^*} \in \mathbf{Z}_b$ such that $\bar{a} \cdot \overline{a^*} = \overline{a^*} \cdot \bar{a} = \bar{1}$. Therefore, (4) is equivalent to

$$\overline{a^*} \cdot \bar{a} \cdot \bar{x} = \overline{a^*} \cdot \bar{c} \Longleftrightarrow \bar{x} = \overline{a^*} \cdot \bar{c}.$$

Therefore, in order to solve Equation (2), it is sufficient to choose x so that

$$x \equiv a^*c \pmod{b}.$$

Then $ax - c$ is divisible by b. Choose y so that $ax - c = -by$. It is clear from the above argument that all solutions of Equation (2) are obtained in this way.

Let us now give an example to show that this whole process is computationally viable. Consider the equation

$$3x + 5y = 12.$$

Then $\overline{3^*} = \bar{2}$ since $3 \cdot 2 \equiv 1 \pmod{5}$. Therefore, we must choose x so that

$$x \equiv 24 \pmod{5},$$

which is equivalent to

$$x \equiv 4 \pmod{5}. \tag{5}$$

For given x satisfying (5), set

$$y = (-3x + 12)/5.$$

For example, some solutions are $x = 4, y = 0$; $x = 9, y = -3$; $x = -1$, $y = 3$.

2.5 *Exercises*

1. Compute $\bar{5} \cdot (\bar{8} + \overline{13})$ among residue classes in $\mathbf{Z}_{17}$.
2. Find all integers x such that $2x + 2 \equiv 7 \pmod{9}$.
3. Find all solutions in integers of the equation $5x + 7y = 17$.

4. Compute $\phi(12)$, $\phi(41)$, and $\phi(22)$.
5. Let p be a prime and r a positive integer. Show that $\phi(p^r) = p^{r-1}(p-1)$.
6. Let m and n be positive integers, $(m, n) = 1$.
   (a) Show that there exists $x \in \mathbf{Z}$ such that
   $$x \equiv 0 \pmod{m},$$
   $$x \equiv 1 \pmod{n}.$$
   (b) Suppose that $a, a' \in \mathbf{Z}$. Show that there exists $x \in \mathbf{Z}$ such that
   $$x \equiv a \pmod{m},$$
   $$x \equiv a' \pmod{n}.$$
   This last assertion is known as the *Chinese remainder theorem.*
7. Let m and n be positive integers, $(m, n) = 1$, and let $a_1, \ldots, a_{\phi(m)}, b_1, \ldots, b_{\phi(n)}$ be integers such that
   $$\mathbf{Z}_m^\times = \{\bar{a}_1, \ldots, \bar{a}_{\phi(m)}\},$$
   $$\mathbf{Z}_n^\times = \{\bar{b}_1, \ldots, \bar{b}_{\phi(n)}\}.$$
   (a) Show that the a_i and b_j may be chosen so that
   $$a_i \equiv 1 \pmod{n} \qquad [1 \leq i \leq \phi(m)],$$
   $$b_j \equiv 1 \pmod{m} \qquad [1 \leq j \leq \phi(n)].$$
   (*Hint:* Use Exercise 6.)
   (b) Show that if the condition of (a) is satisfied, then
   $$\mathbf{Z}_{mn}^\times = \{\overline{a_i b_j} \mid 1 \leq i \leq \phi(m),\ 1 \leq j \leq \phi(n)\}.$$
   (c) Show that if $(m, n) = 1$, then $\phi(mn) = \phi(m)\phi(n)$.
   (d) Let $n = \prod_{i=1}^t p_i^{r_i}$, where the p_i are distinct primes and r_i is a positive integer. Then
   $$\phi(n) = \prod_{i=1}^t p_i^{r_i - 1}(p_i - 1).$$
   [*Hint:* Combine part (c) with Exercise 5.]
8. Let $d = (a, b)$. Show that $(a/d, b/d) = 1$.

2.6 *Equivalence Relations*

In Section 2.5 we introduced the notion of congruence of integers modulo n, and we proved the following basic properties of congruence:

1. $a \equiv a \pmod{n}$.
2. If $a \equiv b \pmod{n}$, then $b \equiv a \pmod{n}$.
3. If $a \equiv b \pmod{n}$ and $b \equiv c \pmod{n}$, then $a \equiv c \pmod{n}$.

Moreover, we showed that the integers are distributed among the residue classes $\{\bar{0}, \bar{1}, \ldots, \overline{n-1}\}$, which satisfy the following properties:

1′. Every integer x belongs to one of the residue classes $\{\bar{0}, \bar{1}, \ldots, \overline{n-1}\}$, so that

$$\mathbf{Z} = \bar{0} \cup \bar{1} \cup \cdots \cup \overline{n-1}.$$

2′. No integer belongs to two of the residue classes $\{\bar{0}, \bar{1}, \ldots, \overline{n-1}\}$, so that

$$\bar{i} \cap \bar{j} = \varnothing \qquad (0 \leq i < j \leq n-1, i \neq j).$$

Let us abstract from this data two very important notions—that of *partition and equivalence relation.*

Let S be a set. A *partition* of S is a collection $\mathfrak{C}$ of subsets of S such that

a. $\bigcup_{A \in \mathfrak{C}} A = S$.
b. If A and B are distinct elements of $\mathfrak{C}$, then $A \cap B = \varnothing$.

For example, 1′ and 2′ assert that $\{\bar{0}, \bar{1}, \ldots, \overline{n-1}\}$ is a partition of $\mathbf{Z}$. Thus, the notion of a partition may be viewed as a generalization of the notion of distributing the integers in residue classes modulo n. By analogy with the example of congruences, let us say that two elements x and y of S are *congruent with respect to the partition* $\mathfrak{C}$ if they lie in the same set of $\mathfrak{C}$. In this case, we will write $x \sim_{\mathfrak{C}} y$. Note that properties (1), (2) and (3) carry over to our more general notion of congruence. Namely, we have

$1_{\mathfrak{C}}$. $x \sim_{\mathfrak{C}} x$ for all $x \in S$.
$2_{\mathfrak{C}}$. If $x \sim_{\mathfrak{C}} y$, then $y \sim_{\mathfrak{C}} x$.
$3_{\mathfrak{C}}$. If $x \sim_{\mathfrak{C}} y$ and $y \sim_{\mathfrak{C}} z$, then $x \sim_{\mathfrak{C}} z$.

The above example suggests the following definition: Let S be a set. Then a *relation* R on S is a subset of $S \times S$. For example, if $S = \{1, 2, 3\}$, then a relation R is given by

$$R = \{(1, 1), (1, 2), (2, 1), (3, 1)\} \subseteq S \times S. \tag{1}$$

If R is a relation on S and if $(x, y) \in R$, then we say that *x is related to y with respect to R*, and we often write $x \sim_R y$. Note that a partition $\mathfrak{C}$ on a set S determines a relation $R_{\mathfrak{C}}$ on S, the relation

$$R_{\mathfrak{C}} = \{(x, y) \in S \times S \,|\, x \sim_{\mathfrak{C}} y\}.$$

In this case, $x \sim_{R_{\mathfrak{C}}} y$ if and only if $x \sim_{\mathfrak{C}} y$. Such a relation has very special properties, $1_{\mathfrak{C}}$–$3_{\mathfrak{C}}$. Lest the reader think that $1_{\mathfrak{C}}$–$3_{\mathfrak{C}}$ are satisfied by any relation, we should examine closely the relation R, which is defined by (1). It is not true that $2 \sim_R 2$, so $1_{\mathfrak{C}}$ is violated. Moreover, $3 \sim_R 1$, but it is not true that $1 \sim_R 3$, so $2_{\mathfrak{C}}$ is violated. Finally, $3 \sim_R 1$ and $1 \sim_R 2$, but it is not true that $3 \sim_R 2$, so that $3_{\mathfrak{C}}$ is violated. A relation satisfying $1_{\mathfrak{C}}$–$3_{\mathfrak{C}}$ is called an *equivalence relation.* More precisely, if S is a set and R is a relation on S, then R is said to be an *equivalence relation* if the following properties hold:

ER1. *For every* $x \in S, x \sim_R x$.
ER2. *If* $x, y \in S$ *and* $x \sim_R y$, *then* $y \sim_R x$.
ER3. *If* $x, y, z \in S$ *and* $x \sim_R y, y \sim_R z$, *then* $x \sim_R z$.

The properties ER1–ER3 are called the *reflexive*, *symmetric*, and *transitive* laws, respectively. They completely characterize an equivalence relation, in the following sense: Suppose there is given a relation R on S which satisfies rules ER1–ER3. Then it is possible to associate to R a partition $\mathcal{C}$ on S such that R is the equivalence relation associated with $\mathcal{C}$. For each $x \in S$, let

$$A_x = \{y \in S \mid x \sim_R y\},$$

and set

$$\mathcal{C} = \{A_x \mid x \in S\},$$

(Note that A_x may be equal to A_y for $x \neq y$, so some sets may be repeated in our description of $\mathcal{C}$.) We claim that

1. $\mathcal{C}$ is a partition on S.
2. R is the equivalence relation associated with $\mathcal{C}$.

It is clear that 2 holds provided 1 does. Thus, it suffices to prove 1. Since $x \sim_R x$, we see that $x \in A_x$, so that

$$\bigcup_{x \in S} A_x = S.$$

Assume that

$$A_x \cap A_y \neq \varnothing.$$

We must show that $A_x = A_y$. Assume that $z \in A_x \cap A_y$, and let $w \in A_x$. Then

$$x \sim_R z, \qquad y \sim_R z, \qquad x \sim_R w.$$

Therefore, by ER2,

$$x \sim_R z, \qquad z \sim_R y, \qquad w \sim_R x,$$

so that by ER3 and ER2,

$$w \sim_R x, \qquad x \sim_R y,$$

which implies that $w \sim_R y$, by ER3. Thus, $y \sim_R w$ by ER2, and we have shown that if $w \in A_x$, then $w \in A_y$. Therefore, $A_x \subseteq A_y$. Interchanging x and y, we see that $A_y \subseteq A_x$. Therefore, $A_x = A_y$.

Note: The student should carefully master the relationship between a partition and its associated equivalence relation, since this idea will be used over and over again throughout this book.

If $\sim$ is an equivalence relation and $\mathcal{C}$ is its associated partition, then the sets contained in $\mathcal{C}$ are called *equivalence classes* with respect to $\sim$. Two elements are equivalent with respect to $\sim$ if and only if they belong to the same equivalence class.

Equivalence relations occur with surprising frequency in even the most innocent of contexts. For example, consider the rational numbers (or fractions). From our days in grade school, we are accustomed to think of a rational number as the quotient a/b of integers a and b ($b \neq 0$). However, strictly speaking, this is not correct, since we do not regard the rational numbers $\frac{1}{3}, \frac{2}{6}$, and $\frac{9}{27}$ as different from one another. However, there is a simple way out of this logical difficulty. Let S denote the set of all quotients a/b, where a and b are integers, $b \neq 0$, and let us define an equivalence relation $\sim$ on S by

$$a/b \sim c/d \longleftrightarrow ad - bc = 0.$$

It is easy to verify that $\sim$ is an equivalence relation and that all the ratios a/b, $2a/2b$, $3a/3b$, etc., are equivalent to one another. Then a rational number can be defined to be an equivalence class belonging to $\sim$! Roughly speaking a rational number is the "common value" of the ratios

$$\{a/b, 2a/2b, (-1)a/(-1)b, 6a/6b, \cdots\}.$$

Suppose that A is a set and $\sim$ is an equivalence relation on A. Moreover, suppose that $*$ is a binary operation on A. We will say that $*$ is *compatible with* $\sim$ if, whenever $a \sim b, c \sim d$ $(a, b, c, d \in A)$, we have $a * c \sim b * d$. For example, if $A = \mathbf{Z}$, and $\sim$ is the equivalence relation on $\mathbf{Z}$ defined by the partition $\{\bar{0}, \bar{1}, \ldots, \overline{n-1}\}$, then $\sim$ is compatible with the binary operations of addition and multiplication. For indeed, if $a,b,c,d \in \mathbf{Z}$, and $a \sim b$, $c \sim d$, then $a \equiv b \pmod{n}$, $c \equiv d \pmod{n}$, so that $a + c \equiv b + d \pmod{n}$, $a \cdot c \equiv b \cdot d \pmod{n}$ by Proposition 2, part (4). Thus, $a + c \sim b + d$ and $a \cdot c \sim b \cdot d$.

If A is a set and $\sim$ is an equivalence relation on A, let $A/\sim\, = \{A_x \mid x \in A\}$ denote the set of equivalence classes of A with respect to $\sim$. If $*$ is a binary operation on A which is compatible with respect to $\sim$, then we can define a binary operation $*'$ on $A/\sim$ as follows:

$$A_x *' A_y = A_{x*y}. \tag{2}$$

Please note that in order for (2) to define a binary operation on $A/\sim$, we must show that the "product" of A_x and A_y depends only on A_x and A_y and not on the choice of x in A_x and y in A_y—that is, we must show that if $A_x = A_{x'}$, $A_y = A_{y'}$, then we have $A_x *' A_y = A_{x'} *' A_{y'}$. But, if $A_x = A_{x'}$, then $x \sim x'$. Similarly, if $A_y = A_{y'}$, then $y \sim y'$. Thus, since $*$ is compatible with $\sim$, we have $x * y \sim x' * y'$, and therefore $A_{x*y} = A_{x'*y'}$, and thus we have shown that $*'$ does, indeed, define, a binary operation on the set of equivalence classes $A/\sim$.

In what follows, we will give many examples of equivalence relations on various algebraic structures (for example, groups, rings) and of binary operations compatible with these equivalence relations, so we will refrain from giving further examples at this point.

2.6 Exercises

1. Let $S = \{a, b, c, d, e\}$ and let $\mathcal{C} = \{\{a, b\}, \{c, d\}, \{e\}\}$.
   (a) Show that $\mathcal{C}$ is a partition of S.
   (b) What is the equivalence relation associated to $\mathcal{C}$?
2. Let $S =$ the Cartesian plane of high school analytic geometry, and let $\sim$ be a relation on S defined by
$$(x, y) \sim (x', y') \text{ if and only if } y = y'.$$
   (a) Show that $\sim$ is an equivalence relation on S.
   (b) What are the equivalence classes of S with respect to $\sim$?
3. Let A and B be sets. Show that $\{A - B, B - A, A \cap B\}$ is a partition of $A \cup B$.
4. Let S be a set and let R be a relation on S. We say that R is a *partial ordering* on S if
   (1) R is reflexive.
   (2) R is transitive.
   (3) If $(x, y) \in R$, and $(y, x) \in R$, then $x = y$.
   If R is a partial ordering on S and if $(x, y) \in R$, we write $x \leq_R y$ (read "x is less than or equal to y with respect to R"). Then (1)–(3) can be rewritten
   (1′) $x \leq_R x$ for all $x \in S$.
   (2′) If $x \leq_R y$, and $y \leq_R z$, then $x \leq_R z$.
   (3′) If $x \leq_R y$ and $y \leq_R x$, then $x = y$.
   (a) Let A be any set and let $S = \mathcal{P}(A)$, where $\mathcal{P}(A)$ denotes the set of all subsets of A. Define a relation R on S by $(x, y) \in R$ if and only if $x \subseteq y$ [Note that since x and y are elements of $\mathcal{P}(A)$, they are both subsets of A, and therefore it makes sense to speak of x being contained in y.] Show that R is a partial ordering on S.
   (b) Using (a), show that it is not necessarily true that if R is a partial ordering on S and $x, y \in S$, then either $x \leq_R y$ or $y \leq_R x$ holds. (Possibly both.)
5. Let A and B be sets and let $f: A \longrightarrow B$ be a function. For each $b \in B$, set
$$A_b = \{a \in A \mid f(a) = b\}.$$
   (a) Show that $\{A_b \mid b \in B\}$ is a partition of A. The sets A_b are called *level sets* of f.
   (b) Describe the equivalence relation associated with the partition of (a).
   (c) Let $f: \mathbf{R} \times \mathbf{R} \longrightarrow \mathbf{R}$ be defined by $f((x, y)) = x^2 - y^2$. Describe the level sets of f.

3 THE THEORY OF GROUPS, I

In this chapter we begin the study of one of the most beautiful branches of mathematics—the theory of groups. A group is one of the most elementary mathematical systems, having only one operation defined among its elements. Therefore, from a logical point of view, groups should be one of the first algebraic systems to be studied. Such study, however, would be wasted if groups did not occur "in nature." We have already hinted at the connection between groups and the solution of equations in radicals. We will see in Section 3.2 that groups occur with surprising frequency in mathematics, and this is precisely the reason that the notion of a group is one of the central notions of modern mathematics.

Our approach to group theory will be "abstract"—that is, we will define a group by a set of axioms and then deduce general properties of groups from the axioms. When the general theorems are applied to particular groups, we are able to determine data about individual groups. The student should recognize that this approach is the reverse of the usual pattern of historical discovery. Mathematicians usually discover theorems by generalizing particular examples or known special cases. The important point for the beginner to recognize is that mathematicians do *not* just write down random systems of axioms and then proceed to deduce theorems based on the axioms. Not all axiom systems are "interesting." In order for an axiom system to be interesting there must be important examples to which the axioms apply.

3.1 *The Concept of a Group*

Our main goals in this section are to carefully define the notion of a group, and to explore some of the elementary logical consequences of our definition.

DEFINITION 1: A group is a nonempty set G together with an associative binary operation $\cdot$ on G which has an identity and such that each element has an inverse with respect to the binary operation: What this means in detail is that the following hold:

G1. *Associative Law.* For all $a, b, c \in G$, we have $a \cdot (b \cdot c) = (a \cdot b) \cdot c$.

G2. *Identity.* There exists an element e of G which has the property

$$e \cdot x = x \cdot e = x,$$

for all $x \in G$. The element e is called an *identity element* of G.

G3. *Inverse.* For every $x \in G$, there exists $y \in G$ such that

$$x \cdot y = y \cdot x = e.$$

The element y is said to be an *inverse* of x.

Remarks: 1. A group can have only one identity element. For if e and e' are both identity elements of G, then $e' = e \cdot e' = e$ by axiom G2 applied with respect to each of the identity elements e and e'. Therefore, we may speak of *the* identity element of G. We will denote the identity element of G by 1_G (or just 1 if G is clear from context). The reader should note the following confusion, which could possibly result from our notation. Note that **Z** is a group with respect to the binary operation of addition. Indeed, we observed in Chapter 2 that addition is associative. The identity element is 0, since $0 + x = x + 0 = x$ for all $x \in \mathbf{Z}$. Finally, $-x$ is an inverse for x, since $x + (-x) = (-x) + x = 0$. Thus, in this particular example, $1_G = 0$. If we were to use the notation 1 for the identity element in this example, we would be misled into thinking that the identity element is the integer 1, which is not the case. Thus, some care must be exercised in using the notation 1 for the identity element.

2. Suppose that G is a group and that $x \in G$. Then x has only one inverse element. For if y and y' are inverses of x, then we have from the definition of an inverse that

$$x \cdot y = y \cdot x = 1_G,$$
$$x \cdot y' = y' \cdot x = 1_G.$$

Therefore, by the first equation,

$$y' \cdot (x \cdot y) = y' \cdot 1_G = y' \qquad \text{(by G2)}.$$

On the other hand, by the second equation and the associative law G3,

$$\begin{aligned} y' \cdot (x \cdot y) &= (y' \cdot x) \cdot y \\ &= 1_G \cdot y \\ &= y \qquad \text{(by G2)}. \end{aligned}$$

Therefore, $y = y'$, and x has only one inverse. Thus, we are entitled to speak of *the* inverse of x, which we will denote by x^{-1}.

3. $1_G^{-1} = 1_G$; for by the definition of the identity, we have $1_G \cdot 1_G = 1_G$. However, 1_G^{-1} is the only element of G which has the property $1_G \cdot 1_G^{-1} = 1_G^{-1} \cdot 1_G = 1_G$.

4. Since the associative law holds, we can omit parentheses in products. For example, we can write $a \cdot b \cdot c$. A priori, this might mean either $a \cdot (b \cdot c)$ or $(a \cdot b) \cdot c$. However, these two products are equal by the associative law. Similar remarks hold for longer products, such as $a \cdot b \cdot c \cdot d$. We will go one step further and omit the dot indicating the binary operation when the binary operation is clear from context. Thus, for example, we will write abc instead of $a \cdot b \cdot c$.

Let us now prove two elementary, but very important, results, which will greatly facilitate doing calculations among elements of a group.

PROPOSITION 2: Let G be a group, and let $a, b \in G$. Then there exists one and only one $x \in G$ such that $ax = b$.

Proof: Set $x = a^{-1}b$. Then $ax = aa^{-1}b = 1 \cdot b = b$. Thus, x exists. If $ax = ax' = b$, then $a^{-1}ax = a^{-1}ax'$, so that $1x = 1x'$, which implies that $x = x'$. Thus, there exists only one x such that $ax = b$.

PROPOSITION 3: If G is a group, and $a, b \in G$, then $(ab)^{-1} = b^{-1}a^{-1}$.

Proof: We know that $(ab)^{-1}$ is the unique element x of G which has the property $(ab)x = x(ab) = 1$. Therefore, it suffices to prove that $b^{-1}a^{-1}$ has this property, But $(ab)(b^{-1}a^{-1}) = a(bb^{-1})a^{-1} = a1a^{-1} = aa^{-1} = 1$. Similarly, $(b^{-1}a^{-1})(ab) = 1$. Thus, we see that $b^{-1}a^{-1}$ has the desired property.

DEFINITION 4: We say that a group G is *abelian* (or *commutative*) if $ab = ba$ for all $a, b \in G$.

If G is an abelian group, we will often denote the group operation by + instead of $\cdot$. The typical example of an abelian group is $\mathbf{Z}$, with respect to addition. We will meet many examples of groups which are not commutative in Section 3.2.

DEFINITION 5: If a group G contains only a finite number of elements, then we say that G is a *finite group*. If G is a finite group, then the number of elements in G is called the *order* of G and is denoted $|G|$.

It is very convenient to use the following power notation for elements of a group. Let G be a group, and let $a \in G, n \in \mathbf{Z}$. Let us define the power a^n of a. First, let us define a^n for $n \geq 0$. First, we define $a^0 = 1_G$. Next, assuming that a^n has been defined, let us set $a^{n+1} = a^n \cdot a$. By the principle of mathematical induction, this suffices to define a^n for all $n \geq 0$. Let us define a^n for $n < 0$ by $a^n = (a^{-1})^{-n}$. [Note that if $n < 0$, then $-n > 0$, so that we have already assigned a meaning to $(a^{-1})^{-n}$.] Then we have the usual laws of exponents:

$$a^n a^m = a^{n+m} \qquad (m, n \in \mathbf{Z}), \tag{1}$$

$$(a^n)^m = a^{nm} \qquad (m, n \in \mathbf{Z}). \tag{2}$$

If G is abelian, then

$$(ab)^n = a^n b^n. \tag{3}$$

Note that if G is nonabelian, then (3) need not hold. It is easy to verify (1)–(3) by induction, and we leave it to the reader to supply the details. [*Hint*: First verify (1)–(3) for m and n both positive.]

Let us make one last remark on our definition of a group. In many books the definition of a group includes an extra axiom, the axiom of *closure under multiplication*, which asserts that if x and y are in G, then $x \cdot y \in G$. This axiom is already built into our definition, since we have assumed that multiplication is a binary operation on G. However, in verifying that a given object is a group, one must not forget to verify that multiplication does, indeed, define a binary operation on G. This point is quietly buried in our definition and is easy to gloss over but is very important in studying subgroups of a group, which we will do in Section 3.3. To summarize: *In order for G to be a group, the product of any two elements of G must be an element of G.*

3.1 *Exercises*

1. Let G be a group and let $a_1, \ldots, a_n \in G$. Show that $(a_1 \ldots a_n)^{-1} = a_n^{-1} \ldots a_1^{-1}$.
2. Prove the laws of exponents (1), (2), and (3).
3. If G is a group such that $x^2 = 1$ for all $x \in G$, show that G is abelian.
4. Let i be a complex number having the property $i^2 = -1$. Show that $\{1, -1, i, -i\}$ is a group with respect to multiplication of complex numbers. For each element of this group, find its inverse.
5. Let G be a group, $x \in G$. Show that $(x^{-1})^{-1} = x$.
6. Let G be a group, and let $a, b, c \in G$. Prove that if $ac = bc$, then $a = b$.
7. Let G be a group. Define a relation $\sim$ on G as follows: We say that $g \sim h$ if and only if there exists $x \in G$ such that $xgx^{-1} = h$.
   (a) Show that $\sim$ is an equivalence relation on G. The equivalence classes of G with respect to $\sim$ are called the *conjugacy classes* of G, and if $g \sim h$, then we say that g is *conjugate* to h.
   (b) Show that every conjugate class of G consists of a single element if and only if G is abelian.

8. Let G be a nonempty set on which there is defined an associative binary operation $*$ which satisfies the following conditions: (1) There exists at least one element e of G such that $e * a = a$ for all $a \in G$. (2) For every e satisfying (1) and for every $a \in G$, there exists $b \in G$ such that $b * a = e$. Show that G is a group with respect to $*$.
9. Let G be a group. Show that G is abelian if and only if $(ab)^2 = a^2b^2$ for all a, b in G.

3.2 Examples of Groups

As we mentioned in the introduction to this chapter, a general theory is not of much value unless there are some interesting examples to which the theory applies. In this sense, the theory of groups is a superb theory! Let us now survey some of the many examples of groups which naturally present themselves in various branches of mathematics.

Examples from Arithmetic

Many of the arithmetic systems which we studied in Chapter 2 were groups, even though we did not call attention to this fact at the time.

EXAMPLE 1: $G = \mathbf{Z}$, with respect to the binary operation of addition. As we mentioned in Section 3.1, the identity element is 0 and the inverse of x is $-x$.

EXAMPLE 2: $G = \mathbf{Q}$, the rational numbers, with respect to the binary operation of addition. The identity element is 0 and the inverse of a/b is $(-a)/b$.

EXAMPLE 3: $G = \mathbf{Q} - \{0\}$, with respect to the binary operation of multiplication of rational numbers. Since the product two nonzero rational numbers is a nonzero rational number, multiplication does, indeed, define a binary operation on $\mathbf{Q} - \{0\}$. The identity element is the rational number 1, and the inverse of the rational number $a/b \neq 0$ is b/a, since $(a/b)(b/a) = 1$.

EXAMPLE 4: $G = \mathbf{N}$, with respect to the binary operation of addition. This is not a group. Axioms G1 and G2 hold, but axiom G3 does not. For example, 1 has no inverse with respect to addition in $\mathbf{N}$.

EXAMPLE 5: $G = \mathbf{Z}_n$ with respect to the operation of addition of residue classes. The identity element is $\bar{0}$, since $\bar{x} + \bar{0} = \overline{x + 0} = \bar{x}$ and similarly $\bar{0} + \bar{x} = \bar{x}$ for all $\bar{x} \in \mathbf{Z}_n$. The inverse of $\bar{x}$ is $\overline{-x}$, since $\bar{x} + \overline{-x} = \overline{x + (-x)} = \bar{0}$ and similarly $\overline{-x} + \bar{x} = 0$. Note that $\mathbf{Z}_n$ is finite abelian group of order n.

EXAMPLE 6: $G = \mathbf{Z}_n^\times$ with respect to the operation of multiplication of residue classes. We saw in Section 2.5 that multiplication of residue classes is associative. Moreover, Proposition 3 of Section 2.5 implies that multiplication of residue classes is actually a binary operation on $\mathbf{Z}_n^\times$—that is, the product of two elements of $\mathbf{Z}_n^\times$ is again an element of $\mathbf{Z}_n^\times$. The identity element is $\bar{1}$. If $\bar{a} \in \mathbf{Z}_n^\times$, then $\bar{a}$ has an inverse with respect to multiplication by Proposition 4 of Section 2.5. Therefore, G is a group. Moreover, by the discussion of Section 2.5, $\mathbf{Z}_n^\times$ is a finite abelian group of order $\phi(n)$, where $\phi(n)$ denotes Euler's function.

The Symmetric Groups

In Section 1.1 we introduced the symmetric group S_n in connection with the solution of equations in radicals. Recall that S_n consists of all permutations

$$\sigma = \begin{pmatrix} 1 & 2 \dots n \\ i_1 & i_2 \dots i_n \end{pmatrix},$$

where $\{i_1, i_2, \dots, i_n\}$ is some rearrangement of $\{1, 2, \dots, n\}$. It is clear that a permutation is just a function σ from $\{1, 2, \dots, n\}$ into $\{1, 2, \dots, n\}$—that function defined by

$$\begin{gathered} \sigma(1) = i_1, \\ \sigma(2) = i_2, \\ \vdots \\ \sigma(n) = i_n. \end{gathered}$$

Moreover, since $\{i_1, \dots, i_n\}$ is a rearrangement of $\{1, 2, \dots, n\}$, it is clear that σ is one to one and onto—that is, σ is a bijection. Conversely, every bijection of $\{1, 2, \dots, n\}$ onto itself defines a permutation. This gives us a clue how to define the product of two permutations σ and η. The product $\sigma\eta$ must be a bijection of $\{1, 2, \dots, n\}$ onto itself. Let it be the function defined by composition:

$$\sigma\eta(i) = \sigma(\eta(i)). \tag{1}$$

Thus, the product $\sigma\eta$ is the permutation gotten by first performing the permutation η and following it by the permutation σ. We will leave to the reader the proof of the fact that (1) actually defines a bijection of $\{1, 2, \dots, n\}$ onto itself. It is easy to compute products of permutations. For example, let us compute the product of $\sigma = \begin{pmatrix} 1 & 2 & 3 \\ 2 & 3 & 1 \end{pmatrix}$, and $\eta = \begin{pmatrix} 1 & 2 & 3 \\ 3 & 2 & 1 \end{pmatrix}$. First, σ maps 1 into 3 and η maps 3 into 1. Therefore, $\eta\sigma$ maps 1 into 1. Since σ maps 2 into 2 and

and η maps 2 into 3, the product $\eta\sigma$ maps 2 into 3. Similarly, $\eta\sigma$ maps 3 into 2. Thus $\eta\sigma = \begin{pmatrix} 1 & 2 & 3 \\ 1 & 3 & 2 \end{pmatrix}$.

Since the composition of functions is associative [see Equation (1) of Section 2.2], we see that multiplication of permutations is associative. Moreover, the permutation $\begin{pmatrix} 1 & 2 & \dots & n \\ 1 & 2 & \dots & n \end{pmatrix}$ is an identity with respect to multiplication of permutations. Finally, if $\sigma = \begin{pmatrix} 1 & 2 & \dots & n \\ i_1 & i_2 & \dots & i_n \end{pmatrix}$, then σ has an inverse with respect to multiplication. Indeed, if we set $\eta = \begin{pmatrix} i_1 & i_2 & \dots & i_n \\ 1 & 2 & \dots & n \end{pmatrix}$, then it is reasonably easy to see that

$$\sigma\eta = \eta\sigma = \begin{pmatrix} 1 & 2 & \dots & n \\ 1 & 2 & \dots & n \end{pmatrix}$$

so that η is an inverse for σ. Thus, we have proved that S_n is a group. Note, however, that S_n is not usually abelian. For example,

$$\begin{pmatrix} 1 & 2 & 3 \\ 3 & 2 & 1 \end{pmatrix}\begin{pmatrix} 1 & 2 & 3 \\ 2 & 3 & 1 \end{pmatrix} = \begin{pmatrix} 1 & 2 & 3 \\ 2 & 1 & 3 \end{pmatrix},$$
$$\begin{pmatrix} 1 & 2 & 3 \\ 2 & 3 & 1 \end{pmatrix}\begin{pmatrix} 1 & 2 & 3 \\ 3 & 2 & 1 \end{pmatrix} = \begin{pmatrix} 1 & 2 & 3 \\ 1 & 3 & 2 \end{pmatrix}.$$

By induction, it is easy to see that the order of S_n equals $1 \cdot 2 \cdot 3 \cdot \cdots \cdot n = n!$ (read "n factorial"). Also, it is easy to see that the inverse of $\sigma \in S_n$ is just the inverse of σ considered as a bijection from $\{1, 2, \dots n\}$ onto itself.

The last comment suggests a generalization of the symmetric group S_n. Let A be an arbitrary set, and let S_A denote the set of bijections of A onto itself. Then S_A can be made into a group by defining multiplication in S_A via $(gf)(x) = g(f(x))$, for $f, g \in S_A$, $x \in A$. The identity element is the identity function on A, and the inverse of the bijection f is the bijection f^{-1}. S_A is called the *symmetric group on the set* A, and the elements of S_A are called *permutations* of A. If $A = \{1, 2, \dots, n\}$, then $S_A = S_n$.

Examples from Geometry

EXAMPLE 7: Let

$$\mathbf{R}^2 = \mathbf{R} \times \mathbf{R} = \{(x, y) \mid x, y \in R\},$$
$$\mathbf{R}^3 = \mathbf{R} \times \mathbf{R} \times \mathbf{R} = \{(x, y, z) \mid x, y, z \in \mathbf{R}\}.$$

Then $\mathbf{R}^2$ is called the *Euclidean plane* and $\mathbf{R}^3$ is called *Euclidean* 3 — *space.* The sets $\mathbf{R}^2$ and $\mathbf{R}^3$ may be visualized as the usual plane and three-dimensional spaces of analytic geometry, respectively. The groups which we consider will consist of various motions of $\mathbf{R}^2$ and $\mathbf{R}^3$. For the moment, let us confine ourselves to $\mathbf{R}^2$.

A *motion* of $\mathbf{R}^2$ is a bijection of $\mathbf{R}^2$ onto itself; that is, a motion of $\mathbf{R}^2$ is just a permutation of $\mathbf{R}^2$. A motion of $\mathbf{R}^2$ is what was classically called a "transformation." From our study of the symmetric groups, we know that the set of all motions of $\mathbf{R}^2$ forms a group under the operation of composition of functions. This group will be denoted $M(\mathbf{R}^2)$. Most motions of $\mathbf{R}^2$ are uninteresting. However, many geometric situations give rise to interesting collections of motions which form groups under the operation of multiplication of permutations. It is useful to think of a motion A of $\mathbf{R}^2$ as moving the point (x, y) to the point $A((x,y))$. The product $A \cdot B$ of the two motions A and B moves the point (x, y) to the point $A(B((x,y)))$—that is, taking the product $A \cdot B$ amounts to first performing the motion B, then the motion A.

EXAMPLE 8: A *linear motion* of $\mathbf{R}^2$ is one which moves the point (x, y) to the point (x', y'), where

$$\begin{aligned} x' &= ax + by, \\ y' &= cx + dy, \end{aligned} \tag{2}$$

and $ad - bc \neq 0$. The last condition guarantees that for every (x', y') there exists a unique (x, y) which satisfies the system of equations (2). For if (x, y) satisfies (2) for a given (x', y'), then by multiplying the first equation by d and the second by b, and then subtracting the second from the first, we see that $dx' - by' = (ad - bc)x$, and thus, since $ad - bc \neq 0$, we have

$$x = (d/(ad - bc))x' + (-b/(ad - bc)y'. \tag{3}$$

Similarly,

$$y = (-c/(ad - bc)x') + (a/(ad - bc))y'. \tag{3'}$$

If A' denotes the linear motion which maps (x', y') into (x'', y''), where

$$\begin{aligned} x'' &= a'x' + b'y', \\ y'' &= c'x' + d'y', \end{aligned}$$

then $A' \cdot A$ maps (x, y) into (x'', y''), where

$$\begin{aligned} x'' &= (a'a + b'c)x + (a'b + b'd)y, \\ y'' &= (c'a + d'c)x + (c'b + d'd)y. \end{aligned}$$

Moreover,

$$(a'a + b'c)(c'b + d'd) - (c'a + dc')(a'b + b'd) = (ad - bc)(a'\,c' - b'c') \neq 0.$$

Therefore, $A' \cdot A$ is a linear motion. The identity motion is clearly a linear motion and the inverse of the linear motion A is described by formulas (3) and (3′). (Check it!) Thus, the set of linear motions of $\mathbf{R}^2$ forms a group, called the two-dimensional *general linear group*, denoted $GL(2,\mathbf{R})$.

EXAMPLE 9: A *translation* of $\mathbf{R}^2$ is a motion which moves the point (x, y) to the point (x', y'), where

$$T: \begin{array}{l} x' = x + a, \\ y' = y + b. \end{array} \tag{4}$$

We can visualize a translation as moving the whole plane along the vector whose initial point is the origin and whose terminal point is (a, b) (Figure 3.1). The product $T_1 \cdot T_2$ of the two translations

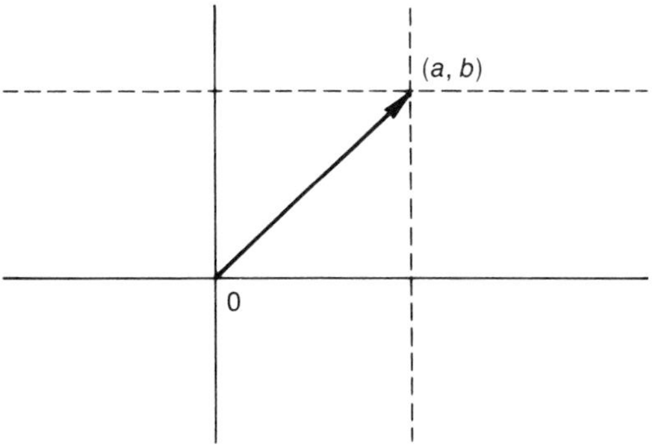

FIGURE 3-1: A Translation in $\mathbf{R}^2$.

$$T_1: \begin{array}{l} x' = x + a_1, \\ y' = y + b_1, \end{array} \qquad T_2: \begin{array}{l} x' = x + a_2, \\ y' = y + b_2 \end{array}$$

is the translation

$$T_3: \begin{array}{l} x' = x + a_1 + a_2, \\ y' = y + b_1 + b_2. \end{array}$$

The identity motion is clearly a translation and the inverse of the translation

$$T: \begin{array}{l} x' = x + a_1, \\ y' = y + b_1 \end{array}$$

is the translation

$$T^{-1}: \begin{array}{l} x' = x - a_1, \\ y' = y - b_1 \end{array}$$

The set of translations of $\mathbf{R}^2$ forms a group, denoted $T(\mathbf{R}^2)$. The translation (4) will be denoted by $T_{a,b}$.

EXAMPLE 10: The motion of $\mathbf{R}^2$ obtained by following a linear motion M by a translation $T_{a,b}$ is called an *affine motion* of $\mathbf{R}^2$. If M corresponds to

$$\begin{pmatrix} \alpha & \beta \\ \gamma & \delta \end{pmatrix},$$

then the affine motion $T_{a,b} \cdot M$ moves the point (x, y) to the point (x', y'), where

$$x' = \alpha x + \beta y + a,$$
$$y' = \gamma x + \delta y + b.$$

The set of affine motions of $\mathbf{R}^2$ forms a group, denoted $A(\mathbf{R}^2)$. We will leave the verification of the group axioms in this case as an exercise.

Thus far, all our examples of groups of motions have been infinite. But many interesting finite groups occur as groups of motions. Let us give some examples.

EXAMPLE 11: A *rigid motion* of $\mathbf{R}^2$ is a motion of $\mathbf{R}^2$ which preserves distances between points. [Recall from analytic geometry that the distance between the points (x, y) and $(x' y')$ is given by

$$\{(x - x')^2 + (y - y')^2\}^{1/2}.$$

An example of a rigid motion of $\mathbf{R}^2$ is the motion which rotates the plane through an angle θ. This motion will be denoted R_θ. Another example is a translation $T_{a,b}$.

Let $S \subseteq \mathbf{R}^2$ be any subset. A *symmetry* of S is a rigid motion which maps S onto itself. At first this might seem like a strange definition. But it really does coincide with our intuitive concept of symmetry! Let us consider an example. Let S be a square, centered at the origin, as in Figure 3.2.

The identity motion I is clearly a symmetry of the square. Let R be the motion of rotation counterclockwise through an angle $\pi/2$. The square of Figure 3.2 is mapped onto itself and is now situated as shown in Figure 3.3(a). Therefore, R is a symmetry of the square. Similarly, $R \cdot R = R^2$ and $R \cdot R \cdot R = R^3$ are symmetries. (R^2 is just counterclockwise rotation

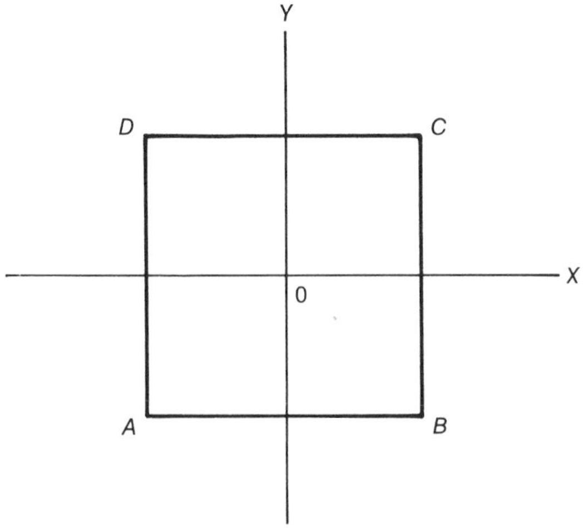

FIGURE 3-2: A Square with Labelled Vertices.

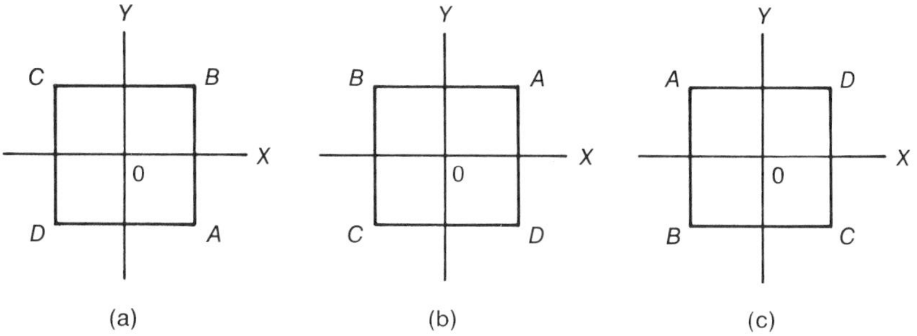

FIGURE 3-3: Symmetries of a Square.

through an angle of π, while R^3 is just counterclockwise rotation through an angle of $3\pi/2$.) Note, however, that R^4 brings the square back to its original position, since R^4 is a counterclockwise rotation through 2π. Therefore, $R^4 = I$.

Another symmetry of the square can be gotten by rotating the square around the X-axis, resulting in the configuration of Figure 3.4. Let us call this symmetry F (for "flip"). Then it is clear that upon repeating F twice, the square is returned to its original position. Therefore, $F^2 = I$. Moreover, a trivial calculation shows that $R \cdot F = F \cdot R^3$.

By combining flips and rotations, we get symmetries of the form

$$I, F, R, R^2, R^3, F \cdot R, F \cdot R^2, F \cdot R^3. \tag{5}$$

It might seem that this list should go on to include many other symmetries, say $F \cdot R^3 \cdot F$. However, since $R \cdot F = FR^3$, $R^4 = I$, $F^2 = I$, we see that

$$F \cdot R^3 \cdot F = R \cdot F \cdot F = R.$$

With a little patience, one can show that the product of any two of the elements (5) is another element of the form (5), and that the elements of (5) form a group. We have assembled the multiplication for the group in Table 3.1. If, in this table, we look up element a on the row index and b on the

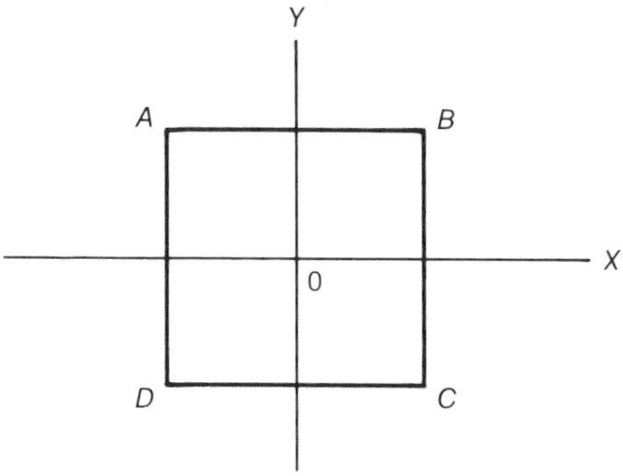

FIGURE 3-4: The Symmetry F.

TABLE 3.1
Multiplication Table for the Group of Symmetries of the Square

	I	R	R^2	R^3	F	FR	FR^2	FR^3
I	I	R	R^2	R^3	F	FR	FR^2	FR^3
R	R	R^2	R^3	I	FR^3	F	FR	FR^2
R^2	R^2	R^3	I	R	FR^2	FR^3	F	FR
R^3	R^3	I	R	R^2	FR	FR^2	FR^3	F
F	F	FR	FR^2	FR^3	I	R	R^2	R^3
FR	FR	FR^2	FR^3	F	R^3	I	R	R^2
FR^2	FR^2	FR^3	F	FR	R^2	R^3	I	R
FR^3	FR^3	F	FR	FR^2	R	R^2	R^3	I

column index, the tabular entry gives the value of $a \cdot b$. The reader is strongly encouraged to verify the entries in the table, using the same reasoning as used above. [The table can easily be used to check that every element has an inverse. Thus, once the table is derived, it is easy to verify that the elements (5) form a group.†]

The group which we have constructed is the *group of symmetries of the square*. It is a group of order 8. There are symmetries of the square which we have not yet mentioned. For example, the symmetry gotten by rotating the square about the Y-axis; or about the line $y = x$; or about the line $y = -x$. The reader should have no difficulty in verifying that these symmetries can be expressed in terms of products of powers of R and F, and are therefore in the group which we have constructed. *Note that the group of symmetries of the square is nonabelian.* Indeed, we have

$$R \cdot F = F \cdot R^{-1} = F \cdot R^3 \neq F \cdot R.$$

EXAMPLE 12: Let us generalize the preceding example. Let $n > 2$ be a positive integer, and let P_n be a regular polygon of n sides with one of its vertices on the X-axis (Figure 3.5). (This last requirement is unnecessary but will make our exposition easier.) We will construct the group of symmetries of P_n. Two obvious symmetries are suggested by Example 11. Let R denote a counterclockwise rotation through an angle $2\pi/n$, and let F denote a rotation about the X-axis. It is then simple to check that R and F satisfy the following relations:

$$R^n = I, \qquad F^2 = I,$$
$$R \cdot F \cdot R = F,$$

where I denotes the identity symmetry. As in Example 11, we can show that the product of any of the two elements

$$I, R, R^2, \ldots, R^{n-1}, F, F \cdot R, F \cdot R^2, \ldots, F \cdot R^{n-1}$$

is again an element of the same form. Of course, in this example, things are

† In this case, the associativity is guaranteed, since permutations are functions, the group operation is composition of functions, and composition of functions is associative.

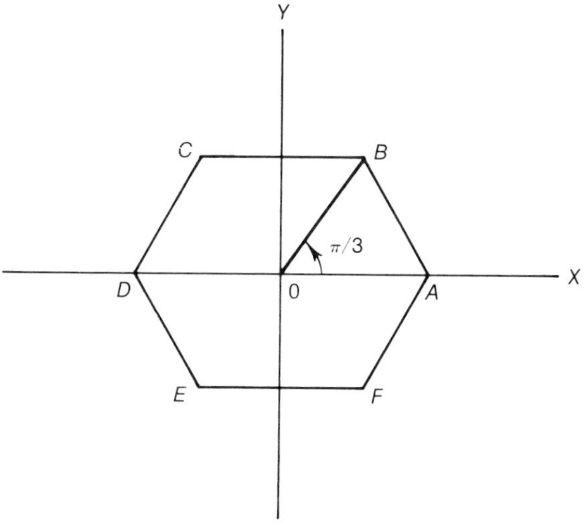

FIGURE 3-5: P_6.

slightly more complicated, so let us outline a more efficient analysis than given in Example 11. First note that since $R^{n-1} \cdot R = R^n = I$, we have $R^{-1} = R^{n-1}$, where we have used the uniqueness of the inverse. Therefore, we see that the relation $R \cdot F \cdot R = F$ is equivalent to $R \cdot F = F \cdot R^{-1} = F \cdot R^{n-1}$. Therefore, if a is any positive integer,

$$R^a \cdot F = F \cdot R^{a(n-1)}.$$

Let us now consider the product of two elements $F^a \cdot R^b$ and $F^c \cdot R^d$, where a, b, c, and d are positive integers. Without loss of generality, we may assume that

$$0 \leq a, \quad c \leq 1, \qquad 0 \leq b, \quad d \leq n-1.$$

Then if $c = 0$,

$$F^a \cdot R^b \cdot F^c \cdot R^d = F^{a+c} \cdot R^{b+d}.$$

On the other hand, if $c = 1$,

$$\begin{aligned} F^a \cdot R^b \cdot F^c \cdot R^d = F^a \cdot (R^b \cdot F) \cdot R^d &= F^a \cdot F \cdot R^{b(n-1)+d} \\ &= F^{a+c} \cdot R^{d-b}. \end{aligned}$$

Therefore, we have proved closure under multiplication. We leave the verification of the remaining group axioms to the reader. The group which is defined in the preceding discussion is called the *dihedral group* and is denoted D_n. In particular, if $n = 4$, D_4 is the group of symmetries of the square. Note that D_n always contains $2n$ elements.

EXAMPLE 13: We can speak of symmetries of subsets of $\mathbf{R}^3$ by taking over piecemeal the definitions for $\mathbf{R}^2$. The three-dimensional analogues of the dihedral groups are gotten by looking at the symmetries of the regular solids. In high school geometry it is proved that there are only five regular solids—the tetrahedron, the cube, the octahedron, the dodecahedron, and the icosahedron. Their groups of symmetries are very complicated.

Examples from Analysis

EXAMPLE 14: Let $\mathfrak{F}$ denote the set of all functions f: $\mathbf{R} \longrightarrow \mathbf{R}$. Define the sum of two functions $f, g \in \mathfrak{F}$ to be the function $f + g$ defined by

$$(f + g)(x) = f(x) + g(x) \qquad (x \in \mathbf{R}).$$

With respect to the operation $+$, $\mathfrak{F}$ becomes an abelian group. The identity element is the zero function $\mathbf{0}$ defined by

$$\mathbf{0}(x) = 0 \qquad (x \in \mathbf{R}).$$

The inverse of the function f is the function $-f$ defined by

$$(-f)(x) = -f(x) \qquad (x \in \mathbf{R}).$$

EXAMPLE 15: Let $\mathfrak{C} \subseteq \mathfrak{F}$ denote the set of all continuous functions. Since the sum of continuous functions is a continuous function, we see that if $f, g \in \mathfrak{C}$, then $f + g \in \mathfrak{C}$. Therefore, the operation $+$ defined in Example 14 also defines a binary operation on $\mathfrak{C}$. With respect to this binary operation, $\mathfrak{C}$ becomes an abelian group.

Miscellaneous Examples

EXAMPLE 16: We may put a group structure on $\mathbf{R}^2$ as follows. Define a binary operation called addition on $\mathbf{R}^2$ setting $(a, b) + (c, d) = (a + c, b + d)$, where $a + c$, $b + d$ denote the sums of a, c and b, d as real numbers. The identity element of $\mathbf{R}^2$ with respect to this binary operation is $(0, 0)$, and the inverse of (a, b) is $(-a, -b)$. The reader should have no difficulty in verifying these assertions.

EXAMPLE 17: This example provides us with a method for manufacturing new groups from given ones. Let G and G' be groups, and let $G \times G'$ denote the Cartesian product of G and G'. Let us define a binary operation on $G \times G'$ by setting $(g, g')\cdot(h, h') = (gh, g'h')$. Then it is easy to see that this binary operation is associative. An identity element is $(1_G, 1_{G'})$, and the inverse of (g, g') is just (g^{-1}, g'^{-1}). The group $G \times G'$ is called the *direct product* of G and G'. If $G = G' = \mathbf{R}$, then $G \times G'$ is just the group of Example 16.

3.2 Exercises

1. Compute the following products of elements of D_4:
   (a) R^7. (b) R^3FR^{12} (c) F^{17}.
   (d) $R^{-1}FR^{-1}$. (e) RFR^3F^{-1}.

2. Compute the following products of elements of S_3:
   (a) $\begin{pmatrix} 1 & 2 & 3 \\ 3 & 2 & 1 \end{pmatrix}\begin{pmatrix} 1 & 2 & 3 \\ 2 & 3 & 1 \end{pmatrix}$.
   (b) $\begin{pmatrix} 1 & 2 & 3 \\ 2 & 3 & 1 \end{pmatrix}\begin{pmatrix} 1 & 2 & 3 \\ 1 & 3 & 2 \end{pmatrix}$.
   (c) $\begin{pmatrix} 1 & 2 & 3 \\ 1 & 2 & 3 \end{pmatrix}\begin{pmatrix} 1 & 2 & 3 \\ 1 & 3 & 2 \end{pmatrix}\begin{pmatrix} 1 & 2 & 3 \\ 3 & 2 & 1 \end{pmatrix}$.
3. Verify the entries in Table 3.1.
4. Show that GL(2, **R**) is a group.
5. Show that $A(\mathbf{R}^2)$ is a group.
6. Show that $T(\mathbf{R}^2)$ is a group.
7. Show that S_n has order $1 \cdot 2 \cdot \ldots n$.
8. For what values of n is S_n abelian?
9. Let G be a group containing an even number of elements. Show that there exists $x \in G$, $x \neq 1$ such that $x = x^{-1}$.
10. Let G be a group and let $a, b \in G$. Suppose that a, b and ab are all elements whose squares are 1. Show that $ab = ba$.
11. A group G of motions of $\mathbf{R}^2$ is said to be *transitive* if any point of $\mathbf{R}^2$ can be mapped into any other point of $\mathbf{R}^2$ by an element of G.
    (a) Show that GL(2, **R**) is not transitive.
    (b) Show that $T(\mathbf{R}^2)$ is transitive.
    (c) Is D_{2n} transitive?
    (d) If G is finite, can G be transitive?
12. Write down the multiplication table for the group of symmetries of a regular tetrahedron.
13. Write down the multiplication table for the group of symmetries of a cube.
14. Write down a multiplication table for the group of symmetries of a parallelogram which is not regular and which is not rectangular.
15. The discussion of linear motions in the plane suggests introducing square arrays of real numbers of the form $\begin{pmatrix} a & b \\ c & d \end{pmatrix}$ $(a,b,c,d \in \mathbf{R})$. Such a square array is called a 2×2 *matrix with real coefficients*. Define addition and multiplication of such matrices by
$$\begin{pmatrix} a & b \\ c & d \end{pmatrix} + \begin{pmatrix} a' & b' \\ c' & d' \end{pmatrix} = \begin{pmatrix} a + a' & b + b' \\ c + c' & d + d' \end{pmatrix}$$
$$\begin{pmatrix} a & b \\ c & d \end{pmatrix}\begin{pmatrix} a' & b' \\ c' & d' \end{pmatrix} = \begin{pmatrix} aa' + bc' & ab' + bd' \\ ca' + dc' & cb' + dd' \end{pmatrix}$$
the latter being motivated by the formula for the product of two linear motions.
    (a) Show that the set of 2×2 matrices forms an abelian group under addition.
    (b) Define the *determinant* of a 2×2 matrix to be the real number $ad - bc$. Show that if A and B are 2×2 matrices, then the determinant of $A \cdot B$ is the product of the determinants of A and of B.

(c) Show that the set of all 2×2 matrices and nonzero determinant form a nonabelian group under multiplication.

3.3 Subgroups

Throughout this section let G denote a group, $H \subseteq G$ a nonempty subset. Our goal is to study the properties of those subsets H which are groups.

DEFINITION 1: We say that H is a *subgroup* of G if H is a group with respect to the multiplication $\cdot$ of G.

Since $\cdot$ is already associative with respect to elements of G, it is automatically associative for elements of H. Therefore, from the group axioms, we deduce that H is a subgroup of G if and only if H has the following properties:

SG1. H is closed under $\cdot$; that is, for $x, y \in H$, we have $x \cdot y \in H$.
SG2. $1_G \in H$.
SG3. If $x \in H$, then $x^{-1} \in H$.

These three properties can be combined into a simple test for H to be a subgroup:

PROPOSITION 2: Let $H \subseteq G$, $H \neq \varnothing$. Then H is a subgroup of G if and only if for $a, b \in H$, we have $a \cdot b^{-1} \in H$.

Proof: There are two statements to prove. First assume that H is a subgroup of G and that $a, b \in H$. Let us show that $a \cdot b^{-1} \in H$. By (SG3), $b^{-1} \in H$, and therefore by (SG1), $a \cdot b^{-1} \in H$.

Conversely, assume that $H \subseteq G$ and that for $a,b \in H$, we have $a \cdot b^{-1} \in H$. It suffices to show that the properties (SG1)–(SG3) are satisfied. Since $H \neq \phi$, there exists $a \in H$. Therefore, $1_G = a \cdot a^{-1} \in H$ and thus (SG2) holds. Moreover, if $b \in H$, $b^{-1} = 1_G \cdot b^{-1} \in H$. Therefore, (SG3) holds. Finally, if $a, b \in H$, then $b^{-1} \in H$ by (SG3), which we have just proved. Note that $(b^{-1})^{-1} = b$ since $b^{-1} \cdot b = b \cdot b^{-1} = 1_G$. Therefore, $a \cdot b = a \cdot (b^{-1})^{-1} \in H$ and (SG1) holds.

EXAMPLE 1: Let G denote the group of symmetries of the square. Two subgroups of G are given by

$$H_1 = \{I, R, R^2, R^3\},$$
$$H_2 = \{I, F\}.$$

EXAMPLE 2: More generally, the dihedral group D_n contains the two subgroups

$$H_1 = \{I, R, R^2, \ldots, R^{n-1}\},$$
$$H_2 = \{I, F\}.$$

EXAMPLE 3: Let $k \in \mathbf{Z}$, and let

$$k\mathbf{Z} = \{k \cdot r \mid r \in \mathbf{Z}\}.$$

Then $k\mathbf{Z}$ is a subgroup of $\mathbf{Z}$, considered as a group under addition.

EXAMPLE 4: The group $\mathcal{C}$ of continuous functions on $\mathbf{R}$ is a subgroup of the group $\mathfrak{F}$ of all functions on $\mathbf{R}$.

EXAMPLE 5: Let f be the polynomial $(x^2 - 2)(x^2 - 3)$ with roots $\pm\sqrt{2}$, $\pm\sqrt{3}$. The set of permutations of the roots forms a group (essentially S_4). The four permutations of this group defined by $\sqrt{2} \to \pm\sqrt{2}$, $-\sqrt{2} \to \mp\sqrt{2}$, $\sqrt{3} \to \pm\sqrt{3}$, and $-\sqrt{3} \to \mp\sqrt{3}$ form a subgroup.

THEOREM 3: If $\mathcal{S}$ is any nonempty collection of subgroups of the group G, then the intersection

$$H^* = \bigcap_{H \in \mathcal{S}} H$$

is a subgroup of G.

Proof: First observe that $H^* \neq \varnothing$ since $1_G \in H$ for all $H \in \mathcal{S}$, so that $1_G \in H^*$. Let us apply Proposition 2. Let $a, b \in H^*$. Then, for every $H \in \mathcal{S}$, we have $a, b \in H$. Thus, since H is a subgroup of G, $a \cdot b^{-1} \in H$. Therefore, $a \cdot b^{-1} \in H^*$ and H^* is a subgroup of G.

We will now consider a general technique for manufacturing subgroups of a group G. Let S be any subset of G. Let $\mathcal{S}$ be the collection of all subgroups containing S. $\mathcal{S}$ is nonempty because $G \in \mathcal{S}$. By Theorem 3

$$[S] = \bigcap_{H \in \mathcal{S}} H$$

is a subgroup of G and by the definition of $\mathcal{S}$, $[S]$ contains S. *Actually* $[S]$ *is the smallest subgroup of* G *which contains* S. For if H is a subgroup of G containing S, then $H \in \mathcal{S}$, so that $[S] \subseteq H$.

DEFINITION 4: $[S]$ is called the subgroup of G *generated* by S. If $[S] = G$, then S is said to be a *set of generators* for G.

We can give another description of $[S]$. It is clear that $[S]$ contains all products of the form

$$a_1 \cdot a_2 \cdot a_3 \cdot \ldots a_n, \tag{1}$$

where either $a_i \in S$ or $a_i^{-1} \in S$ $(1 \leq i \leq n)$. But the set of all these products forms a subgroup H of G. And this subgroup contains S. Therefore, $[S] \subseteq H$.

But we also have $H \subseteq [S]$. Therefore, $H = [S]$. In other words, $[S]$ is the subgroup of G consisting of all products of the form (1). For example, D_n is generated by $S = \{R, F\}$.

DEFINITION 5: A group G is said to be *cyclic* if G is generated by a single element. That is, G is cyclic if there exists $a \in G$ such that $G = [S]$, where $S = \{a\}$.

From our above characterization of $[S]$ we see that G is cyclic if and only if there exists $a \in G$ such that

$$G = \{\ldots, a^{-2}, a^{-1}, 1_G, a, a^2, \ldots\}.$$

The element a is not usually unique. Moreover, the various powers of a need not all be distinct. We will study this possibility more closely below.

Cyclic groups are the simplest groups. Later we will prove the fundamental theorem of abelian groups, which states that the cyclic groups are the building blocks from which abelian groups with a finite number of generators can be built.

We have already met several examples of cyclic groups.

EXAMPLE 6: $G = \mathbf{Z}$ is a cyclic group. As generator we may take 1.

EXAMPLE 7: $G = \mathbf{Z}_n$ is a finite cyclic group of order n. As generator we may take $\bar{1}$.

EXAMPLE 8: Let G be an arbitrary group, $a \in G$. Set $S = \{a\}$. Then $[S]$ is a cyclic subgroup of G. We will usually denote this subgroup by $[a]$.

DEFINITION 6: Let G be a group and let $a \in G$. If $[a]$ is finite, then we define the *order* of a to be the number of elements in $[a]$. If $[a]$ is infinite, we define the order of a to be infinity.

PROPOSITION 7: Suppose that a has finite order r. Then r is the smallest positive integer such that $a^r = 1_G$, and $[a] = \{1_G, a, \ldots, a^{r-1}\}$.

Proof: Since $[a]$ is finite, not all the powers a^m ($m \in \mathbf{N}$) are distinct. Suppose that $a^m = a^n$ ($m, n \geq 0$, $m > n$). Then $a^{m-n} = 1_G$ and $m - n > 0$. Therefore, there exist positive integers r such that $a^r = 1_G$. Let r be the smallest such. We will prove that $[a] = \{1_G = a^0, a, a^2, \ldots, a^{r-1}\}$, where the powers of a mentioned are all distinct. Indeed, if $a^i = a^j$ ($0 \leq i < j \leq r - 1$), then $a^{j-i} = 1_G$ and $0 < j - i < r$, which is a contradiction to the definition of r. Therefore, the powers a^j ($0 \leq j \leq r - 1$) are all distinct. Let $\alpha \in [a]$. In order to prove our assertion, it suffices to show that $\alpha = a^j$ for some j ($0 \leq j \leq r - 1$). But $\alpha = a^n$ for some $n \in \mathbf{Z}$. Assume for the moment that $n \geq 0$. By the division algorithm, there exist $p, q \in \mathbf{N}$ such that

$$n = pr + q, \qquad 0 \leq q \leq r - 1.$$

Then

$$\alpha = a^n = a^{pr+q} = (a^r)^p a^q$$
$$= a^q \qquad (\text{since } a^r = 1).$$

Thus, we are done in case $n \geq 0$. If $n < 0$, then $-n > 0$, so that by what we have already proved, we see that $\alpha^{-1} = a^{-n} = a^j$ for some j $(0 \leq j \leq r - 1)$. If $j = 0$, then we may take $n = 0$, in which case the assertion has already been proved. Thus, we may assume that $j > 0$. Then

$$\alpha = a^{-j} = a^r a^{-j} = a^{r-j}$$

and $0 \leq r - j \leq r - 1$. Thus, the assertion is proved for $n < 0$.

PROPOSITION 8: Suppose that a has infinite order. Then $a^r \neq 1_G$ for all $r \in \mathbf{Z} - \{0\}$, and $[a] = \{\ldots, a^{-2}, a^{-1}, 1_G, a, a^2, \ldots\}$, where all the powers of a are distinct.

Proof: It suffices to consider the case $r > 0$, for if $a^r = 1_G$, then $a^{-r} = 1_G$. By exactly the same reasoning as was used to prove Proposition 7, we see that if, for some $r > 0$, we have $a^r = 1_G$, then

$$[a] = \{1_G = a^0, a^1, \ldots, a^{r-1}\}.$$

(The powers of a may not all be distinct, however.) But this contradicts the fact that $[a]$ is infinite. Moreover, if $a^i = a^j$, then $a^{i-j} = 1_G$, so that $i - j = 0$ by what we just proved. Thus, all powers of a are distinct.

Combining Propositions 7 and 8 we can say something about cyclic groups. For let G be a cyclic group with generator a. There are two cases to consider:

Case 1: G is finite. Suppose that G has order r. Then, by Proposition 7,

$$G = \{1_G = a^0, a^1, a^2, \ldots, a^{r-1}\},$$

and $a^r = 1_G$.

Case 2: G is infinite. In this case,

$$G = \{\ldots, a^{-2}, a^{-1}, a^0 = 1_G, a^1, a^2, \ldots\},$$

where all powers of a are distinct.

Note that in case 1, G closely resembles $\mathbf{Z}_r$, while in case 2, G closely resembles $\mathbf{Z}$. In fact, in Section 3.4 we will show that for purposes of group theory, any cyclic group is either $\mathbf{Z}$ or $\mathbf{Z}_r$.

Let us now return to the consideration of the subgroups of an arbitrary group G. One very important, but very difficult, problem is to determine all subgroups of G. When G is finite, this problem can be solved by brute force. For we can list all the subsets of G and then test each subset to determine whether it is a subgroup. But this solution to the problem is not very practical. (If any reader doubts this assertion, let him try to determine all subgroups of S_5 by exhaustion!) Therefore, we must look for some simple criteria which can rule out most subsets of G from being subgroups. The simplest criterion is due to Lagrange and is one of the most fundamental theorems in the theory of groups.

THEOREM 9 (Lagrange): Let G be a finite group, H a subgroup of G. Then the order of H divides the order of G.

Before proving Lagrange's theorem, it will be necessary for us to build up some machinery. This machinery may seem very ad hoc at this time, but the ideas we develop here will be quite natural and will be extremely important later in this chapter when we define quotient groups.

Let $a \in G$ and set

$$Ha = \{ha \mid h \in H\}.$$

The set Ha is called a *right coset of G with respect to H*. We could just as easily consider left cosets of the form

$$aH = \{ah \mid h \in H\}.$$

But throughout this argument, we will stick to right cosets. We will denote the set of all right cosets of G with respect to H by $H \backslash G$. The set of all left cosets will be denoted G/H.

LEMMA 10: All right cosets contain $|H|$ elements.

Proof: It is clear that Ha contains at most $|H|$ elements. In order to prove that the number is exactly $|H|$, it suffices to show that all the elements $h \cdot a$ ($h \in H$) are different. But if

$$h_1 \cdot a = h_2 \cdot a \qquad (h_1, h_2 \in H),$$

we have $h_1 = h_2$ on multiplying on the right by a^{-1}.

LEMMA 11: If H$a \cap Hb \neq \varnothing$, then $Ha = Hb$.

Proof: Let $g \in Ha \cap Hb$. Then

$$g = h_1 \cdot a = h_2 \cdot b \qquad (h_1, h_2 \in H)$$

Therefore, $a = h_1^{-1} \cdot h_2 \cdot b$, and

$$Ha = Hh_1^{-1} \cdot h_2 \cdot b. \tag{1}$$

Since H is a subgroup, $h_1^{-1} \cdot h_2 \in H$, so that $Hh_1^{-1} \cdot h_2 \subseteq H$. Moreover, if $h \in H$, then $h = (h \cdot h_2^{-1} \cdot h_1) \cdot (h_1^{-1} \cdot h_2) \in Hh_1^{-1} \cdot h_2$. Therefore, $H \subseteq Hh_1^{-1} \cdot h_2$ and we have shown that $H = Hh_1^{-1} \cdot h_2$. Thus, by (1),

$$Ha = Hb.$$

Every element of G is contained in some right coset. For example, a is contained in Ha. Let the number of distinct right cosets of G with respect to H be k. Then, if the right cosets are

$$Ha_1, Ha_2, \ldots, Ha_k,$$

we can write

$$G = Ha_1 \cup Ha_2 \cup \cdots \cup Ha_k. \tag{2}$$

By Lemma 11, the union (2) defines a partition of G and, by Lemma 10, each subset of the partition contains $|H|$ elements.

Proof of Lagrange's Theorem: Let $|G| = n$, $|H| = m$. Then by (2) and the remarks just made, we see that

$$n = km \Longrightarrow |H| \,\big|\, |G|.$$

In Chapter 2 we showed that with every partition of a set there is associated an equivalence relation such that the subsets in the partition are precisely the equivalence classes of the equivalence relation. In the case of the partition (2), the equivalence relation is

$$a \sim b \Longleftrightarrow b \cdot a^{-1} \in H. \tag{3}$$

We leave the verification that (3) defines an equivalence relation as well as the fact that this equivalence relation gives the partition (2) as an exercise.

Remark: Lagrange's theorem does *not* assert that if m is a positive integer which divides the order of G, then there exists a subgroup H of G having order m. In fact, this statement is false. (See Exercise 12 of Section 3.8.)

Lagrange's theorem has a number of very important consequences. To mention but a few:

COROLLARY 12: Let G be a finite group, $a \in G$. Then the order of a divides $|G|$.

Proof: The order of a equals the order of the subgroup $[a]$ of G. Therefore, the assertion follows from Lagrange's theorem.

COROLLARY 13: Let G be a finite group of order n, $a \in G$. Then $a^n = 1_G$.

Proof: Let m denote the order of a. Then by Corollary 12, $m \mid n \Rightarrow n = m \cdot k$ for some positive integer k. Therefore,

$$a^n = a^{mk} = (a^m)^k = 1_G^k = 1_G.$$

COROLLARY 14: A group of prime order is cyclic.

Proof: Let G have prime order p. Then the only positive divisors of p are 1 and p. Let $a \in G$, $a \neq 1_G$. By Corollary 12, the order of $[a]$ divides p, so that $[a]$ has either 1 or p elements. But $[a]$ contains 1_G and a, so that $[a]$ has p elements. Therefore, $G = [a]$ and G is cyclic.

COROLLARY 15: Let n be a positive integer, $(a, n) = 1$. Then

$$a^{\phi(n)} \equiv 1 \pmod{n},$$

where $\phi(n)$ denotes Euler's function. In particular, if p is a prime and $p \nmid a$, then

$$a^{p-1} \equiv 1 \pmod{p}.$$

Proof: Let us apply Corollary 13 to the element $\bar{a}$ of the group $\mathbf{Z}_n^\times$ of order $\phi(n)$. We get that

$$\bar{a}^{\phi(n)} = \overline{a^{\phi(n)}} = \bar{1}.$$

Therefore,

$$a^{\phi(n)} \equiv 1 \pmod{n}.$$

Corollary 15 was first discovered by Pierre Fermat in case $n =$ a prime and is known as "Fermat's little theorem." The general form of Corollary 15 is due to Euler.

Let m be a positive integer. Since $[a^m]$ is a subgroup of $[a]$, Corollary 12 implies that the order of a^m divides the order of a. However, we can be much more specific:

THEOREM 16: Let a have order n. Then a^m has order $n/(n, m)$. In particular, if $(m, n) = 1$, then a^m has order n.

Proof: Let us first consider two special cases of the theorem:

Case 1: $m \mid n$. Suppose that $n = k \cdot m$. Then $(n, m) = m$, so that we must prove that a^m has order $n/m = k$. First, it is clear that $(a^m)^k = a^n = 1_G$. Suppose that r is a positive integer less than k such that $(a^m)^r = 1_G$. Then $a^{mr} = 1_G$ and $m \cdot r < n$. But this contradicts the assumption that n is the smallest positive integer such that $a^n = 1_G$. Thus, k is the smallest positive integer such that $(a^m)^k = 1_G$; that is, a^m has order k.

Case 2: $(m, n) = 1$. In this case we must prove that a^m has order n. Let r be the smallest positive integer such that $(a^m)^r = 1_G$. Since $(m, n) = 1$, there exist integers s, t such that $ms + nt = 1$. Therefore,

$$msr + ntr = r$$

and

$$\begin{aligned} a^r &= a^{msr} \cdot a^{ntr} \\ &= (a^m)^{rs} \cdot (a^n)^{tr} \\ &= 1_G. \end{aligned}$$

Therefore, since a has order n, we see that $r \geq n$. But since $(a^m)^n = 1_G$, we have $r = n$; that is, a^m has order n.

General case: Suppose that $(m, n) = p$. Set

$$m = m_0 p, \qquad n = n_0 p.$$

Then $(m_0, n_0) = 1$, since if $(m_0, n_0) > 1$, we have $(m, n) > p$. By case 1, a^p has order $n/p = n_0$. Therefore, by case 2, $a^m = (a^p)^{m_0}$ has order $n_0 = n/(n, m)$.

DEFINITION 17: Let G be a group, and let H be a subgroup of G. The number of right cosets in $H\backslash G$ is called the *index of H in G* and is denoted $[G: H]$.

The mapping $f: H\backslash G \longrightarrow G/H$ which maps the right coset Ha onto the left coset $a^{-1}H$ is easily seen to be a bijection, so $[G:H]$ could just as well be defined as the number of left cosets in G/H. From our proof of Lagrange's theorem [see Equation (2)], we see that $|G| = |H|\cdot[G:H]$, in case G is a finite group.

3.3 Exercises

1. Show that $\left\{\begin{pmatrix} 1 & 2 & 3 \\ 1 & 2 & 3 \end{pmatrix}, \begin{pmatrix} 1 & 2 & 3 \\ 1 & 3 & 2 \end{pmatrix}\right\}$ is a subgroup of S_3.
2. Find all subgroups of S_2; of S_3.
3. Find all subgroups of D_4.
4. Determine the orders of the following elements of S_4:

   (a) $\begin{pmatrix} 1 & 2 & 3 & 4 \\ 2 & 1 & 3 & 4 \end{pmatrix}$. (c) $\begin{pmatrix} 1 & 2 & 3 & 4 \\ 4 & 3 & 2 & 1 \end{pmatrix}$.

   (b) $\begin{pmatrix} 1 & 2 & 3 & 4 \\ 2 & 1 & 4 & 3 \end{pmatrix}$. (d) $\begin{pmatrix} 1 & 2 & 3 & 4 \\ 4 & 2 & 1 & 3 \end{pmatrix}$.
5. Determine the subgroup of S_4 generated by each of the elements in Exercise 4.
6. Show that $\mathbf{Z}_7^\times$ is cyclic, and exhibit a generator for it.
7. Prove that $\mathbf{Z}_8^\times$ is not cyclic. (*Hint:* Show that $\mathbf{Z}_8^\times$ does not contain an element of order 4.)
8. Compute the order of each element of D_6.
9. Let G be a cyclic group of order n and let m be a positive integer dividing n. Show that G contains one and only one subgroup of order m.
10. Let G be an abelian group. Show that the set of elements of G having finite order forms a subgroup of G. This subgroup is called the *torsion subgroup* of G.

*11. Let $G = \mathbf{Z}_p^\times$, where p is a prime, and let $G_2 = \{\bar{x}^2 \mid \bar{x} \in G\}$. Show that G_2 is a subgroup of G of order $(p - 1)/2$ if p is odd and of order 1 if $p = 2$. G_2 is called the group of *quadratic residues modulo p*. This group was studied in great detail by Gauss, in connection with the law of quadratic reciprocity mentioned in Chapter 2.

12. Let G be a group and set $Z(G) = \{x \in G \mid xg = gx \text{ for all } g \in G\}$.
    (a) Show that $Z(G)$ is a subgroup of G, called the *center* of G.
    (b) Show that G is abelian if and only if $Z(G) = G$.

13. (a) Show that (3) defines an equivalence relation on G.
    (b) Show that the partition associated with the equivalence relation (3) is given by (2).

14. Let G be an arbitrary group, $a,b \in G$. Show that ab and ba have the same order.

15. Let a and b be elements of finite order in the group G. Is it necessarily true that ab has finite order?

16. Suppose that G is a group with exactly one element of order 2, say g. Prove that $gx = xg$ for all $x \in G$.

In Exercises 17–22, we will develop the notion of a group acting on a set. Throughout, let G be a group, S a set. We say that G *acts on* S *as a group of transformations* (or just G *acts on* S) if, for every $g \in G$, there is given a bijection ρ_g of S such that $(\rho_g \rho_h)(s) = \rho_{gh}(s)$ $(s \in S)$. Very often, we write gs instead of $\rho_g(s)$. We say that G acts *transitively* on S if, for every pair of elements $s,t \in S$, there exists $g \in G$ such that $gs = t$.

17. Let H be a subgroup of G, and let $S = G/H$. Set $\rho_g(xH) = gxH$. Show that G acts as a transitive group of transformations on S.

18. Fix $s \in S$, and let $H_s = \{g \in G \mid gs = s\}$. Then H_s is called the *stabilizer* of s. Show that H_s is a subgroup of G.

19. Show that for $s \in S$, $g \in G$, we have $H_{gs} = gH_sg^{-1}$.

20. Suppose that G acts transitively on S. Fix $s \in S$. Define the mapping $\psi: G/H_s \longrightarrow S$ by $\psi(gH_s) = gs$.
    (a) Show that ψ is well defined.
    (b) Show that ψ is bijective.
    (c) Show that ψ is compatible with the action of G on G/H_s and S—that is, $\psi(g \cdot xH_s) = g \cdot \psi(xH_s)$.
    (d) Show that if $g \in G$, then $\{gt \mid t \in S\}$ consists of all elements in S, each repeated $|H_s|$ times.
    (e) To what extent are (a)–(d) true if G does not act transitively on S?

21. Let $s \in S$. A set of the form $T_s = \{gs \mid g \in G\}$ is called a *G-orbit* of S.
    (a) Show that $s \in T_s$.
    (b) Show that if $\sim$ denotes the relation on S defined by $s_1 \sim s_2$ if and only if $s_1 = gs_2$ for some $g \in G$, then $\sim$ is an equivalence relation on S. The equivalence classes are just the G-orbits of S.
    (c) Show that G acts transitively on S if and only if there exists $s \in S$ such that $T_s = S$.

22. Let G be any group, $S =$ the set of all subgroups of G. If $g \in G$, $H \in S$, define $g \cdot H = g^{-1}Hg$. Show that this defines an action of G on S. Does G act transitively?

3.4 Isomorphism

Let us consider the group G of symmetries of an equilateral triangle. As we saw in Section 3.3, this group is the dihedral group D_3. Recall that D_3 is a group of order 6, generated by a counterclockwise rotation R through an angle of $2\pi/3$ and a flip F about the X-axis. The elements of D_3 are then

$$\{I, R, R^2, F, FR, FR^2\}.$$

The multiplication table for D_3 was seen to be as follows:

	I	R	R^2	F	FR	FR^2
I	I	R	R^2	F	FR	FR^2
R	R	R^2	I	FR^2	F	FR
R^2	R^2	I	R	FR	FR^2	F
F	F	FR	FR^2	I	R	R^2
FR	FR	FR^2	F	R^2	I	R
FR^2	FR^2	F	FR	R	R^2	I

On the other hand, the symmetric group on three letters is also a group of order 6, since

$$S_3 = \left\{\begin{pmatrix}1 & 2 & 3\\1 & 2 & 3\end{pmatrix}, \begin{pmatrix}1 & 2 & 3\\3 & 1 & 2\end{pmatrix}, \begin{pmatrix}1 & 2 & 3\\2 & 3 & 1\end{pmatrix}, \begin{pmatrix}1 & 2 & 3\\2 & 1 & 3\end{pmatrix}, \begin{pmatrix}1 & 2 & 3\\1 & 3 & 2\end{pmatrix}, \begin{pmatrix}1 & 2 & 3\\3 & 2 & 1\end{pmatrix}\right\}.$$

Let us set

$$A = \begin{pmatrix}1 & 2 & 3\\3 & 1 & 2\end{pmatrix}, \qquad B = \begin{pmatrix}1 & 2 & 3\\2 & 1 & 3\end{pmatrix}.$$

Then,

$$A^2 = \begin{pmatrix}1 & 2 & 3\\2 & 3 & 1\end{pmatrix}, \qquad BA = \begin{pmatrix}1 & 2 & 3\\3 & 2 & 1\end{pmatrix}, \qquad BA^2 = \begin{pmatrix}1 & 2 & 3\\1 & 3 & 2\end{pmatrix}.$$

Therefore, if we denote the identity of S_3 by I, we see that

$$S_3 = \{I, A, A^2, B, BA, BA^2\}.$$

Moreover, a mildly tedious calculation shows that the multiplication table for S_3 is given by

	I	A	A^2	B	BA	BA^2
I	I	A	A^2	B	BA	BA^2
A	A	A^2	I	BA^2	B	BA
A^2	A^2	I	A	BA	BA^2	B
B	B	BA	BA^2	I	A	A^2
BA	BA	BA^2	B	A^2	I	A
BA^2	BA^2	B	BA	A	A^2	I

We now observe a curious phenomenon. Consider the function $\phi: S_3 \longrightarrow D_3$ defined by

$$\begin{aligned} \phi(I) &= I, & \phi(B) &= F, \\ \phi(A) &= R, & \phi(BA) &= FR, \\ \phi(A^2) &= R^2, & \phi(BA^2) &= FR^2. \end{aligned} \tag{1}$$

If we apply ϕ to every element in the multiplication table for S_3, we get the multiplication table for D_3! Let us pause for a moment and ponder the significance of the existence of the function ϕ. The function ϕ just replaces all A's by R's and all B's by F's. Therefore, the structures of the multiplication tables for S_3 and D_3 are identical for the two tables differ only in which letters we have chosen to call elements. But the multiplication table of a group tells us everything there is to know about the group. Therefore, since D_3 and S_3 have the "same" multiplication table, we should regard them as the same group. For every property of D_3, there is a corresponding property of S_3, and for every property of S_3, there is a corresponding property of D_3.

Since it is clumsy to speak of two groups having the "same" multiplication table, let us examine more closely just what this entails. We started out with the function $\phi: S_3 \longrightarrow D_3$ which transformed the multiplication table of S_3 into the multiplication table of D_3. In order for ϕ to accomplish this, it must preserve multiplication. That is, if $a, b \in S_3$, then

$$\phi(a \cdot b) = \phi(a) \cdot \phi(b). \tag{2}$$

Note that the product $a \cdot b$ is taken in S_3, whereas the product $\phi(a) \cdot \phi(b)$ is taken in D_3. Equation (2) does not reflect the only important property of ϕ. Indeed, it is clear that ϕ is also an injection. Therefore, we are led to make the following definition:

DEFINITION 1: Let G_1 and G_2 be groups. An *isomorphism* from G_1 to G_2 is an injection $\phi: G_1 \longrightarrow G_2$ such that

$$\phi(a \cdot b) = \phi(a) \cdot \phi(b) \qquad (a, b \in G_1).$$

If there exists an isomorphism $\phi: G_1 \longrightarrow G_2$ which is surjective, then we say that G_1 and G_2 are *isomorphic* (denoted $G_1 \approx G_2$).

It is clear that the two groups S_3 and D_3 are isomorphic with respect to the isomorphism ϕ which is defined by (1). Moreover, isomorphic groups have the "same" multiplication table, in the sense which we meant in our example, and therefore, in the theory of groups, isomorphic groups can be regarded as the same. Usually, it is not sufficient to say that two groups are isomorphic; rather, one has to specify the isomorphism. For it is the isomorphism which allows one to translate, directly, properties of one group onto the other.

The following properties of isomorphism are more or less obvious:

(a) $G \approx G$; (b) if $G_1 \approx G_2$, then $G_2 \approx G_1$; (c) if $G_1 \approx G_2$ and $G_2 \approx G_3$, then $G_1 \approx G_3$. Thus, isomorphism is an equivalence relation.

An extremely difficult problem in group theory is to determine all finite groups having given order n. Of course, without the notion of isomorphism, such a problem would be hopeless. For there are infinitely many groups isomorphic to a given group. (Just keep changing the names of the elements.) Thus, there are infinitely many groups having a given order n. However, let us rephrase the problem to read: Determine all nonisomorphic groups having order n. There are only finitely many for given n. For let us choose fixed names for the elements of the group, say $A_1, \ldots, A_n$. Then the group will be completely determined by its multiplication table, which is an $n \times n$ array whose entries are chosen from among $A_1, \ldots, A_n$. It is clear that there are at most n^{n^2} such arrays. Of course very few of these arrays will satisfy the group axioms. But nevertheless, we can conclude:

THEOREM 2: There are at most n^{n^2} nonisomorphic groups of order n.

Using the next result, we will be able to list all nonisomorphic groups of order n for $n \leq 7$. As far as the general question of determining all finite groups of a given order, it seems to be far beyond our present state of knowledge. In fact, the determination of a satisfactory answer to this question can be viewed as one of the basic objectives of the theory of groups. For the moment, let us be content with the following modest result:

THEOREM 3: Let n be a positive integer. Then every cyclic group G of order n is isomorphic to $\mathbf{Z}_n$. Therefore, any two cyclic groups of order n are isomorphic.

Proof: Let G be a cyclic group of order n and let a be a generator of G. Then $a^n = 1$, and $G = \{1, a, \ldots, a^{n-1}\}$. Let $f\colon \mathbf{Z}_n \to G$ be defined by $f(\bar{i}) = a^i$ $(0 \leq i \leq n-1)$. It is clear that f is a bijection. We assert that f is an isomorphism. Suppose that i and j both satisfy $0 \leq i, j \leq n-1$, and let p and q be positive integers such that

$$i + j = np + q \qquad (0 \leq q \leq n-1).$$

p and q exist by the division algorithm. Then $f(\bar{i} + \bar{j}) = f(\bar{q}) = a^q$, by the definition of f, and thus

$$\begin{aligned} f(\overline{i+j}) &= a^{i+j-np} \\ &= a^i a^j (a^n)^{-p} \\ &= a^i a^j \qquad (\text{since } a^n = 1) \\ &= f(\bar{i}) f(\bar{j}) \end{aligned}$$

Thus, f is an isomorphism of $\mathbf{Z}_n$ onto G, so that G is isomorphic to $\mathbf{Z}_n$.

Let us now determine a list of all nonisomorphic groups of order at most 7. For each positive integer n, there exists a group of order n, the cyclic group of order n. Morever, if n is prime, then a group of order n is cyclic, so that up to isomorphism, there is only one group of order n if n is prime, the cyclic group $\mathbf{Z}_n$. This takes care of all groups of order 2, 3, 5, and 7. There is only one group of order 1, the group $G = \{1\}$. Thus, it remains to consider the cases $n = 4$ and $n = 6$. It will turn out that there are two non-isomorphic groups for each of these orders.

Let $G = \{1, a, b, c\}$ be a noncyclic group of order 4. Then G cannot contain an element of order 4, since otherwise G would be cyclic. But the orders of a, b, and c must divide 4 and must be greater than 1, so that a, b, and c are all of order 2: $a^2 = b^2 = c^2 = 1$. We assert that $ab = c$. If $ab = 1$, then $a = b^{-1}$; but $b^{-1} = b$, since $b^2 = 1$. Therefore, if $ab = 1$, we see that $a = b$, which contradicts the fact that G is of order 4. Thus, $ab \neq 1$. If $ab = a$, then $b = 1$, a contradiction. Thus, $ab \neq a$. Similarly, $ab \neq b$. The only possibility is $ab = c$. Thus, we have shown that if G is a noncyclic group of order 4, then G is of the form $G = \{1, a, b, ab,\}$ where $a^2 = b^2 = (ab)^2 = 1$. It is easy to see that from these facts, the product of any two elements of G is completely specified, and G has the multiplication table pictured in Table 3.2.

TABLE 3.2
Multiplication Table of the Group V_4

	1	a	b	ab
1	1	a	b	ab
a	a	1	ab	b
b	b	ab	1	a
ab	ab	b	a	1

Thus, we see that if G is a noncyclic group of order 4, then G is uniquely determined up to isomorphism, and its multiplication table must be given by Table 3.2. One could check that Table 3.4 defines a group. But the verification of associativity is somewhat laborious. It is much easier to observe that $\mathbf{Z}_2 \times \mathbf{Z}_2$ is a group of order 4 which is not cyclic, and is thus an example of a group having a multiplication table as in Table 3.2. In particular, Table 3.2 does define a group. This group is called *Klein's 4-group* and is denoted V_4. Thus, we have shown that a group of order 4 is isomorphic to either $\mathbf{Z}_4$ or V_4.

Next , let G be a noncyclic group of order 6. The order of an element of G is either 1, 2, 3, or 6. Since G is noncyclic, G does not contain an element of order 6. Thus, if x is an element of G other than 1_G, then x has order 2 or 3. Assume for a moment that G contains no elements of order 3 and let $a, b \neq 1_G$. Then by reasoning as in the case of groups of order 4, we can see that $\{1_G, a, b, ab\}$ is a subgroup of G of order 4 (Exercise; referred to as fact 1). But since $4 \nmid 6$, we see that this contradicts Lagrange's theorem. Thus, G

has at most one element of order 2, and therefore G has an element a of order 3. Choose $b \in G$, so that $b \neq 1_G, a, a^2$. Then it is easy to see that $G = \{1_G, a, a^2, b, ab, a^2b\}$ (Exercise; referred to as fact 2). The order of b must be 2 or 3. Let us show that it must be 2. If not, then $b^3 = 1_G$, and $b^2 \neq 1_G$. Therefore, $b^2 = a, a^2, b, ab, a^2b$. The last three possibilities are impossible, since $b \neq 1_G, a, a^2$. Thus, either $b^2 = a$ or $b^2 = a^2$. In the first case, $b^3 = ab = 1_G$, which is a contradiction. In the second case, $b^3 = a^2b = 1_G$, which is a contradiction. Thus, we see that the order of b cannot be 3 and b must have order 2. Now what are the possibilities for ba? It is trivial to see that ba cannot be 1_G, a, a^2, or b, since a has order 3 and b has order 2. Thus, either $ba = ab$ or $ba = a^2b$. In the first case, G is abelian. Moreover, it is easy to check that ab has order 6, so G is cyclic, a contradiction. Thus, we see that $ba = a^2b$. The reader should have no difficulty in convincing himself that once ba is determined, the entire multiplication table for G is completely determined. Thus, if a noncyclic group of order 6 exists, then it is uniquely determined up to isomorphism. Does such a group exist? Of course. We have seen that S_3 is a non-abelian group of order 6. In particular, S_3 is not cyclic. Thus, we have proved that if G is a group of order 6, then G is isomorphic to either $\mathbf{Z}_6$ or S_3.

3.4 *Exercises*

1. Show that if the finite groups G and G' are isomorphic, then G and G' have the same order.
2. Let $\phi: G \longrightarrow G'$ be an isomorphism, H a subgroup of G, $a \in G$.
   (a) $\phi(H)$ is a subgroup of G'.
   (b) The order of $\phi(a)$ ($a \in G$) is equal to the order of a.
3. When is D_{2k} isomorphic to one of the groups S_n?
4. An isomorphism of G onto itself is called an *automorphism* of G.
   (a) Since an automorphism σ of G is a bijection of G, the inverse function σ^{-1} is well defined. Show that σ^{-1} is an automorphism of G.
   (b) Show that with respect to the law of multiplication defined by composition of functions, the set of all automorphisms form a group, denoted Aut(G).
5. Show that every automorphism of $\mathbf{Z}$ under addition is of the form $x \longrightarrow x$ or $x \longrightarrow -x$.
6. Show that every automorphism of $\mathbf{Z}_n$ is of the form $\bar{x} \longrightarrow \bar{a}\bar{x}$ ($\bar{a} \in \mathbf{Z}_n^\times$).
7. Consider the following functions of $z \in \mathbf{R} - \{0, 1\}$: z, $1/(1-z)$, $(z-1)/z$, $1/z$, $1-z$, $z/(z-1)$.
   (a) Show that with respect to the operation of composition of functions, these six functions form a group.
   (b) Show that this group is isomorphic to S_3.
8. Prove facts 1 and 2 used in the classification of groups of order 6.

9. Let G be a group, $a \in G$. Show that the mapping $x \longrightarrow axa^{-1}$ $(x \in G)$ is an automorphism of G (see Exercise 4). Such an automorphism is called an *inner automorphism* of G.

10. Let notation be as in Exercise 9.
 (a) Show that the set of inner automorphisms forms a subgroup of Aut(G).
 (b) Show that all inner automorphisms of G are trivial if and only if G is abelian.

3.5 *Normal Subgroups*

Let us now return to the study of the properties of the subgroups of a group. Our major task will be to introduce the extremely important concept of a *normal subgroup.*

Throughout this section let G be a group. If A, B are arbitrary subsets of G, let us set

$$AB = \{a \cdot b \mid a \in A, b \in B\}.$$

Since multiplication in G is associative, we see that for any subsets A, B, and C of G, we have

$$A(BC) = (AB)C.$$

We will denote each of these sets by ABC. If $A = \{a\}$, we will write aB (respectively, Ba) instead of $\{a\}B$ (respectively, $B\{a\}$). Note that this use of notation coincides with our previous use of aB and Ba for left and right cosets in case B is a subgroup of G. However, the two uses of the same notation are consistent with one another.

PROPOSITION 1: Let H be a subgroup of G, $a \in G$. Then aHa^{-1} is a subgroup of G.

Proof: From the definition of our notation above, we see that

$$aHa^{-1} = \{aha^{-1} \mid h \in H\}.$$

Let $\alpha, \beta \in aHa^{-1}$. By Proposition 2 of Section 3.3 it suffices to show that $\alpha \cdot \beta^{-1} \in aHa^{-1}$. Since $\alpha, \beta \in aHa^{-1}$, there exist $x, y \in H$ such that

$$\alpha = axa^{-1}, \qquad \beta = aya^{-1}.$$

Now $\beta^{-1} = (aya^{-1})^{-1} = ay^{-1}a^{-1}$. Therefore,

$$\alpha \cdot \beta^{-1} = (axa^{-1})(ay^{-1}a^{-1}) = a(xy^{-1})\, a^{-1} \in aHa^{-1}.$$

We say that the subgroup aHa^{-1} is *conjugate* to H. Usually, $aHa^{-1} \neq H$. That is, most subgroups H have conjugates other than themselves. (See below for an example.) However, the subgroups which coincide with all their conjugates are a very important class of subgroups. They play a very special role in group theory. Let us single them out in the following:

DEFINITION 2: *Let H be a subgroup of G.* If for every $a \in G$, we have $aHa^{-1} = H$, then H is called a *normal subgroup* of G. (The terms *invariant subgroup* and *self-conjugate subgroup* are used interchangeably with *normal subgroup*.)

Before giving examples of subgroups which are and are not normal, it is convenient to establish a simple criterion which can be used to determine whether a given subgroup is normal.

PROPOSITION 3: Let H be a subgroup of G. Then H is normal if and only if for every $a \in G$, we have $aHa^{-1} \subseteq H$.

Proof: Assume first that H is normal. Then for every $a \in G$, we have $aHa^{-1} = H$, so that we certainly have $aHa^{-1} \subseteq H$. Conversely, assume that $aHa^{-1} \subseteq H$ for all $a \in G$. Let us prove that H is normal. Let $b \in G$. Then, by assumption,

$$bHb^{-1} \subseteq H. \tag{1}$$

However, since $(b^{-1})^{-1} = b$,

$$b^{-1}Hb = b^{-1}H(b^{-1})^{-1} \subseteq H.$$

Therefore, if $h \in H$, $h' = b^{-1}hb \in H$, so that

$$h = bh'b^{-1} \in bHb^{-1}.$$

Since h is any element of H, we see that

$$H \subseteq bHb^{-1}. \tag{2}$$

Therefore, by (1) and (2), we deduce that

$$H = bHb^{-1}.$$

Since $b \in G$ was arbitrary, we have proved that H is normal.

EXAMPLE 1: If G is any group, $\{1\}$ and G are normal subgroups. These normal subgroups are said to be *trivial*.

EXAMPLE 2: If G is abelian, every subgroup H is normal. For if $a \in G$, $h \in H$, we have $aha^{-1} = h$.

EXAMPLE 3: Let $G = D_n$, $H = [R]$. We assert that H is a normal subgroup of G. For we certainly have

$$R^aHR^{-a} \subseteq H \qquad (a = 0, 1, 2, \ldots, n-1).$$

Further, since $FRF^{-1} = R^{-1} = R^{n-1}$, we have

$$FR^aF^{-1} = (FRF^{-1})^a = R^{a(n-1)} \qquad (a = 0, 1, \ldots, n-1),$$

$$\Longrightarrow FHF^{-1} \subseteq H.$$

Therefore,

$$(FR^a)H(FR^a)^{-1} = F(R^aHR^{-a})F^{-1} \subseteq FHF^{-1} \subseteq H$$

for $a = 0, 1, \ldots, n-1$. But every element of G is of the form R^a or FR^a $(a = 0, 1, \ldots, n-1)$. Therefore, by Proposition 3, H is normal.

EXAMPLE 4: Let $G = D_n$, $H = [F]$. Since $FRF = R^{-1}$, we have

$$RFR^{-1} = F^{-1}R^{-2} = FR^{n-2} \qquad (\text{since } F = F^{-1},\ R^{-1} = R^{n-1},\ R^n = I).$$

Since $n \geq 3$, we see that $0 < n-2 < n$, so that $R^{n-2} \neq I$. Therefore, $FR^{n-2} \notin H$, so that $RFR^{-1} \notin H$ so that $RHR^{-1} \nsubseteq H$. Thus, H is not a normal subgroup of G.

Let H be a subgroup of G and let us consider the sets $H\backslash G$ and G/H of right and left cosets, respectively. A typical left coset is a set of the form aH $(a \in G)$; a typical right coset is a set of the form Ha $(a \in G)$. It is not generally true that every left coset is also a right coset. The reader is referred to the exercises for a counterexample. The important thing to note, however, is that if H is normal, then for every $a \in G$ we have $aHa^{-1} = H$. Therefore,

$$H \cdot a = aHa^{-1} \cdot a = aH \cdot 1_G = aH. \tag{3}$$

Thus we have proved

PROPOSITION 4: If H is a normal subgroup, then every right coset Ha of G with respect to H is also a left coset—namely aH; and, conversely, every left coset is a right coset. In other words, if H is normal, then $G/H = H\backslash G$.

We will now go through a construction which will appear in various guises a number of times in this book. So let us try to get some sort of broad insight into the general philosophy which underlies the construction. The primary mathematical objects we have been discussing so far have all been sets. Not just arbitrary sets, however. Our sets in this chapter have been endowed with some extra structure—they are groups. We have already considered the problem of translating this extra structure from the original set to one of its subsets. What we arrived at were subgroups. Now we will consider a similar problem. Suppose that we have an equivalence relation on our original set. Can we translate the structure on the original set to the set of equivalence classes? To be more specific, let us consider a group G and a subgroup H. In our proof of Lagrange's theorem, we considered the equivalence relation on G defined by

$$a \sim b \text{ if and only if } ba^{-1} \in H. \tag{*}$$

The equivalence classes turned out to be the left cosets of G with respect to H. The set of equivalence classes is therefore G/H. In reference to our general question raised above, we can now ask: Can we define a group structure on G/H? Of course, we aren't interested in just any group structure on G/H. We are interested only in a group structure which is naturally connected with the group structures on G and H. The first law of multiplication on G/H which comes to mind is the one defined by

$$aH \cdot bH = abH. \tag{4}$$

The very simplicity of this definition leads us to suspect that it should be the "natural" multiplication on G/H. However, we run into problems. It is not in general true that (4) defines a binary operation on G/H. What can (and sometimes does) go wrong is that the "product" defined by (4) may depend on the choice of a and b and not merely on the cosets aH and bH. *However, this difficulty goes away if H is a normal subgroup of G*! Let us prove this nontrivial statement.

What we claim is that (4) defines a binary operation on G/H in case H is a normal subgroup of G. We remarked above that G/H is the set of equivalence classes of G with respect to the equivalence relation $\sim$ defined by (*). Provided that $\sim$ is compatible with multiplication in G, the results of Section 2.6 imply that (4) defines a binary operation on G/H. Thus, let $a,b,c,d \in G$, and assume that $a \sim b$, $c \sim d$. Let us show that $ac \sim bd$. This will imply the compatibility of $\sim$. But $ac \sim bd$ if and only if $(ac)(bd)^{-1} \in H$. However, since $a \sim b$ and $c \sim d$, we see that $b = ah_1$, $d = ch_2$ $(h_1, h_2 \in H)$. Therefore, $(ac)(bd)^{-1} = acd^{-1}b^{-1} = ah_2^{-1}b^{-1}$. However, since H is normal in G, $Hb^{-1} = b^{-1}H$, so that $h_2^{-1}b^{-1} = b^{-1}h_3$ for some $h_3 \in H$. Thus, $(ac)(bd)^{-1} = ab^{-1}h_3 = h_1^{-1}h_3$, and thus $(ac)(bd)^{-1} \in H$, which implies that $ac \sim bd$, which was what we desired to prove.

THEOREM 5: Let H be a normal subgroup of G. Then with respect to the rule of multiplication (4), G/H is a group. The identity element is the coset $H = 1H$, while the inverse of aH is $a^{-1}H$.

Proof: It suffices to verify the group axioms G1–G3.

Associativity: $aH \cdot (bH \cdot cH) = aH \cdot bcH = (a \cdot bc)H = (ab \cdot c)H = abH \cdot cH$ $(aH \cdot bH) \cdot cH$, where we have made essential use of the associativity of multiplication in G.

Identity: $1H \cdot aH = (1 \cdot a)H = aH$. Similarly, $aH \cdot 1H = aH$.

Inverse: $aH \cdot a^{-1}H = aa^{-1}H = 1H$. Similarly, $a^{-1}H \cdot aH = 1H$.

One further comment about the definition (4) of multiplication in G/H is in order. It is possible to interpret the product $aH \cdot bH$ as the product of two sets—aH and bH—which we defined as $aH \cdot bH = \{m \cdot n \mid m \in aH, n \in bH\} = \{ahbh' \mid h, h' \in H\}$. If H is normal, then $Hg = gH$ for all $g \in G$, so that $aH \cdot bH$, when interpreted as a product of two sets, equals $a \cdot Hb \cdot H = a \cdot bH \cdot H = ab \cdot H \cdot H = abH$. Therefore, the product of aH and bH as subsets of G coincides with the product defined by (4), and our notation cannot lead to any confusion.

DEFINITION 6: Let H be a normal subgroup of the group G. Then the set of cosets G/H, with the group structure defined above, is called the *quotient group of G modulo H*. The term *quotient group* is used interchangeably with *factor group*.

EXAMPLE 5: Let G denote the group $\mathbf{Z}$ with respect to addition, and let $H = n\mathbf{Z}$. It is clear that H is a subgroup of G, and H is normal since G is abelian. A typical element of $\mathbf{Z}/n\mathbf{Z}$ is a coset $a + n\mathbf{Z}$, which consists of all integers of the form $a + nz$ ($z \in \mathbf{Z}$). Thus, we see that an element of $\mathbf{Z}/n\mathbf{Z}$ is a residue class modulo n. The law of addition in the quotient group $\mathbf{Z}/n\mathbf{Z}$ is just the addition of residue classes which we defined in Chapter 2. Thus, $\mathbf{Z}/n\mathbf{Z}$ is the group $\mathbf{Z}_n$ of residue classes modulo n.

EXAMPLE 6: Let G be the group $\mathfrak{F}$ of all functions $f: \mathbf{R} \longrightarrow \mathbf{R}$, where the group operation is addition of functions; let

$$H = \{f \in \mathfrak{F} \mid f(0) = 0\}.$$

Then H is a normal subgroup of G. The elements of G/H are cosets $f + H$ ($f \in \mathfrak{F}$). Of course, not all these cosets are distinct. Therefore, let us now describe the elements of G/H more explicitly. For each real number a, let us choose a function $f_a \in \mathfrak{F}$ such that $f(0) = a$. Then the cosets $f_a + H$ belong to G/H. Let us prove that every element of G/H is a coset of this form and that all these cosets are distinct. If $f + H \in G/H$, let us choose $a \in \mathbf{R}$ such that $f(0) = a$. Then

$$\begin{aligned}(f - f_a)(0) &= f(0) - f_a(0) \\ &= a - a \\ &= 0.\end{aligned}$$

Therefore, $f - f_a \in H \Rightarrow f + H = f_a + H$, so that every element of G/H is of the form $f_a + H$ for some $a \in \mathbf{R}$. If $f_a + H = f_b + H$, then $f_a = f_b + h$ for some $h \in H$. But since $h(0) = 0$, $f_a(0) = f_b(0)$, and thus $a = b$. Therefore, all the cosets $f_a + H$ are distinct and

$$G/H = \{f_a + H \mid a \in \mathbf{R}\}.$$

It is easy to see that $(f_a + H) + (f_b + H) = f_{a+b} + H$. Therefore, the mapping

$$k: \mathbf{R} \longrightarrow G/H$$

defined by $k(a) = f_a + H$ is a surjective isomorphism, so that $G/H \approx \mathbf{R}$.

3.5 Exercises

1. Find all normal subgroups of S_3.
2. Let $G = \{(a, b) \mid a, b \in \mathbf{R}, a \neq 0\}$. Define multiplication on G by

$$(a, b)\cdot(c, d) = (ac, ad + b).$$

(a) Show that with respect to the multiplication defined, G is a group.
(b) Let $T = \{(a, 0) \mid a \in \mathbf{R}^\times\}$, $U = \{(1, b) \mid b \in \mathbf{R}\}$. Show that T and U are subgroups of G. ($\mathbf{R}^\times = \mathbf{R} - \{0\}$.)
(c) Show that U is a normal subgroup of G, while T is not.
(d) Describe G/U.

3. Let G and G' be groups. On the Cartesian product $G \times G'$, define the multiplication

$$(g_1, g_1')\cdot(g_2, g_2') = (g_1 g_2, g_1' g_2').$$

(a) Show that with respect to this multiplication, $G \times G'$ is a group, called the *direct product* of G and G'.
(b) Let $\bar{G} = \{(g, 1_{G'}) \mid g \in G\}$, $\bar{G}' = \{(1_G, g') \mid g' \in G'\}$. Show that $\bar{G}$ and $\bar{G}'$ are normal subgroups of $G \times G'$.
(c) Show that $G \approx \bar{G}$, $G' \approx \bar{G}'$.
(d) Describe $(G \times G')/\bar{G}$ and $(G \times G')/\bar{G}'$.

*4. (a) Show that there exists a group Q which is generated by the two elements A and B, which satisfy the relations

$$A^4 = 1, \qquad A^2 = B^2, \qquad BA = A^3 B, \qquad A^2 \neq 1.$$

(b) Show that Q has order 8. (Q is called the *quaternionic group.*)
(c) Find all subgroups of Q.
(d) Show that every subgroup of Q is normal, but nevertheless Q is nonabelian.
[*Hint:* Show that the relations of (a) determine the multiplication table of Q uniquely. The major difficulty in proving that the multiplication table defines a group consists of verifying the associative law. Show that we may take A and B to be the matrices

$$A = \begin{pmatrix} 0 & i \\ i & 0 \end{pmatrix}, \qquad B = \begin{pmatrix} 0 & -1 \\ 1 & 0 \end{pmatrix},$$

where the group operation is matrix multiplication, which is known to be associative (Exercise 15 of Section 3.2).]

5. Let G be a group, H a subgroup of G. Define the *normalizer of H in G*, denoted $N_G(H)$, by

$$N_G(H) = \{g \in G \mid gHg^{-1} = H\}.$$

(a) Show that $N_G(H)$ is a subgroup of G which contains H.
(b) Show that $N_G(H)$ is the largest subgroup of G in which H is normal.

6. Let G be a group, $a, b \in G$. The *commutator* of a and b is the element of G defined by $[a, b] = aba^{-1}b^{-1}$. The *commutator subgroup* $[G, G]$ *of* G is the subgroup of G generated by all commutators.
(a) Show that $[G, G]$ is a normal subgroup of G.
(b) Show that $G/[G, G]$ is abelian.
(c) Show that G is abelian if and only if $[G, G] = \{1\}$.
(d) Suppose H is normal in G and G/H is abelian. Show that $H \supseteq [G, G]$.

7. Let G denote the group of all translations $(x, y) \longrightarrow (x + a, y + b)$. Let S denote the set of translations for which $b = 0$. Show that S is a subgroup of G. Is S normal? Describe the sets of right and left cosets G/S and $S\backslash G$.

8. Let $\mathbf{Q}$ be the additive group of rational numbers, $\mathbf{Z}$ = the integers.
   (a) Show that $\mathbf{Z}$ is a normal subgroup of $\mathbf{Q}$.
   (b) Show that $\mathbf{Q}/\mathbf{Z}$ is an infinite group in which every element has finite order.
9. Let G be a group, H a subgroup such that $[G: H] = 2$. Show that H is a normal subgroup.

3.6 Homomorphisms

Recall that an isomorphism f from a group G into a group H is a function $f: G \rightarrow H$ which is injective and which satisfies

$$f(g \cdot g') = f(g) \cdot f(g') \tag{1}$$

for all $g, g' \in G$. There are many interesting examples of functions $f: G \rightarrow H$ which satisfy (1) but which are not injective. Such functions are fundamental in the theory of groups. Therefore, let us make the following

DEFINITION 1: Let G and H be groups. A function $f: G \rightarrow H$ which satisfies (1) for all $g, g' \in G$ is called a *homomorphism* of G into H.

Note that the product $g \cdot g'$ is taken with respect to multiplication in G, whereas the product $f(g) \cdot f(g')$ is taken with respect to multiplication in H.

EXAMPLE 1: An isomorphism $f: G \rightarrow H$ is a homomorphism.

EXAMPLE 2: Let $f: G \rightarrow H$ be defined by

$$f(g) = 1_H.$$

for all $g \in G$. Then f is a homomorphism.

EXAMPLE 3: Let $G = D_n$, $H = [F]$. Then every element of D_n is of the form $F^a R^b$ ($a = 0, 1$; $b = 0, 1, \ldots, n-1$). Let us define $f: G \rightarrow H$ by

$$f(F^a R^b) = F^a.$$

To prove that (1) holds, we must show that

$$f([F^a R^b] \cdot [F^c R^d]) = f(F^a R^b) \cdot f(F^c R^d). \tag{2}$$

However, we proved in Section 3.2 that

$$[F^a R^b] \cdot [F^c R^d] = \begin{cases} F^{a+c} R^{b+d} & \text{if } c = 0, \\ F^{a+c} R^{d-b} & \text{if } c = 1. \end{cases}$$

Therefore, the left-hand side of (2) equals F^{a+c} in both cases, and

$$F^{a+c} = f(F^a R^b) \cdot f(F^c R^d), \tag{3}$$

which proves (2). Thus, f is a homomorphism.

EXAMPLE 4: Let $f: \mathbf{Z} \longrightarrow \mathbf{Z}$ be defined by

$$f(a) = 2a \qquad (a \in \mathbf{Z}).$$

Since

$$f(a + b) = 2(a + b) = 2a + 2b = f(a) + f(b),$$

we see that f is a homomorphism.

EXAMPLE 5: We may generalize Example 4 as follows: Let G be an abelian group, and let n be an integer. Then let us define $f_n: G \longrightarrow G$ by

$$f_n(a) = a^n \qquad (a \in G).$$

Since

$$f_n(ab) = (ab)^n = a^n b^n = f_n(a) f_n(b),$$

we see that f_n is a homomorphism.

PROPOSITION 2: Let $f: G \longrightarrow H$ be a homomorphism.

(1) $f(1_G) = 1_H$.

(2) $f(a^{-1}) = f(a)^{-1} \quad (a \in G)$.

(3) If K is a subgroup of G, then $f(K)$ is a subgroup of H.

Proof: (1) Since $1_G \cdot 1_G = 1_G$ and since f is a homomorphism, we see that

$$f(1_G) = f(1_G \cdot 1_G) = f(1_G) \cdot f(1_G).$$

Multiplying both sides by $f(1_G)^{-1}$, we get $f(1_G) = 1_H$.

(2) It suffices to show that

$$f(a^{-1}) \cdot f(a) = f(a) f(a^{-1}) = 1_H. \tag{4}$$

For then, by the uniqueness of inverses in H, we have $f(a^{-1}) = f(a)^{-1}$. However,

$$a^{-1}a = a \cdot a^{-1} = 1_G.$$

Therefore, by part (1) and the fact that f is a homomorphism, we get (4).

(3) By Proposition 2 of Section 3.3 it suffices to show that if $a, b \in K$ then $f(a) \cdot f(b)^{-1} \in f(K)$. But, by part (2), $f(b)^{-1} = f(b^{-1})$, so that by the fact that f is a homomorphism, $f(a) \cdot f(b)^{-1} = f(a \cdot b^{-1}) \in f(K)$.

Suppose that $f: G \longrightarrow H$ is a homomorphism. Set

$$\ker(f) = \{x \in G \mid f(x) = 1_H\}.$$

We will call $\ker(f)$ the *kernel* of f. If f is an isomorphism, then $\ker(f) = \{1_G\}$. The larger the kernel of f, the farther f is from an isomorphism.

PROPOSITION 3: If $f: G \longrightarrow H$ is a homomorphism, then $\ker(f)$ is a normal subgroup of G.

Proof: First let us show that $\ker(f)$ is a subgroup of G. Let $a, b \in \ker(f)$. It suffices to show that $a \cdot b^{-1} = \in \ker(f)$. Now

$$\begin{aligned} a, b \in \ker(f) &\Longrightarrow f(a) = f(b) = 1_H \\ &\Longrightarrow f(a) = f(b^{-1}) = 1_H \text{ [by proposition 2, part (2)]} \\ &\Longrightarrow f(a \cdot b^{-1}) = f(a)f(b^{-1}) = 1_H \\ &\Longrightarrow a \cdot b^{-1} \in \ker(f). \end{aligned}$$

Let us now show that $\ker(f)$ is normal. Let $x \in \ker(f)$, $a \in G$. Then

$$\begin{aligned} f(axa^{-1}) &= f(a) \cdot f(x) \cdot f(a)^{-1} \\ &= f(a) \cdot 1_H \cdot f(a)^{-1} \\ &= 1_H, \end{aligned}$$

so that $axa^{-1} \in \ker(f)$. Therefore, $\ker(f)$ is normal.

If one knows the kernel of a homomorphism, then one has a good deal of information about the homomorphism, as the next two results show.

PROPOSITION 4: Let $f: G \longrightarrow H$ be a homomorphism, $h \in H$. Let $g \in G$ satisfy $f(g) = h$. Then

$$f^{-1}(h) = g \cdot \ker(f).$$

Proof: If $g' \in g \cdot \ker(f)$, then $g' = g \cdot k$ for some $k \in \ker(f)$. Therefore,

$$\begin{aligned} f(g') = f(g \cdot k) &= f(g)f(k) \\ &= f(g) \quad [\text{since } k \in \ker(f)] \\ &= h. \end{aligned}$$

Therefore, $g' \in f^{-1}(h)$ and $g \cdot \ker(f) \subseteq f^{-1}(h)$. In order to prove the reverse inclusion, note that if $g' \in f^{-1}(h)$, then $f(g') = h \Rightarrow f(g') \cdot f(g)^{-1} = h \cdot h^{-1} = 1_H$. But $f(g') \cdot f(g)^{-1} = f(g' \cdot g^{-1})$ by Proposition 2, part (2). Therefore,

$$\begin{aligned} f(g' \cdot g^{-1}) = 1_H &\Longrightarrow g' \cdot g^{-1} \in \ker(f) \\ &\Longrightarrow g' \in g \ker(f) \end{aligned}$$

Thus, $f^{-1}(h) \subseteq g \cdot \ker(f)$.

COROLLARY 5: Let $f: G \longrightarrow H$ be a homomorphism. Then f is an injection $\Longleftrightarrow \ker(f) = \{1_G\}$.

Proof: $\Rightarrow$ clear.

$\Leftarrow$ Suppose that $\ker(f) = \{1_G\}$. By Proposition 4, if $h \in f(G)$, then $f^{-1}(h)$ consists of exactly one element. Therefore, f is an injection.

Proposition 3 tells us that with every homomorphism f of G into a group H is associated a normal subgroup of G, $\ker(f)$. Conversely, suppose that we are given a normal subgroup N of G. Then there is an associated homomorphism, which we will now construct. We have already defined the quotient group G/N. Let us associate to N the homomorphism

$$i_N: G \longrightarrow G/N,$$
$$i_N(g) = gN.$$

Indeed, i_N is a homomorphism, since

$$\begin{aligned} i_N(g_1g_2) &= g_1g_2N \\ &= g_1N \cdot g_2N \\ &= i_N(g_1)i_N(g_2) \qquad (g_1, g_2 \in G) \end{aligned}$$

by the definition of multiplication in G/N.

PROPOSITION 6: The homomorphism $i_N: G \longrightarrow G/N$ is surjective and $\ker(i_N) = N$.

Proof: It is clear that i_N is surjective. Moreover,

$$\begin{aligned} g \in \ker(i_N) &\Longleftrightarrow i_N(g) = 1_{G/N} \\ &\Longleftrightarrow gN = N \\ &\Longleftrightarrow g \in N. \end{aligned}$$

Therefore, $\ker(i_N) = N$.

The homomorphism $i_N: G \longrightarrow G/N$ is called the *canonical homomorphism* (associated with N) Note that by Proposition 6, every normal subgroup N of G appears as the kernel of a homomorphism of G, the canonical homomorphism.

Suppose that G, H, and J are groups, and suppose that $a: G \longrightarrow H$, $b: H \longrightarrow J$, and $c: G \longrightarrow J$ are homomorphisms. It is easy to check that the composite function $ba: G \longrightarrow J$ is a homomorphism. We can picture this situation by drawing the following diagram:

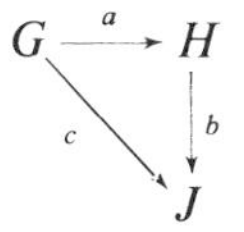

We see from the diagram that there are two homomorphisms defined from G into J, ba and c. If $c = ba$, then we say that the *diagram is commutative*. In other words, if the diagram is commutative, then the functions defined by the two paths from G to J are the same.

Let $f: G \longrightarrow H$ be a homomorphism. Then we can construct the following diagram:

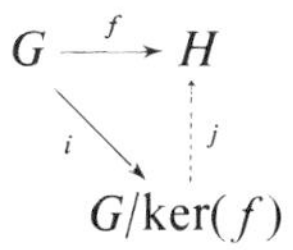

(5)

Here i is the canonical homomorphism $i_{\ker(f)}$. A common sort of question which arises in many contexts is: Does there exist a homomorphism $j: G/\ker(f) \longrightarrow H$ [dashed arrow in (5)] which makes the diagram commute? We will prove the existence and uniqueness of j in the following theorem.

THEOREM 7 (FIRST ISOMORPHISM THEOREM): Let $f: G \longrightarrow H$ be a homomorphism. Then there exists a homomorphism $j: G/\ker(f) \longrightarrow H$ which makes the diagram (5) commute. Such a j is unique. Moreover, j is an injection, so that j is an isomorphism and $\text{im}(f) \approx G/\ker(f)$, where *im*($f$) denotes the image of f. If, in addition, f is surjective, then j is surjective and $H \approx G/ker(f)$.

Remarks: 1. In many books, the first isomorphism theorem consists of the statement that if $f: G \longrightarrow H$ is a surjective homomorphism, then $H \approx G/\ker(f)$. Indeed, this is how the first isomorphism theorem is usually applied. However, for many purposes, it is important to know not only that H and $G/\ker(f)$ are isomorphic, but that there is an isomorphism which makes the diagram (5) commute, which makes them isomorphic.

2. The real utility of Theorem 7 (as well as the other isomorphism theorems) is that they allow one to conclude that two groups are isomorphic to one another without actually constructing the isomorphism. Thus, these isomorphism theorems allow one to build a calculus for computing with groups.

3. The fundamental theorem asserts that $G/\ker(f)$ can be identified with the image of f under the isomorphism j. Moreover, once this identification is carried out, the commutativity of the diagram implies that the homomorphism $f: G \longrightarrow H$ is identified with the canonical isomorphism. Thus, roughly speaking, all homomorphisms $f: G \longrightarrow H$ are "the same as" canonical homomorphisms. This is the real impact of the fundamental theorem.

EXAMPLE 6: Let $f: G \longrightarrow \{1_G\}$ be the homomorphism defined by $f(g) = 1_G$ for all $g \in G$. Then f is surjective and $\ker(f) = G$, and from the theorem we conclude that $G/G \approx \{1_G\}$.

EXAMPLE 7: Let $f: D_n \longrightarrow [F]$ be the homomorphism of Example 3 of Section 3.2 Then f is surjective and $\ker(f) = [R]$. Then from the theorem we conclude that $D_n/[R] \approx [F]$, a fact which could be easily verified directly.

Proof of Theorem: Let $K = \ker(f)$. Then a typical element of G/K is of the form gK ($g \in G$). Since $i(g) = gK$, if (5) commutes, then we must have

$$j(gK) = f(g). \tag{6}$$

Therefore, if j exists, it is unique and must be given by (6). Let us try to define j by (6). First, we must check that the definition of $j(gK)$ depends only on the coset gK and not on the choice of g. But if $gK = g'K$, then $gg'^{-1} \in K$, say

$gg'^{-1} = k$, so that $f(gg'^{-1}) = 1_H$, since $K = \ker(f)$. But then, since f is a homomorphism, $1_H = f(gg'^{-1}) = f(g)f(g')^{-1}$, which implies that $f(g) = f(g')$. Therefore, the definition (6) makes sense. It is clear that j makes the diagram (5) commute. (This was how j was chosen!) Moreover, j is a homomorphism since

$$\begin{aligned} j(gK\cdot g'K) &= j(gg'K) = f(gg') = f(g)f(g') \\ &= j(gK)j(g'K). \end{aligned}$$

Finally, j is an isomorphism, since if $gK \in \ker(j)$, then $f(g) = 1_H$, which implies that $g \in K$ and thus $gK = K$. Therefore, $\ker(j) = \{1_{G/K}\}$, and j is an isomorphism. The other assertions of the theorem follow immediately.

We will return to the subject of isomorphism theorems in Chapter 8. For the time being, it will be sufficient for our purposes to use the first isomorphism theorem.

3.6 *Exercises*

1. Show that the image of a cyclic group under a homomorphism is cyclic.
2. Show that the image of an abelian group under a homomorphism is abelian.
3. Let $\mathfrak{F}$ denote the group of all functions f: $\mathbf{R} \longrightarrow \mathbf{R}$. Let us define ϕ_a: $\mathfrak{F} \longrightarrow \mathbf{R}$ by

$$\phi_a(f) = f(a) \qquad (a \in \mathbf{R}).$$

(a) Show that ϕ_a is a surjective homomorphism.
(b) Find $\ker(\phi_a)$.
(c) Describe the elements of $\mathfrak{F}/\ker(\phi_a)$.
(d) Verify that the first isomomorphism theorem holds in this example.

4. Let G be a group and let $H,K \subseteq G$ be subgroups. Show that HK is a subgroup of G if and only if $HK = KH$.
5. Refer to Exercise 4. Assume that K is normal in G.
(a) Show that $HK = KH$.
(b) Show that K is a normal subgroup of HK.
(c) Show that $H \cap K$ is a normal subgroup of H.
(d) Show that the mapping ψ: $H \longrightarrow HK/K$ defined by $\psi(h) = hK$ is a surjective homomorphism and that $\ker(\psi) = H \cap K$.
(e) Show that $HK/K \approx H/H \cap K$ (Second Isomorphism theorem).
6. Let G be a group, $K \subseteq H \subseteq G$, where H and K are subgroups, with H, K normal in G.
(a) Show that K is normal in H and H/K is normal in G/K.
(b) Show that $(G/K)/(H/K) \approx G/H$.
7. Let G be an abelian group, and let f: $G \longrightarrow G$ be defined by $f(x) = x^n$. Show that f is a homomorphism.

8. Let $\mathbf{R}$ denote the additive group of real numbers, and let $\mathbf{R}_+$ denote the group of positive real numbers with respect to multiplication. Show that the function $f: \mathbf{R} \longrightarrow \mathbf{R}_+$ defined by $f(x) = e^x$ is a surjective isomorphism. What is the inverse of f?

9. Let m and n be positive integers. Prove that $\mathbf{Z}_{mn}/m\mathbf{Z}_{mn} \approx \mathbf{Z}_m$.

10. Let m and n be positive integers.
   (a) Show that if m and n are relatively prime, then $\mathbf{Z}_{mn}^\times/m\mathbf{Z}_m^\times \approx \mathbf{Z}_n^\times$.
   (b) Does the conclusion of (a) remain true if m and n are not relatively prime?

11. Let G_1 and G_2 be two groups, H_i a normal subgroup of G_i $(i = 1, 2)$. Give examples to show that none of the following statements are true:
   (a) $H_1 \approx H_2$ and $G_1/H_1 \approx G_2/H_2 \Longrightarrow G_1 \approx G_2$.
   (b) $G_1 \approx G_2$ and $G_1/H_1 \approx G_2/H_2 \Longrightarrow H_1 \approx H_2$.

*12. Let G be a group and let H be a subgroup of G such that $[G: H] = n < \infty$. Show that H contains a normal subgroup K of G such that $[G: K] < n!$, where $n! = n(n - 1) \ldots 1$.
   [*Hint:* Let G/H be the set of left cosets of H in G. Define a mapping $k: G \longrightarrow S_{G/H}$ by $(k(g))(g_1H) = gg_1H$. Show that k is a homomorphism. What is $\ker(k)$?]

13. Let m, n be positive integers, $(m, n) = 1$. Show that $\mathbf{Z}_{mn}^\times \approx \mathbf{Z}_m^\times \times \mathbf{Z}_n^\times$.
   [*Hint:* There is an obvious homomorphism $f: \mathbf{Z}_{mn}^\times \longrightarrow \mathbf{Z}_m^\times \times \mathbf{Z}_n^\times$. This map is surjective by Exercise 7 of Section 2.5.]

3.7 Cayley's Theorem

We have stated that one of the main objectives of group theory is to write down a complete list of nonisomorphic groups. At first, such a task appears hopeless. For, as we have seen, groups pop up in some very unexpected places and, therefore, if we set out to compile a list of all nonisomorphic groups, we would hardly begin to know where to look. The following theorem of Cayley solves this dilemma.

THEOREM 1: Every group is isomorphic to a subgroup of a permutation group.

What Cayley's theorem tells us is that the permutation groups and their subgroups are, roughly speaking, all the groups that can exist. Unfortunately, the problem of classifying the subgroups of a permutation group is fantastically complicated, even in the case of a finite permutation group. Therefore, Cayley's theorem does not allow us to easily write down a complete list of groups, as one might at first suspect.

Proof of Theorem: Let G be a group, $g \in G$. Define

$$\rho_g: G \longrightarrow G$$
$$\rho_g(x) = xg^{-1} \qquad (x \in G).$$

If $\rho_g(x) = \rho_g(y)$, then $xg^{-1} = yg^{-1}$, so that $x = y$. Therefore, ρ_g is an injection. If $y \in G$, then $\rho_g(yg) = yg \cdot g^{-1} = y$. Therefore, ρ_g is a surjection. Thus, since ρ_g is a bijection, ρ_g is a permutation of the elements of (the set) G, and $\rho_g \in S_G$. Let us consider the mapping

$$G \longrightarrow S_G$$

defined by

$$g \longrightarrow \rho_g. \tag{1}$$

Since $\rho_{gg'}(x) = x(gg')^{-1} = \rho_g \rho_{g'}(x)$, mapping (1) is a homomorphism. But $\rho_g = 1_{S_G}$ if and only if $x \cdot g = x$ for all $x \in G$, which occurs if and only if $g = 1_G$. Therefore, the kernel of the homomorphism (1) is 1_G, and therefore the mapping (1) is an injection. Thus, we have shown that G is isomorphic to a subgroup of S_G.

The above argument actually proves somewhat more than we claimed. For if G is finite, having order n, then G is isomorphic to a subgroup of S_G. Therefore, we have

COROLLARY 2: If G has finite order n, then G is isomorphic to a subgroup of S_n.

3.8 *The Symmetric Groups*

In Section 3.7 we proved that every finite group is isomorphic to a subgroup of S_n for some n. Therefore, the symmetric groups play a much more fundamental role in group theory than would appear at the outset. In this section we will begin an in-depth study of these groups.

Before we begin discussing the properties of S_n, let us introduce some new notation for permutations, which is more amenable to computation than our previous notation. Suppose that $i_1, \ldots, i_r$ are distinct integers, $1 \leq i_j \leq n$. Let

$$(i_1 i_2 \ldots i_r)$$

denote the permutation of S_n which maps i_1 onto i_2, i_2 onto $i_3, \ldots,$ and i_r onto i_1, and which leaves fixed the integers not expressly listed. Such a permutation is called an r-*cycle*. For example, the permutation (123) of S_5 would be written in our old notation as

$$\begin{pmatrix} 1 & 2 & 3 & 4 & 5 \\ 2 & 3 & 1 & 4 & 5 \end{pmatrix}.$$

Any 1-cycle is the identity permutation. Moreover,

$$(i_1 i_2 \ldots i_r)^{-1} = (i_r i_{r-1} \ldots i_1). \tag{1}$$

We will say that two cycles are *disjoint* if they have no elements in common. For example, (123) and (456) are disjoint. It is clear that disjoint cycles commute with one another. Note, however, that the product of cycles is not necessarily a cycle, but only a permutation. We say that an r-cycle is *nontrivial* if $r \neq 1$.

LEMMA 1: Let $\sigma \in S_n$, $\sigma \neq 1$. Then σ can be written as a product of nontrivial disjoint cycles. This representation is unique up to rearrangement of the cycles.

We will postpone the proof of Lemma 1 for a moment. The following example should suggest a proof to the reader. Let $\sigma \in S_8$ be given by

$$\sigma = \begin{pmatrix} 1 & 2 & 3 & 4 & 5 & 6 & 7 & 8 \\ 7 & 1 & 4 & 5 & 8 & 6 & 2 & 3 \end{pmatrix}.$$

Then, σ maps 1 into 7, 7 into 2, and 2 back to 1. Thus, (172) is a component of σ. Moreover, σ maps 3 into 4, 4 into 5, 5 into 8, and 8 back into 3. Thus, a second component cycle of σ is (3458). Finally, 6 is mapped into 6, so that $\sigma = (172)(3458)(6)$. Moreover, since a 1-cycle is the identity, we have

$$\sigma = (172)(3458),$$

an expression of σ as a product of nontrivial, disjoint cycles.

Proof of Lemma 1: We will say that an integer i is fixed by σ if $\sigma(i) = i$. Let us show how to express σ as a product of nontrivial, disjoint cycles by using induction on the number of integers not left fixed by σ. Since $\sigma \neq 1$, there is an integer i_1 not fixed by σ. Thus, $\sigma(i_1) = i_2$, $i_2 \neq i_1$. If $\sigma(i_2) \neq i_1, i_2$, define i_3 by $i_3 = \sigma(i_2)$. If $\sigma(i_3) \neq i_1, i_2, i_3$, define i_4 by $i_4 = \sigma(i_3)$. And so on. Thus, we define a sequence of integers $i_1, i_2, \ldots, i_s$, all distinct, and such that $\sigma(i_k) = i_{k+1}$. Eventually, this process must stop, and $\sigma(i_s)$ will be one of $i_1, \ldots, i_s$. But then $\sigma(i_s) = i_1$. For if $\sigma(i_s) = i_k$ $(1 < k \leq s)$, then σ maps both i_s and i_{k-1} onto i_k, which contradicts the fact that σ is a bijection. Set $\eta = (i_1 i_2 \ldots i_s)$. Then $\sigma\eta^{-1}$ fixes precisely $i_1, \ldots, i_s$ and all those integers fixed by σ. Therefore, $\sigma\eta^{-1}$ has fewer integers not left fixed than σ. Thus, by induction, $\sigma\eta^{-1} = c_1 \ldots c_t$, where $c_1', \ldots, c_t$ are nontrivial, disjoint cycles. Moreover, the integers appearing in $c_1, \ldots, c_t$ are just those not fixed by $\sigma\eta^{-1}$. Thus, $c_1, \ldots, c_t$ are disjoint from η and $\sigma = \eta\, c_1 \ldots c_t$ is an expression of σ as a product of nontrivial, disjoint cycles. The uniqueness statement is left as an exercise for the reader.

DEFINITION 2: A *transposition* is a 2-cycle (ij), $i \neq j$.

A simple inductive argument suffices to show that S_n contains $n(n-1)/2$ transpositions.

PROPOSITION 3: S_n is generated by the set of transpositions.

Proof: Let $\sigma \in S_n$, $\sigma \neq 1$. We will show that σ can be written as a product of transpositions. Let

$$\sigma = (i_1 i_2 \ldots i_r)(j_1 \ldots j_s)(k_1 \ldots k_t) \ldots \tag{2}$$

be the expression of σ as a product of disjoint, nontrivial cycles. Then

$$\sigma = (i_1 i_r)(i_1 i_{r-1}) \ldots (i_1 i_2)(j_1 j_s) \ldots (j_1 j_2)(k_1 k_t) \ldots (k_1 k_2) \ldots . \tag{3}$$

We see from the proof of Proposition 3 that if $\sigma \in S_n$, $\sigma \neq$ the identity, then σ can be written as a product of $N(\sigma)$ transpositions, where

$$N(\sigma) = (r - 1) + (s - 1) + (t - 1) + \ldots,$$

and σ is given by (2). Let us set

$$N(1) = 0.$$

Note that although $\sigma \neq 1$ can always be expressed as a product of $N(\sigma)$ transpositions, it will usually be possible to write σ as a product of some other number of transpositions. The next result will be used to prove that in any two such representations of σ, the number of transpositions is always even or always odd.

THEOREM 4: Let $\eta, \sigma \in S_n$. Then $N(\eta\sigma) \equiv N(\eta) + N(\sigma) \pmod 2$.

Proof: First we assert that it suffices to prove the theorem in the case η = a transposition. For assume that we know this special case, and let η and σ be arbitrary permutations; let

$$\sigma = \sigma_1 \ldots \sigma_a, \eta = \eta_1 \ldots \eta_b$$

be expressions of σ and η, respectively, as products of transpositions. Then, by induction and the assumed special case of the theorem, it is easy to verify that

$$\begin{aligned} N(\sigma\eta) &\equiv (N(\sigma_1) + N(\sigma_2) + \cdots + N(\sigma_a)) + (N(\eta_1) + \cdots \\ &\quad + N(\eta_b)) \pmod 2 \\ &\equiv N(\sigma) + N(\eta) \pmod 2. \end{aligned}$$

Thus, let us asume that $\eta = (ij)$, and let us assume that σ is given by (2). There are four cases to consider:

Case 1: Neither i nor j appear in the cycle decomposition (2). In this case, $N(\eta\sigma) = (2 - 1) + (r - 1) + (s - 1) + (t - 1) + \cdots = 1 + N(\sigma) = N(\eta) + N(\sigma)$.

Case 2: Exactly one of i and j appears in the cycle decomposition (2). Without loss of generality, let us asume that $i = i_1$. Then

$$(ij)\sigma = (jii_2 \ldots i_r)(j_1 \ldots j_s)(k_1 \ldots k_t) \ldots .$$

Therefore,

$$N(\eta\sigma) = 1 + N(\sigma) = N(\eta) + N(\sigma).$$

Case 3: Both i and j appear in the cycle decomposition (2), but they appear in different cycles. Without loss of generality, let us assume that $i = i_1$ and $j = j_1$. Then

$$(ij)\sigma = (jj_2j_3 \dots j_s ii_2 \dots i_r)(k_1 \dots k_t) \dots ,$$

so that

$$N(\eta\sigma) = (r + s - 1) + (t - 1) + \dots = N(\eta) + N(\sigma).$$

Case 4: Both i and j appear in the same cycle. Without loss of generality, let $i_1 = i$, $i_k = j$. Then

$$(ij)\sigma = (ii_2 \dots i_{k-1})(ji_{k+1} \dots i_r)(j_1 \dots j_s)(k_1 \dots k_t) \dots ,$$

so that

$$\begin{aligned} N(\eta\sigma) &= (k - 2) + (r - k) + (s - 1) + (t - 1) + \cdots \\ &= N(\eta) + N(\sigma) - 2 \\ &\equiv N(\eta) + N(\sigma) \pmod{2}. \end{aligned}$$

DEFINITION 5: Let $\sigma \in S_n$. Then σ is *even* (respectively, *odd*) if $N(\sigma)$ is even (respectively, odd).

Thus, a permutation σ is even if, when σ is written in the standard way [as in (3)] as a product of transpositions, the number of transpositions is even. However, if σ is even and if $\sigma = \sigma_1 \dots \sigma_a$ is *any* expression of σ as a product of transpositions, Theorem 4 and a simple induction imply that

$$\begin{aligned} N(\sigma) &\equiv N(\sigma_1) + \cdots + N(\sigma_a) \pmod{2} \\ &\equiv a \pmod{2}, \end{aligned}$$

and thus a is even. Thus, *a permutation is even if and only if it can be expressed (in any manner) as a product of an even number of transpositions.*

By Theorem 4 and the fact that $N(\sigma^{-1}) = N(\sigma)$, we see that the set of even permutations of S_n form a subgroup. This subgroup is called the *alternating group on n letters* and is denoted A_n.

THEOREM 6:

(1) *A_n is a normal subgroup of S_n.*

(2) *If $n \geq 2$, the index of A_n in S_n is 2.*

Proof: Let us consider the mapping

$$\psi : S_n \longrightarrow \mathbf{Z}_2$$

$$\psi(\sigma) = N(\sigma) \pmod{2}.$$

By Theorem 4, this mapping is a homomorphism. Its kernel is clearly A_n. Thus, part (1) is proved by Proposition 3 of Section 3-6. If $n \geq 2$, S_n contains the 2-cycle (12) and $N((12)) = 1$. Therefore, if $n \geq 2$, ψ is surjective. By the first isomorphism theorem, $S_n/A_n \approx \mathbf{Z}_2$. Therefore, the index of A_n in S_n is 2.

In Proposition 3 we proved that S_n is generated by the set of all transpositions. Let us now prove an analogous statement for A_n.

PROPOSITION 7: A_n is generated by the set of all 3-cycles (ijk).

Proof: Let S denote the set of all 3-cycles, $G = [S]$. If $n \leq 2$, $S = \varnothing$, so that $G = \{1\}$. (The smallest subgroup of S_n containing S is just $\{1\}$.) But if $n \leq 2$, then $A_n = \{1\}$. Thus, we may assume that $n \geq 3$. Since a 3-cycle is even, it is clear that $G \subseteq A_n$. Let us prove the reverse inclusion. Since a permutation is even if and only if it can be written as a product of an even number of transpositions, we see that A_n is generated by all products of the form $(i_1i_2)(i_3i_4)$ $(i_1 \neq i_2,\ i_3 \neq i_4)$. Therefore, it suffices to show that every such product is contained in the group G. There are three cases:

Case 1: Only two of the integers $i_j (j = 1, 2, 3, 4)$ are distinct. Without loss of generality, we may assume that $i_1 = i_3$, $i_2 = i_4$. Then

$$(i_1i_2)(i_3i_4) = 1 \in G.$$

Case 2: Three of the integers i_1, i_2, i_3, and i_4 are distinct. Without loss of generality, assume that $i_1 = i_3$. Then

$$(i_1i_2)(i_3i_4) = (i_1i_4i_2) \in G.$$

Case 3: All four of the integers i_1, i_2, i_3, and i_4 are distinct. Then

$$(i_1i_2)(i_3i_4) = (i_1i_3i_2)(i_1i_3i_4) \in G.$$

DEFINITION 8: A group G is said to be *simple* if G has no nontrivial normal subgroups. Equivalently, G is simple if the only normal subgroups N of G are $N = G$ and $N = \{1\}$.

By Lagrange's theorem, if $G = \mathbf{Z}_p$, p a prime, then G is simple. On the other hand, if $G = \mathbf{Z}_n$, $n \neq 1$, prime, then G is not simple. For if p is a prime dividing n, then $N = \{\bar{0}, \bar{p}, \overline{2p}, \ldots, \overline{(n-1)\cdot p}\}$ is a nontrivial normal subgroup of G.

Another example of a simple group is given by the following important result.

THEOREM 9 (ABEL): Assume that $n \neq 4$. Then A_n is simple.

In order to prove this result, we require

LEMMA 10: Let N be a normal subgroup of A_n $(n \geq 5)$. If N contains a 3-cycle, then $N = A_n$.

Proof: By Proposition 7, it suffices to show that every 3-cycle belongs to N. Let $(ijk) \in N$ and let $(i'\,j'\,k')$ be an arbitrary 3-cycle. Let $\alpha \in S_n$ be any permutation having the property

$$\alpha(i) = i',$$
$$\alpha(j) = j',$$
$$\alpha(k) = k'.$$

Then, by Exercise 9 of this section,

$$\alpha(ijk)\alpha^{-1} = (i'j'k'). \tag{4}$$

If $\alpha \in A_n$, then (4) asserts that $(i'j'k') \in N$, since N is normal. Therefore, assume that $\alpha \notin A_n$. Since $n \geq 5$, there exist integers a, a' $(1 \leq a, a' \leq n)$ such that

$$\{a, a'\} \cap \{i', j', k'\} = \varnothing.$$

Then, since $\alpha \notin A_n$, $\beta = (aa')\alpha \in A_n$. However,

$$\begin{aligned}\beta(ijk)\beta^{-1} &= (aa')\alpha(ijk)\alpha^{-1}(aa') \\ &= (aa')(i'j'k')(aa') \\ &= (i'j'k').\end{aligned}$$

Thus, again $(i'j'k') \in N$ and the lemma is proved.

Proof of Theorem 9: A_n has order 1, 1, 3 for $n = 1, 2, 3$, respectively. In these cases, A_n has no proper subgroups by Lagrange's theorem. Thus, let us assume that $n \geq 5$. Let $N \subseteq A_n$ be a normal subgroup $N \neq \{1\}$. We must show that $N = A_n$. By Lemma 10 it suffices to show that N contains a 3-cycle. For each $\sigma \in S_n$, let $f(\sigma) =$ the number of integers $j\,(1 \leq j \leq n)$ such that $\sigma(j) = j$. Let us choose $\sigma \in N - \{1\}$ for which $f(\sigma)$ is a maximum. Since $\sigma \neq 1$, $f(\sigma) \leq n - 1$. However, if $\sigma(j) = j$ for $n - 1$ values of j, then $\sigma = 1$, so that $f(\sigma) \leq n - 2$. If $f(\sigma) = n - 2$, then σ is a transposition, which contradicts the fact that $\sigma \in A_n$. Therefore, $f(\sigma) \leq n - 3$. We assert that $f(\sigma) = n - 3$. Assume the contrary. Then $f(\sigma) \leq n - 4$, and there are at least four integers not mapped into themselves by σ. Let us examine the expression of σ as a product of nontrivial disjoint cycles. There are two possibilities. Either there exists an r-cycle $(r \geq 3)$ in the expression or only transpositions occur. In the former case, σ looks like

$$(i_1i_2i_3\ldots)\ldots, \tag{5}$$

while in the latter σ is of the form

$$(i_1i_2)(i_3i_4)\ldots. \tag{6}$$

In the first case, if $f(\sigma) = n - 4$, then σ is a 4-cycle, which contradicts the fact that σ is even. Therefore, in the first case, $f(\sigma) \leq n - 5$. Therefore, there exist integers i_4, i_5 $(1 \leq i_4, i_5 \leq n)$, distinct from i_1, i_2, i_3, and not left fixed by σ. Let $\eta = (i_3i_4i_5)$. Then

$$\sigma' = \eta\sigma\eta^{-1} = (i_1i_2i_4\ldots)\ldots,$$

and any integer left fixed by σ is also left fixed by σ'. Since N is normal, we have $\sigma' \in N$. Moreover, $\sigma\sigma'^{-1} \neq 1$, so that $f(\sigma\sigma'^{-1}) \leq f(\sigma)$ by the definition of σ. But $\sigma\sigma'^{-1}$ leaves fixed all integers which are left fixed by σ and $\sigma\sigma'^{-1}$ leaves i_2 fixed as well. Therefore, $f(\sigma\sigma'^{-1}) \geq f(\sigma) + 1$. Thus, a contradiction is reached if σ is of the form (5). Assume that σ is of the form (6). Let i_5 be an integer left fixed by σ and set $\eta = (i_3 i_4 i_5)$. Then

$$\sigma' = \eta\sigma\eta^{-1} = (i_1 i_2)(i_4 i_5) \ldots$$

is an element of N, which leaves fixed every integer left fixed by σ, except for i_5. Moreover, $\sigma \neq \sigma'$, so that $\sigma\sigma'^{-1} \neq 1$. Further, $\sigma\sigma'^{-1}$ leaves fixed the $f(\sigma) - 1$ integers which are left fixed both by σ and σ' and it leaves i_1 and i_2 fixed as well. Therefore, $\sigma\sigma'^{-1}$ leaves $f(\sigma) + 1$ integers fixed, which is a contradiction to the choice of σ. We have finally proved that the assumption $f(\sigma) \leq n - 4$ leads to a contradiction, and therefore $f(\sigma) = n - 3$. But then there are exactly three integers not left fixed by σ, so σ is a 3-cycle. Thus, N contains a 3-cycle as claimed.

3.8 *Exercises*

1. (a) For every element σ of S_4, decompose σ into a product of disjoint cycles and compute $N(\sigma)$.
   (b) Find A_4.
   (c) Write every element of A_4 as a product of 3-cycles.
2. Show that the decomposition of a permutation into a product of disjoint cycles (omitting 1-cycles) is unique up to the order of the cycles.
3. Show that the number of transpositions in S_n is $n(n-1)/2$.
4. Generalize Exercise 3 and show that the number of r-cycles in S_n is $n(n-1)\ldots(n-r+1)/r$.
5. In Theorem 6 we showed that A_n is of index 2 in S_n ($n \geq 2$) and that A_n is a normal subgroup of S_n. Show that if G is a group and H is a subgroup of index 2, then H is normal in G.
6. (a) Let $n_1, n_2, \ldots, n_t$ be positive integers. The *least common multiple* of $n_1, \ldots, n_t$ is the smallest positive integer m such that $n_i | m$ for all i ($1 \leq i \leq t$). Let

   $$n_i = \prod_p p^{a_i(p)}$$

   be the decomposition of n_i into a product of distinct prime powers. Show that the least common multiple is given by

   $$\prod_p p^{b(p)}$$

   where $b(p) = \max\{a_1(p), \ldots, a_t(p)\}$.
   (b) Let $\sigma_1, \ldots, \sigma_t$ be disjoint cycles having orders $n_1, \ldots, n_t$, respectively. Then the order of $\sigma_1 \ldots \sigma_t$ is the least common multiple of $n_1, \ldots, n_t$.
   (c) Is (b) true if the cycles are not disjoint?

7. Let G be a group, $a \in G$. An element of the form gag^{-1} for some $g \in G$ is called a *conjugate of* a. Define the relation $\sim$ on G by: $a \sim b \Leftrightarrow a$ is a conjugate of b. Show that $\sim$ is an equivalence relation. The equivalence classes of G with respect to $\sim$ are called the *conjugacy classes* of G.

8. Find all conjugacy classes of S_4.

9. Let $(i_1 \ldots i_r) \in S_n$ be an r-cycle and let $\sigma \in S_n$. Show that

$$\sigma(i_1 \ldots i_r)\sigma^{-1} = (\sigma(i_1)\sigma(i_2) \ldots \sigma(i_r)).$$

10. If $\sigma \in S_n$ is written in the standard manner as a product of disjoint cycles, say

$$\sigma = (i_1 \ldots i_r)(j_1 \ldots j_s)(k_1 \ldots k_t) \ldots ,$$

then the set of integers $\{r, s, t, \ldots\}$ is called the *cycle structure* of σ.
   (a) Show that σ is conjugate to σ' if and only if σ and σ' have the same cycle structures.
   (b) How many conjugacy classes does S_3 have? S_4? S_7? S_n?
   *(c) Find a formula for the number of elements of S_n having a given cycle structure.

11. Show that A_4 is not simple. (*Hint:* Look for normal subgroups of order 4 in A_4.)

12. Prove that A_4 is a group of order 12 which does not have a subgroup of order 6.

*13. Prove that A_n can be generated by two elements.

14. Let G be a simple group, H an arbitrary group, $f: G \longrightarrow H$ a homomorphism. Prove that either (a) f is an isomorphism, or (b) $f(g) = 1_H$ for all $g \in G$.

3.9 *Historical Perspectives The Theory of Groups*

At the end of the eighteenth century, Lagrange noticed that if the polynomial

$$f = X^n + a_1 X^{n-1} + \ldots + a_n$$

with complex coefficients has zeros $\alpha_1, \alpha_2, \ldots, \alpha_n$, then

$$f = (X - \alpha_1)(X - \alpha_2) \ldots (X - \alpha_n),$$

and therefore

$$\begin{aligned} a_1 &= -(\alpha_1 + \cdots + \alpha_n) \\ a_2 &= \alpha_1\alpha_2 + \alpha_1\alpha_3 + \cdots + \alpha_{n-1}\alpha_n \\ &\vdots \\ a_n &= (-1)^{n-1}\alpha_1 \cdots \alpha_n. \end{aligned}$$

Therefore, Lagrange observed, if $\alpha_1, \ldots, \alpha_n$ are subjected to any permutation, then the coefficients $a_1, \ldots, a_n$ are unaltered. To put it another way, the coefficients are symmetric functions of the zeros. It is this observation which is at the heart of Lagrange's method for solving equations of degree ≤ 4 in radicals. Moreover, it is this observation which makes the connection between groups and equations. We will pursue this theme in Chapter 9.

Elementary properties of permutations were found by Lagrange, Galois, and Abel, but it was the great French mathematician Cauchy who first made an exhaustive study of the elementary properties of permutation groups. Cauchy's work was published in a series of memoirs in the 1840s. It was Cauchy who first clearly enunciated the group concept, although Cauchy studied only groups of permutations. The abstract notion of a group was formulated by Arthur Cayley in 1853. It was Galois, a short-lived genius (1811–1832), who first recognized the importance of normal subgroups of a group, in connection with the theory of equations. We will come to this important connection in Chapter 9, when we prove the fundamental theorem of Galois theory.

The theory of finite groups has been an active field of research for nearly two centuries now, with the main problem being the classification of all finite groups into an organized listing of some form. It is possible to break the classification of all finite groups into two subproblems:

PROBLEM A: Determine all finite simple groups.

PROBLEM B (THE EXTENSION PROBLEM): Let H and K be groups. Determine, up to isomorphism, all finite groups G which contain H as a normal subgroup and such that $G/H \approx K$.

Problems A and B are both extremely difficult and no early solution of either is expected in spite of extensive efforts made during the last 80 years. In Chapter 8 we will prove that if problems A and B are solved, then we could determine all finite groups up to isomorphism.

The problem of determining all finite, simple groups has received a great deal of attention, especially in recent years, when many new examples of simple groups have been discovered.

The work on the extension problem has led to a field of mathematics called the *cohomology of groups*, which is a field of current interest.

In the latter part of the nineteenth century, a number of mathematicians, in particular Sophus Lie, began to study the class of infinite groups known as *continuous groups*. In modern terminology, these groups are called Lie groups. Typical examples of Lie groups are the groups of all translations, or of all rigid motions of the plane. The original motivation for studying Lie groups was provided by the theory of differential equations and their applications to mechanics. The theory of Lie groups is an entire discipline in itself and is currently of great importance to people interested in particle physics.

Another class of infinite groups is the class of *discontinuous groups*, which were extensively studied by Poincaré, Klein, and Fricke, as well as many others, in connection with certain functions of a complex variable, called *automorphic functions*. Today, the theory of automorphic functions is an immense field in which research is currently being done.

Thus, the reader can see that, although the idea of a group was first studied in the rather restricted setting of permutation groups, it has come to be one of the fundamental unifying concepts of mathematics.

4 THE THEORY OF RINGS

In Chapter 1 we introduced the notion of a group of permutations, in connection with our study of the solution of cubic equations in radicals. Then, in Chapter 3, we abstracted from this notion the general idea of a group and proceeded to establish many properties of groups. The net result of our efforts was a theory which applied to groups of permutations and to many other interesting situations as well. In this chapter we will pursue a similar plan to that carried out in Chapters 1 and 3 and will develop the theory of a type of algebraic system called a *ring*. The reader will recall that in Chapter 2 we devoted considerable energy to studying the properties of the integers **Z**. We subsequently found that with respect to the operation of addition, **Z** is an abelian group. However, **Z** is much more. In **Z**, there are two operations: $+$ and $\cdot$. Each of these operations is associative and commutative and has an identity. Each element a of **Z** has an inverse $-a$ with respect to addition. Finally, the two operations are connected via the distributive laws:

$$a\cdot(b+c) = a\cdot b + a\cdot c, \quad (b+c)\cdot a = b\cdot a + c\cdot a \qquad (a,b,c \in \mathbf{Z}).$$

In this chapter we will abstract from these properties of **Z** the notion of a ring. We will then study the properties of rings. Many of the phenomena which we have already observed in **Z** will carry over to other rings. And this is precisely the power of the "abstract approach" to the theory of rings.

4.1 *The Concept of a Ring*

Without further ado, let us formulate the definition of a ring.

DEFINITION 1: A *ring* is a nonempty set R on which there are defined two binary operations $+$ and $\cdot$, which satisfy the following axioms:

R1. With respect to $+$, R is an abelian group.

R2. $\cdot$ is associative: $a\cdot(b\cdot c) = (a\cdot b)\cdot c$ for all $a, b, c \in R$.

R3. The following distributive laws are satisfied:

$$\left.\begin{aligned} a\cdot(b + c) &= a\cdot b + a\cdot c \\ (b + c)\cdot a &= b\cdot a + c\cdot a \end{aligned}\right\} \text{for all } a, b, c \in R.$$

The operation $+$ is called *addition*, and the operation $\cdot$ is called *multiplication*.

In a ring R, we will denote the identity with respect to addition by 0, and we will denote the inverse of a with respect to $+$ by $-$a. We will call the identity with respect to addition the *zero element of R* or just *zero*.

Note that although we have assumed that R contains an identity with respect to addition, we have made no such assumption as regards multiplication. An identity element i with respect to multiplication is an element of R such that

$$i\cdot a = a\cdot i = a$$

for all $a \in R$. A given ring may or may not contain such an element. We will give examples of both phenomena below. However, if i and i' are two identity elements with respect to multiplication, then we must have

$$i = i\cdot i' = i'.$$

Therefore, any two identities of R are equal and R contains at most one identity element with respect to multiplication.

DEFINITION 2: Let R be a ring. If R contains an identity with respect to multiplication, then we say that R is a *ring with identity*. The unique identity with respect to multiplication will be denoted 1_R (or just 1 if R is clear from context) and will be called the *identity* of R. (Note that some authors require that all rings have an identity. We will not adopt this convention until the end of Chapter 4.)

We will see below that the operation of multiplication need not satisfy the commutative law. Therefore, let us make the following definition.

DEFINITION 3: Let R be a ring. We say that R is a *commutative* ring if

$$a\cdot b = b\cdot a$$

for all $a, b \in R$.

Let us now get some feel for the variety of possible rings by studying a number of examples.

EXAMPLE 1: Let $R = \mathbf{Z}$, with respect to the usual operations of addition and multiplication. Then as was mentioned at the outset, all the ring axioms are verified. The ring $\mathbf{Z}$ is commutative and has identity 1.

EXAMPLE 2: Let $R = \mathbf{Q}$, the set of all rational numbers, and let addition and multiplication, respectively, be addition and multiplication of fractions. With respect to these operations, $\mathbf{Q}$ becomes a ring. The ring $\mathbf{Q}$ is commutative and has identity 1.

EXAMPLE 3: Let $R = \mathbf{R}$, the set of all real numbers. Let addition and multiplication be, respectively, addition and multiplication of real numbers. Then $\mathbf{R}$ becomes a commutative ring with identity 1.

EXAMPLE 4: Let $R = \{2n \mid n \in \mathbf{Z}\} = \{\cdots, -4, -2, 0, 2, 4, \cdots\}$, and let the operations of addition and multiplication be the same as in $\mathbf{Z}$. Since the sum, difference, and product of even integers is even, the addition and multiplication in $\mathbf{Z}$ define binary operations on R, with respect to which R becomes a ring. R is clearly commutative, but has no identity. For if $2n \in R$ is an identity, then $(2n)\cdot 2 = 2$, which leads to a contradiction, since the left-hand side is divisible by 4, but the right-hand side is not. The ring R is sometimes denoted $2\mathbf{Z}$. It is possible to generalize this example as follows: Let r be an integer, and let

$$r\mathbf{Z} = \{rn \mid n \in \mathbf{Z}\}.$$

Then, with respect to the usual operations of addition and multiplication of integers, $r\mathbf{Z}$ is a commutative ring, having no identity unless $r = \pm 1$ (see Exercise 9 of this section).

EXAMPLE 5: Let $R = \{0\}$, and let us define addition and multiplication by

$$0 + 0 = 0,$$
$$0 \cdot 0 = 0.$$

Then the axioms R1–R3 are satisfied in a trivial way. Thus, R is a ring called the *trivial ring*. Note that R is commutative with identity 0. Thus, in this example, $1_R = 0_R$.

EXAMPLE 6: Let $R = \mathbf{Z}_n$ for some positive integer n, and let addition and multiplication on R be defined as addition and multiplication of residue classes, as introduced in Section 2.5. By the results which we established in Section 2.5, axioms R1–R3 are satisfied, so that R becomes a ring. Note that R is a finite ring with n elements. Moreover, R is commutative with identity $\bar{1}$. The most peculiar property exhibited by some of the rings $\mathbf{Z}_n$ is that it often happens that the product of two nonzero elements of $\mathbf{Z}_n$ equals $\bar{0}$, which is the zero element of $\mathbf{Z}_n$. For example, consider $\bar{2}, \bar{3} \in \mathbf{Z}_6$. Then $\bar{2} \neq \bar{0}$, $\bar{3} \neq \bar{0}$, but nevertheless $\bar{3}\cdot\bar{2} = \bar{6} = \bar{0}$. This situation should be viewed in contrast with the situation which occurs in Examples 1–5. In each of these examples, the product of two nonzero ring elements is nonzero.

EXAMPLE 7: Let R denote the set of functions $f: \mathbf{R} \longrightarrow \mathbf{R}$. If $f, g \in R$, let us define the sum $f + g$ and the product $f \cdot g$ by

$$(f + g)(x) = f(x) + g(x) \qquad (x \in \mathbf{R}),$$
$$(f \cdot g)(x) = f(x) \cdot g(x).$$

For example, if $f(x) = x$, $g(x) = 2x^2$, we have $(f + g)(x) = x + 2x^2$, and $(f \cdot g)(x) = 2x^3$. Note that we have defined the sum and product of functions by specifying the values of the sum and product for all values $x \in \mathbf{R}$. Thus, the sum and product of functions in R also belong to R, and so we have defined two binary operations on R. With respect to these operations, R is a ring. We have already verified in Section 3.2 that with respect to $+$, R is an abelian group. Thus, it suffices to check axioms R2 and R3. We leave this verification as an exercise. It is fairly easy to see that R is a commutative ring with identity. The identity element is the function I defined by $I(x) = 1$ for all $x \in R$. Note that as in Example 6, it can happen that the product of two nonzero elements of R is the zero element of R. For the zero element of R is the function $\mathbf{0}$ defined by $\mathbf{0}(x) = 0$ for all $x \in \mathbf{R}$. Let $f, g \in R$ be defined by

$$f(x) = 0 \quad (x \neq 0), \qquad f(0) = 1,$$
$$g(x) = 1 \quad (x \neq 0), \qquad g(0) = 0.$$

Then $f \neq \mathbf{0}$, $g \neq \mathbf{0}$, but $f \cdot g = \mathbf{0}$.

EXAMPLE 8: Let R be the set of all functions $f: \mathbf{R} \longrightarrow \mathbf{R}$ of the form

$$f(x) = a_0 + a_1 x + a_2 x^2 + \cdots + a_n x^n \qquad (a_0, \ldots, a_n \in \mathbf{R}).$$

Such a function is called a *polynomial function*. Let addition and multiplication of functions be defined as in Example 7. Then the sum (respectively, product) of two polynomial functions is again a polynomial function, so that we do indeed have two binary operations defined on R. With respect to these operations, R is a commutative ring with identity.

EXAMPLE 9: Let R denote the set of all 2×2 matrices with real entries. A typical element of R looks like

$$\begin{pmatrix} a & b \\ c & d \end{pmatrix},$$

where $a, b, c, d \in \mathbf{R}$. Let us define addition and multiplication of matrices by

$$\begin{pmatrix} a & b \\ c & d \end{pmatrix} + \begin{pmatrix} a' & b' \\ c' & d' \end{pmatrix} = \begin{pmatrix} a + a' & b + b' \\ c + c' & d + d' \end{pmatrix},$$
$$\begin{pmatrix} a & b \\ c & d \end{pmatrix} \cdot \begin{pmatrix} a' & b' \\ c' & d' \end{pmatrix} = \begin{pmatrix} aa' + bc' & ab' + bd' \\ ca' + dc' & cb' + dd' \end{pmatrix}.$$

With respect to these operations, R becomes a ring with identity (Exercise). The zero element and the identity of R are given by

$$0 = \begin{pmatrix} 0 & 0 \\ 0 & 0 \end{pmatrix}, \qquad 1 = \begin{pmatrix} 1 & 0 \\ 0 & 1 \end{pmatrix}.$$

Note that

$$\begin{pmatrix} 1 & 0 \\ 0 & 0 \end{pmatrix} \cdot \begin{pmatrix} 0 & 1 \\ 0 & 0 \end{pmatrix} = \begin{pmatrix} 0 & 1 \\ 0 & 0 \end{pmatrix},$$

$$\begin{pmatrix} 0 & 1 \\ 0 & 0 \end{pmatrix} \cdot \begin{pmatrix} 1 & 0 \\ 0 & 0 \end{pmatrix} = \begin{pmatrix} 0 & 0 \\ 0 & 0 \end{pmatrix}.$$

Therefore, R is *not* a commutative ring. We will henceforth denote the ring of 2×2 matrices with real entries by $M_2(\mathbf{R})$. Note that our last computation also shows that the product of two nonzero matrices may be zero.

EXAMPLE 10: Let $\mathbf{R}$ be the set of all real numbers with respect to the usual operations of addition and multiplication. Then the usual rules for operating with real numbers imply that the real numbers form a commutative ring with identity 1. Later in this chapter we will give a precise construction of the real numbers, using only the rational numbers. From this construction, the fact that the real numbers form a ring will be proved in detail.

EXAMPLE 11: Let $\mathbf{C}$ be the set of complex numbers $a + bi$, where a and b are real numbers and $i^2 = -1$. Then with respect to the usual rules for adding and multiplying complex numbers, the set of all complex numbers is a commutative ring with identity 1. Later in this chapter we will give a careful construction of the ring of complex numbers proceeding from the real numbers. Thus, we will refrain from discussing the complex numbers further here.

EXAMPLE 12: Let d be an integer, and let R consist of the set of all complex numbers of the form $a + b\sqrt{d}$, where $a, b \in \mathbf{Z}$. Let addition and multiplication in R be the usual addition and multiplication for complex numbers, respectively. Since

$$(a + b\sqrt{d}) + (a' + b'\sqrt{d}) = (a + a') + (b + b')\sqrt{d},$$

$$(a + b\sqrt{d})(a' + b'\sqrt{d}) = (aa' + bb'd) + (ab' + ba')\sqrt{d},$$

we see that addition and multiplication define binary operations on R. The fact that R forms a ring is now trivial to verify from the properties of the complex numbers. We leave the details as an exercise. The ring R will be denoted $\mathbf{Z}[\sqrt{d}]$.

EXAMPLE 13: Let R be the set of all complex numbers of the form $a + b\sqrt{d}$, where d is given, and $a, b \in \mathbf{Q}$. Then R is a ring which contains $\mathbf{Z}[\sqrt{d}]$. We will denote the ring of this example by $\mathbf{Q}(\sqrt{d})$.

EXAMPLE 14: Let R and S be any two rings, and let $R \times S$ denote the Cartesian product of R and S (as sets). Let us define addition and multiplication in $R \times S$ by

$$(r, s) + (r', s') = (r + r', s + s'),$$
$$(r, s)(r', s') = (rr', ss').$$

Then it is easy to verify that $R \times S$ is a ring, called the *Cartesian product* of R and S. Sometimes, the Cartesian product is called the *direct sum* of R and S and is denoted $R \oplus S$.

Now that we have surveyed the wide variety of possibilities which the ring axioms admit, let us establish some elementary properties of rings. The following proposition is a modest start in this direction.

PROPOSITION 4: Let R be a ring, $a, b \in R$. Then

(1) $a \cdot 0 = 0 \cdot a = 0$.
(2) $a \cdot (-b) = (-a) \cdot b = -(a \cdot b)$.
(3) $(-a) \cdot (-b) = a \cdot b$.

Note that these rules of arithmetic in an arbitrary ring R are just generalizations of well-known facts from the algebra of **Z**.

Proof of Proposition 4: (1) Since 0 is the identity with respect to addition, $0 + 0 = 0$. Therefore, by the distributive law,

$$a \cdot 0 = a \cdot (0 + 0) = a \cdot 0 + a \cdot 0.$$

Therefore, by adding $-(a \cdot 0)$ to both sides of this last equation, we get $a \cdot 0 = 0$. Similarly, $0 \cdot a = 0$.

(2) Let $y = -(a \cdot b)$. Then $a \cdot b + y = 0$. Therefore,

$$\begin{aligned}(-a) \cdot b &= (-a) \cdot b + 0 \\ &= (-a) \cdot b + [a \cdot b + y] \\ &= (-a + a) \cdot b + y \qquad \text{(associative and distributive laws)} \\ &= 0 \cdot b + y \qquad (\text{since } -a + a = 0) \\ &= 0 + y \qquad \text{[by Part (1) of the present proposition]} \\ &= y \qquad \text{[by Axiom R1 in Section 4.1].}\end{aligned}$$

This completes the proof of part (2).

(3) By part (2),

$$\begin{aligned}(-a) \cdot (-b) &= -a \cdot (-b) \\ &= -[-a \cdot b] \\ &= a \cdot b \text{ (since } R \text{ is a group with respect to addition, and} \\ &\quad \text{Exercise 5 of Section 3.1 applies).}\end{aligned}$$

Let us now examine some of the phenomena which were exhibited by our examples. In Examples 6 and 7 we found that there exist nonzero ring elements a and b such that $a \cdot b = 0$. Note that if a or b is zero, then $a \cdot b = 0$ by Proposition 4, part (1). However, it is somewhat surprising that nonzero elements can have zero product. Therefore, we are prompted to make the following definition.

DEFINITION 5: Let R be a ring, $a \in R, a \neq 0$. We say that a is a *right zero divisor* (respectively, *left zero divisor*) if there exists $b \in R, b \neq 0$, such that $a \cdot b = 0$ (respectively, $b \cdot a = 0$). An element of R is called a *zero divisor* of R if it is either a right or left zero divisor.

If R is a commutative ring, then the distinction between right and left zero divisors disappears. An element of a commutative ring R is a right zero divisor if and only if it is a left zero divisor.

DEFINITION 6: A commutative ring R is said to be an *integral domain*† if R contains no zero divisors. In other words, R is an integral domain if the product of any two nonzero elements of R is nonzero.

For example, $\mathbf{Z}_6$ is not an integral domain because $\bar{2}$ is a zero divisor. Similarly, the discussion of Example 7 shows that the ring of functions $f: \mathbf{R} \to \mathbf{R}$ is not an integral domain. On the other hand, $\mathbf{Z}$ is an integral domain. For if a and b are nonzero integers, then $a \cdot b \neq 0$ (Theorem 4 of Section 2.3). Moreover, if p is prime, then $\mathbf{Z}_p$ is an integral domain. (Exercise.)

PROPOSITION 7: Suppose that R is an integral domain, $a, b, c \in R, a \neq 0$. Suppose that $a \cdot b = a \cdot c$. Then $b = c$.

Proof: By Proposition 4 and the distributive law, we have

$$\begin{aligned} 0 &= a \cdot b - a \cdot c = a \cdot b + a \cdot (-c) \\ &= a \cdot (b - c). \end{aligned}$$

However, since $a \neq 0$ and R is an integral domain, $b - c = 0$, and thus $b = c$.

If R is a ring with identity 1, it makes sense to ask whether an element a of R has an inverse with respect to multiplication. That is, does there exist $b \in R$ such that

$$a \cdot b = b \cdot a = 1?$$

If such an element b exists, it is unique. We have already worked through this sort of argument several times, and therefore we will leave the verification that b is unique as an exercise for the reader. If a has an inverse with respect to multiplication, then the unique inverse will be denoted a^{-1}.

DEFINITION 8: Let R be a ring with identity 1, and let $a \in R$. If a has an inverse with respect to multiplication, then we say that a is a *unit* of R.

† For convenience in stating various results, we will always assume that an integral domain always contains an identity.

EXAMPLE 15: Let R be any ring with identity. Then 1 and -1 are units. In fact, since $1 \cdot 1 = 1, (-1) \cdot (-1) = 1 \cdot 1 = 1$ (Proposition 4), we see that

$$1^{-1} = 1, \qquad (-1)^{-1} = -1.$$

EXAMPLE 16: Let $R = \mathbf{Z}$. Then the only units of R are 1 and -1. For if $a \in R$ is a unit, there exists $b \in R$ such that $a \cdot b = 1$. But then, by Exercise 12 of this section, $a = b = 1$ or $a = b = -1$.

EXAMPLE 17: Let $R = \mathbf{Q}$. Then 0 is not a unit of R since $0 \cdot b = 0 \neq 1$ for all $b \in R$. However, if $a/b \in \mathbf{Q} - \{0\}$, then a/b is a unit of R. In fact, since

$$\frac{a}{b} \cdot \frac{b}{a} = 1,$$

we see that $(a/b)^{-1} = b/a$.

EXAMPLE 18: Let $R =$ the trivial ring. Then, since $1 = 0$ in this ring, 0 is a unit.

EXAMPLE 19: Let R be any ring with identity which is not the trivial ring. In this ring, $1 \neq 0$. For if R is not the trivial ring, there exists a nonzero element in R. Pick one such element a. Then, if $1 = 0$, we would have

$$1 \cdot a = 0 \cdot a = 0.$$

However, $1 \cdot a = a \neq 0$. Thus, we reach a contradiction, so that $1 \neq 0$. Let us now show that 0 is not a unit of R. Indeed, for all $b \in R$, we have $0 \cdot b = 0 \neq 1$. Thus in any ring which is not the trivial ring, 0 is not a unit.

EXAMPLE 20: Let $R = \{a + b\sqrt{d} \mid a, b \in \mathbf{Z}\}$ be the ring of Example 12. The units of R are not so easy to write down, as the following special cases show:

$d = -1$: $1 + 0\sqrt{-1}, -1 + 0\sqrt{-1}, 0 + \sqrt{-1}, 0 - \sqrt{-1}$ are units with respective inverses $1 + 0\sqrt{-1}, -1 + 0\sqrt{-1}, 0 - \sqrt{-1}, 0 + \sqrt{-1}$.

$d = 2$: $1 + 0\sqrt{2},\ -1 + 0\sqrt{2},\ 1 + \sqrt{2},\ 1 - \sqrt{2},\ 3 + 2\sqrt{2},\ 3 - 2\sqrt{2}, -3 + 2\sqrt{2}, -3 - 2\sqrt{2}$ are units with respective inverses $1 + 0\sqrt{2}, -1 + 0\sqrt{2}, -1 + \sqrt{2}, -1 - \sqrt{2}, 3 - 2\sqrt{2}, 3 + 2\sqrt{2}, -3 - 2\sqrt{2}, -3 + 2\sqrt{2}$. [All these units are powers of $\pm(1 + \sqrt{2})^n$ for various values of n.]

It is worthwhile to record the observation made in Example 19 as a proposition.

PROPOSITION 21: Let R be a ring with identity which is not the trivial ring. Then $1_R \neq 0_R$.

For any ring with identity, let us denote by U_R the set of all units of R. From our above examples, we know that $1, -1 \in U_R$, and that if R is not the trivial ring, then $0 \notin U_R$.

PROPOSITION 22: Let R be a ring with identity, and let $a, b \in U_R$.

(1) $a^{-1} \in U_R$.
(2) $a \cdot b \in U_R$ and $(a \cdot b)^{-1} = b^{-1} \cdot a^{-1}$.
(3) U_R is a group with respect to multiplication.

Proof: (1) Since $a \in U_R$, a has an inverse a^{-1} and $a \cdot a^{-1} = a^{-1} \cdot a = 1$. But these last equations imply that a^{-1} is a unit of R and that its inverse is is a. Thus, we have part (1).

(2) Since $a \cdot a^{-1} = a^{-1} \cdot a = 1$, $b \cdot b^{-1} = b^{-1} \cdot b = 1$, we see that

$$(a \cdot b) \cdot (b^{-1} \cdot a^{-1}) = a \cdot 1 \cdot a^{-1} = 1,$$
$$(b^{-1} \cdot a^{-1}) \cdot (a \cdot b) = b^{-1} \cdot 1 \cdot b = 1.$$

These last two equations show that $a \cdot b$ is a unit of R and that its inverse is $b^{-1} \cdot a^{-1}$. This proves part (2).

(3) By part (2), U_R is closed with respect to multiplication. Moreover, we observed above that $1 \in U_R$, so that U_R has an identity with respect to multiplication. Multiplication in U_R is associative since multiplication in R is associative. Finally, by part (1), if $a \in U_R$, then a has an inverse with respect to multiplication in U_R, namely a^{-1}. Thus, U_R satisfies the group axioms.

The group U_R is called the *group of units* of R.

In order to introduce the last of the special classes of rings which we will examine in this section, let us confine our discussion to a ring R with identity, which is not the trivial ring. Let us denote the subset $R - \{0\}$ of R by $R^\times$. We showed in Example 19 that 0 is not a unit of R. Therefore, $U_R \subseteq R^\times$. Note that it is not always true that $U_R = R^\times$. For example, we saw above that if $R = \mathbf{Z}$, then $U_R = \{1, -1\}$. However, we also found that if $R = \mathbf{Q}$, then $U_R = \mathbf{Q} - \{0\} = R^\times$. Thus, for some rings R we indeed have $U_R = R^\times$. Such rings play an exceedingly important role in algebra, for reasons which we will discuss below. Before we give this class of rings a name, let us observe that $U_R = R^\times$ if and only if every nonzero element of R has an inverse with respect to multiplication. Thus, let us make the following definition.

DEFINITION 23: A *field* is a nontrivial commutative ring with identity in which every nonzero element has an inverse with respect to multiplication.

Note that a field must contain at least two elements, since it is a nontrivial ring. Therefore, in particular, in a field, we have $1 \neq 0$. Note also that we have assumed that a field is commutative. Some texts define a field without this condition. And, indeed, there are many systems which satisfy the axioms for a field, with the exception of the commutativity. Such systems are often called *skew fields*, *division algebras*, or *division rings*, all these terms being used interchangeably. But we will not meet any of these objects in this book, so that it seems most convenient to incorporate the commutativity in the definition of a field.

EXAMPLE 24: As a result of our discussion, we see that $\mathbf{Q}$ is a field.

EXAMPLE 25: $\mathbf{R}$, the set of all real numbers, is a field which we will examine at some length later in this chapter.

EXAMPLE 26: Let p be a prime and let $\mathbf{Z}_p$ denote the ring of residue classes modulo p. Recall that $\mathbf{Z}_p = \{\bar{0}, \bar{1}, \ldots, \overline{p-1}\}$. Since p is a prime, we have $(i, p) = 1$ for $i = 1, 2, \ldots, p-1$. Therefore, each of the residue classes $\bar{1}, \bar{2}, \ldots, \overline{p-1}$ is reduced. Therefore, since $\bar{0}$ is clearly not reduced, we have $\mathbf{Z}_p^\times = \{\bar{1}, \bar{2}, \ldots, \overline{p-1}\}$. That is, every nonzero element of $\mathbf{Z}_p$ is a reduced residue class. However, in Proposition 4 of Section 2.5 we showed that every reduced residue class has an inverse with respect to multiplication in $\mathbf{Z}_p$. Therefore, $\mathbf{Z}_p$ is a field. In fact, $\mathbf{Z}_p$ is a field with p elements. A field with a finite number of elements is called a *finite field*. Therefore, we can summarize this example by saying that $\mathbf{Z}_p$ is a finite field.

PROPOSITION 27: Let F be a nontrivial commutative ring with identity. Then F is a field if and only if $F^\times$ is a group with respect to multiplication.

Proof: $\Rightarrow$ Assume that F is a field. Then by the definition of a field, $F^\times = U_F$. However, we have shown that U_F is a group. Therefore, $F^\times$ is a group.
$\Leftarrow$ Assume that $F^\times$ is a group. Then every element of $F^\times$ has an inverse with respect to multiplication. Therefore, F is a field.

COROLLARY 28: Let F be a field. Then F is an integral domain.

Proof: If F is a field, then $F^\times$ is a group by Proposition 27. In particular, $F^\times$ is closed with respect to multiplication. Therefore, F is an integral domain.

The extreme importance of fields in mathematics is due to the fact that in a field it is possible to combine elements using all the operations which one meets in elementary arithmetic. That is, given any two elements a and b of a field, it is possible to form their sum $(a + b)$, difference $(a - b)$, and product $(a \cdot b)$ and, if $b \neq 0$, their quotient $(a \cdot b^{-1})$.

If R is a ring with identity, then it is convenient to introduce the language of exponents in R. Let $a \in R$, n a nonnegative integer. Then we define a^n inductively by

$$a^0 = 1,$$

$$a^{n+1} = a \cdot a^n.$$

Then the following laws of exponents hold:

$$a^n a^m = a^{n+m}, \tag{1}$$

$$(a^n)^m = a^{nm}. \tag{2}$$

Moreover, if R is commutative, then

$$(ab)^n = a^n b^n. \tag{3}$$

4.1 Exercises

1. Fill in the details necessary to verify that Examples 1–11 define rings.
2. Find all zero divisors in $\mathbf{Z}_{16}$.
3. Prove that $\mathbf{Z}_n$ is not an integral domain if n is not prime, $n > 1$.
4. Which of the following rings is an integral domain (respectively, field)?
   (a) $\mathbf{Z}$. (b) $\mathbf{Z}_7$. (c) $\mathbf{Z}_4$.
   (d) $\{f: \mathbf{R} \longrightarrow \mathbf{R} \mid f(0) = 0\}$.
   (e) $\{f: \mathbf{R} \longrightarrow \mathbf{R} \mid f \text{ is a polynomial function}\}$.
5. Let R be a ring $a, b, c, d \in R$. Show that
$$(a + b)(c + d) = ac + ad + bc + bd.$$
6. Let R be any ring, n a positive integer, $a \in R$. Define $n \cdot a$ by
$$n \cdot a = 0 \quad \text{if } n = 0$$
$$= \underbrace{a + a + \cdots + a}_{n \text{ times}} \quad \text{if } n > 1$$
   (a) Show that if R is commutative, and $a, b \in R$, then
$$(a + b)^2 = a^2 + 2 \cdot ab + b^2,$$
$$(a + b)^3 = a^3 + 3 \cdot a^2 b + 3 \cdot ab^2 + b^3.$$
   (b) Do the results of (a) hold if R is not commutative?
   (c) Assume that R is commutative. Let $n \geq m \geq 0$ and let $\binom{n}{m}$ denote the binomial coefficient
$$\binom{n}{m} = \frac{n(n-1) \cdots (n-m+1)}{1 \cdot 2 \cdot \cdots \cdot m}.$$
   Prove the binomial theorem in R:
$$(a + b)^n = a^n + \binom{n}{1} a^{n-1} b + \binom{n}{2} a^{n-2} b^2 + \cdots + \binom{n}{n} b^n.$$
7. Determine U_R in case $R = \mathbf{Z}[\sqrt{-1}]$; $R = \mathbf{Z}[\sqrt{-3}]$; $R = \mathbf{Z}[\sqrt{-5}]$.

8. Determine U_R in case $R = M_2(\mathbf{R})$.
9. Show that if $r > 1$, then $r\mathbf{Z}$ does not have an identity.
10. Let R be a ring with identity, $a \in R$. Show that there exists at most one $b \in R$ such that $ab = ba = 1$.
11. Prove the laws of exponents (1)–(3).
12. Let $a,b \in \mathbf{Z}$ be such that $ab = 1$. Show that either $a = b = 1$ or $a = b = -1$.
13. Let R be any ring, and let (a_{ij}) $(i, j = 1, 2, \ldots, n)$ be a set of n^2 elements of R. Such a collection of elements of R (with appropriate indexing) is called an $n \times n$ matrix with entries in the ring R. Define addition and multiplication of such matrices by

$$(a_{ij}) + (b_{ij}) = (c_{ij}), \quad c_{ij} = a_{ij} + b_{ij}, \qquad i, j = 1, 2, \ldots, n,$$

$$(a_{ij}) \cdot (b_{ij}) = (d_{ij}), \quad d_{ij} = \sum_{k=1}^{n} a_{ik} b_{kj}.$$

Show that with respect to these operations, the set of $n \times n$ matrices with entries in R form a ring, denoted $M_n(R)$.

14. Refer to Exercise 13. Suppose that $n > 1$ and that R has an identity. Show that $M_n(R)$ is not commutative and has divisors of zero.
15. Refer to Exercise 13 for notation. Let us define the following 2×2 matrices belonging to $M_2(\mathbf{C})$, where $\mathbf{C}$ = the complex numbers.

$$\mathbf{1} = \begin{pmatrix} 1 & 0 \\ 0 & 1 \end{pmatrix}, \quad i = \begin{pmatrix} \sqrt{-1} & 0 \\ 0 & -\sqrt{-1} \end{pmatrix}, \quad j = \begin{pmatrix} 0 & 1 \\ -1 & 0 \end{pmatrix}, \quad k = \begin{pmatrix} 0 & \sqrt{-1} \\ \sqrt{-1} & 0 \end{pmatrix},$$

and let $\mathbf{H}$ denote the set of all matrices of the form $a_1\mathbf{1} + a_2 i + a_3 j + a_4 k$, where $a_i \in \mathbf{R}$, and where the product aA of a real number a and a matrix $A = (a_{ij})$ is defined to be the matrix (aa_{ij}) $(a_{ij} \in \mathbf{R})$.

(a) Show that $ij = -ji = k$; $jk = -kj = i$; $ki = -ik = j$.

(b) Show that with respect to the operations of addition and multiplication of matrices, $\mathbf{H}$ forms a noncommutative ring.

*(c) Show that $\mathbf{H}$ is a division algebra. $\mathbf{H}$ is called the division ring of *Hamilton's quaternions*.

16. Let A be any set and let R be any commutative ring, and let $R_A = \{f\colon A \longrightarrow R\}$. Define addition and multiplication on R_A as follows: If $f, g \in R_A$, set $(f + g)(a) = f(a) + g(a)$ $(a \in A)$; $(fg)(a) = f(a)g(a)$. Show that with respect to these operations, R_A is a ring.

4.2 *Polynomial Rings*

As every student of high school algebra knows,

$$X + X^2, \qquad 5 + X^3, \qquad 17 + XY + Z^2W, \qquad 3 + 2X$$

are all examples of polynomials. In these examples the coefficients of the polynomials all belong to the field of real numbers. In this section we will generalize the notion of a polynomial to allow for coefficients in an arbitrary ring. In doing so, we will show that the set of all polynomials in X having coefficients in the ring R is itself a ring, with respect to suitably defined addition and multiplication of polynomials. These rings of polynomials provide us with yet another interesting class of rings. Moreover, rings of polynomials are important in their own right, since they will provide us with a technical tool by means of which we can study the properties of arbitrary rings.

Let R be a ring and let X be a symbol. (X is an "indeterminate.") Then a *polynomial in X with coefficients in R* is a formal sum

$$a_0 + a_1X + a_2X^2 + \ldots,$$

where $a_i \in R$ for all i and $a_i = 0$ for i sufficiently large. The set of all polynomials in X with coefficients in R will be denoted $R[X]$. Note that every element of R is contained in $R[X]$. The elements of R are called *constant polynomials.*

Let us adopt a notational convention for writing polynomials. In specifying a polynomial, we will omit all terms which appear with a zero coefficient. Thus, instead of

$$1 + 2X + 0\,X^2 + 0X^3 + \ldots \in \mathbf{Z}[X],$$

we will write

$$1 + 2X.$$

By this convention, the typical polynomial for which $a_i = 0$ for $i > n$ will be denoted

$$a_0 + a_1X + \ldots + a_nX^n.$$

The one exception to this convention will be the polynomial

$$0 + 0X + 0X^2 + \ldots,$$

which will be denoted simply by $\mathbf{0}$. The polynomial $\mathbf{0}$ will be called the *zero polynomial.*

Let

$$f = a_0 + a_1X + a_2X^2 + \cdots, \tag{1}$$

$$g = b_0 + b_1X + b_2X^2 + \cdots \tag{2}$$

be two polynomials. We say that *f is equal to g*, denoted $f = g$, if

$$a_0 = b_0, a_1 = b_1, a_2 = b_2, \ldots.$$

Thus, two polynomials are equal if their corresponding coefficients are equal.

Addition of Polynomials

Let f and g be given by (1) and (2), respectively. The sum $f + g$ is defined by

$$f + g = (a_0 + b_0) + (a_1 + b_1)X + (a_2 + b_2)X^2 + \cdots. \tag{3}$$

Thus, to add two polynomials, we merely add corresponding coefficients. Note that (3) actually defines a polynomial—that is, $a_i + b_i = 0$ for all sufficiently large i. Indeed, suppose that $a_i = 0$ for $i \geq N$ and that $b_i = 0$ for $i \geq M$. Then $a_i + b_i = 0$ for all $i \geq \max(M, N)$, where $\max(M, N)$ denotes the larger of M and N. It is trivial to see that with respect to the operation $+$, $R[X]$ becomes an abelian group. The identity element is $\mathbf{0}$ and the inverse of f under addition is

$$-f = (-a_0) + (-a_1)X + (-a_2)X^2 + \cdots.$$

Multiplication of Polynomials

Let f and g be defined by (1) and (2), respectively, We define the product $f \cdot g$ by

$$f \cdot g = c_0 + c_1 X + c_2 X^2 + \cdots,$$

where

$$\begin{aligned} c_0 &= a_0 \cdot b_0 \\ c_1 &= a_0 \cdot b_1 + a_1 \cdot b_0 \\ c_2 &= a_0 \cdot b_2 + a_1 \cdot b_1 + a_2 \cdot b_0 \\ &\cdot \\ &\cdot \\ &\cdot \\ c_n &= a_0 \cdot b_n + a_1 \cdot b_{n-1} + \cdots + a_{n-1} \cdot b_1 + a_n \cdot b_0. \end{aligned}$$

Let us see what this definition really says. Let us make the agreement that the polynomial $a_0 X^0$ is just the polynomial $a_0 + 0X + 0X^2 + \cdots$. Then the definition of multiplication essentially amounts to the following: To form $f \cdot g$, take the product of each term of f by each term of g using the rule $a_i X^i \cdot b_j X^j = a_i \cdot b_j X^{i+j}$ $(i, j \geq 0)$, and collect all terms containing the same power of X. Thus, our definition of multiplication coincides with the familiar definition of multiplication of polynomials with coefficients in $\mathbf{R}$.

Note that $f \cdot g$ is a polynomial in X with coefficients in R—that is, $c_i = 0$ for i sufficiently large. Indeed, let M and N be so large that $a_i = 0$ for all $i \geq M$ and $b_i = 0$ for all $i \geq N$. Assume that $i \geq M + N$. Then

$$c_i = a_0 \cdot b_i + a_1 \cdot b_{i-1} + \cdots + a_i \cdot b_0.$$

The typical term in this sum is $a_j \cdot b_{i-j}$ $(0 \leq j \leq i)$. Since $i \geq M + N$, we must have either $j \geq M$ or $i - j \geq N$. [For if both of $j < M, i - j < N$ hold, then

$$i = j + (i - j) < M + N.]$$

Therefore, for $i \geq M + N$, we must have either $a_j = 0$ or $b_{i-j} = 0$ $(0 \leq j \leq i)$, so that every term in the expression for c_i is zero. Therefore, $c_i = 0$ for $i \geq M + N$, and $f \cdot g$ is a polynomial.

It is not obvious that multiplication of polynomials is associative or that the distributive laws are satisfied. By way of illustration, let us verify the right distributive law. Let f and g be given by (1) and (2), respectively, and let

$$h = c_0 + c_1X + \cdots \in R[X], \qquad c_i \in R. \tag{4}$$

Then

$$(f + g)\cdot h = d_0 + d_1X + d_2X^2 + \cdots,$$

where

$$d_i = (a_0 + b_0)c_i + (a_1 + b_1)c_{i-1} + \cdots + (a_i + b_i)c_0 \qquad (i = 0, 1, \ldots) \tag{5}$$

Moreover,

$$f\cdot h = e_0 + e_1X + e_2X^2 + \cdots,$$
$$g\cdot h = f_0 + f_1X + f_2X^2 + \cdots,$$

where

$$e_i = a_0\cdot c_i + a_1\cdot c_{i-1} + \cdots + a_i\cdot c_0 \quad (i = 0, 1, \ldots), \tag{6}$$

$$f_i = b_0\cdot c_i + b_1\cdot c_{i-1} + \cdots + b_i\cdot c_0 \quad (i = 0, 1, \ldots). \tag{7}$$

Finally,

$$f\cdot h + g\cdot h = (e_0 + f_0) + (e_1 + f_1)X + \cdots.$$

However, by (5), (6), (7), and the distributive law in R, we see that $e_i + f_i = d_i$ $(i = 0, 1, \ldots)$, so that

$$(f + g)\cdot h = f\cdot h + g\cdot h.$$

This proves the right distributivite law in $R[X]$. Similarly, it is possible to prove the left distributive law and the associativity of multiplication. We leave these proofs as exercises for the reader. Modulo these exercises, we have

THEOREM 1: $R[X]$ is a ring.

The ring $R[X]$ is called the *ring of polynomials in X over R*. The next result is a result which is typical in the theory of polynomial rings. It asserts that certain of the properties of the ring R carry over to the ring $R[X]$.

THEOREM 2: Let R be a ring, X an indeterminate over R.

(1) If R is a ring with identity 1, then $R[X]$ is a ring with identity

$$\mathbf{1} = 1 + 0X + 0X^2 + \cdots.$$

(2) If R is commutative, then $R[X]$ is commutative.

(3) If R is an integral domain then $R[X]$ is an integral domain.

Proof: (1) Let $f \in R[X]$ be given by Equation (1). Then

$$f\cdot\mathbf{1} = c_0 + c_1X + c_2X^2 + \cdots,$$

where

$$c_0 = a_0 \cdot 1 = a_0$$
$$c_1 = a_1 \cdot 1 + a_0 \cdot 0 = a_1$$
$$c_2 = a_2 \cdot 1 + a_1 \cdot 0 + a_2 \cdot 0 = a_2$$
$$\vdots$$
$$c_n = a_n \cdot 1 + a_{n-1} \cdot 0 + \cdots + a_0 \cdot 0 = a_n.$$

Therefore, $f \cdot \mathbf{1} = f$. Similarly, $\mathbf{1} \cdot f = f$. Therefore, $R[X]$ is a ring with the identity **1**.

(2) Let f and g be given by Equations (1) and (2), respectively, and let

$$f \cdot g = c_0 + c_1 X + c_2 X^2 + \cdots,$$
$$g \cdot f = d_0 + d_1 X + d_2 X^2 + \cdots.$$

Then, by the definition of multiplication of polynomials,

$$c_n = a_n \cdot b_0 + a_{n-1} \cdot b_1 + \cdots + a_0 \cdot b_n,$$
$$d_n = b_n \cdot a_0 + b_{n-1} \cdot a_1 + \cdots + b_0 \cdot a_n.$$

But if R is commutative, $a_i \cdot b_j = b_j \cdot a_i$ for all i, j and therefore, $c_n = d_n$ for all n. Thus, $f \cdot g = g \cdot f$. Since f and g are arbitrary elements of $R[X]$, we see that $R[X]$ is commutative.

(3) By the definition of an integral domain it suffices to show that, if R is an integral domain, then the product of two nonzero polynomials is nonzero. Let

$$f = a_0 + a_1 X + \cdots + a_m X^m, \qquad a_m \neq 0,$$
$$g = b_0 + b_1 X + \cdots + b_n X^n, \qquad b_n \neq 0.$$

We may assume that f and g have this form since both are assumed to be nonzero. Then the coefficient of X^{m+n} in $f \cdot g$ is given by $a_m \cdot b_n$. However, since $a_m \neq 0$, $b_n \neq 0$, and R is an integral domain, we have $a_m \cdot b_n \neq 0$. Thus, $f \cdot g$ has a nonzero coefficient and hence is nonzero.

COROLLARY 3: Let F be a field. Then $F[X]$ is a commutative integral domain.

Proof: F is a commutative integral domain with identity.

Note, however, that $F[X]$ is not a field since, for example, X has no multiplicative inverse.

Let R be a ring, X an indeterminate over R and f a polynomial in $R[X]$. If f is not the zero polynomial, then

$$f = a_0 + a_1 X + \cdots + a_n X^n \qquad a_n \neq 0.$$

In this case, we say that f has *degree n*, and we write $\deg(f) = n$. Defining the degree of the zero polynomial presents something of a problem. Strictly as a matter of convenience, let us introduce a symbol $-\infty$ called *minus infinity*. We will perform arithmetic with $-\infty$ according to the following rules:

$$-\infty + n = -\infty, \qquad -\infty + (-\infty) = -\infty, \qquad -\infty < n,$$

for every integer n. Then, let us set the degree of the zero polynomial equal to $-\infty$. Thus, the zero polynomial has smaller degree than any nonzero polynomial. The reason for the very wierd choice of degree for the zero polynomial is explained by the following result:

PROPOSITION 4: Let R be an integral domain, X an indeterminate over R, $f, g \in R[X]$. Then $\deg(fg) = \deg(f) + \deg(g)$.

Proof: If either f or g is zero, then $fg = \mathbf{0}$ and both sides of the above equation are $-\infty$. Therefore, it suffices to assume that f and g are both nonzero. Assume that $\deg(f) = m$, $\deg(g) = n$. Then f and g have the form

$$f = a_0 + a_1X + \cdots + a_mX^m, \qquad a_m \neq 0,$$
$$g = b_1 + b_1X + \cdots + b_nX^n, \qquad b_n \neq 0.$$

Then $f \cdot g$ is of the form

$$f \cdot g = c_0 + c_1X + \cdots + c_{m+n}X^{n+m},$$

where, in particular, $c_{m+n} = a_m \cdot b_n$. But since R is an integral domain and a_m and b_n are nonzero, we see that $c_{m+n} \neq 0$. Therefore, $\deg(f + g) = m + n$.

Since we made the agreement that $-\infty < n$ for all integers n, we may state the following additive analogue of Proposition 4.

PROPOSITION 5: Let R be a ring, X an indeterminate over $R, f, g \in R[X]$. Then $\deg(f + g) \leq \max(\deg(f), \deg(g))$, where $\max(\deg(f), \deg(g))$ denotes the larger of $\deg(f)$ and $\deg(g)$. If $\deg(f) \neq \deg(g)$, then the inequality may be replaced by equality.

Proof: Exercise. (Consider separately the cases $f + g = \mathbf{0}, f + g \neq \mathbf{0}$.)

We have thus far considered the addition, subtraction, and multiplication of polynomials. Let us now begin to consider the problem of division of one polynomial by another. Let us recall what happens in the case of polynomials with real coefficients. Let f and g be polynomials with real coefficients, $g \neq \mathbf{0}$. In high school algebra we considered the problem of "dividing" f by g. We learned a procedure called "long division," whereby f/g could be written as a quotient q plus a remainder of the form r/g, where q and r are polynomials. For example, if $f = X^5 + 2X^3 + X + 1$ and $g = 2X^3 + 2$, we have $q = \frac{1}{2}X^2 + 1, r = -X^2 + X - 1$. The reader is urged to carry out this computation in order to recall the procedure. By inspection of the long-division process, we can see that $\deg(r) < \deg(g)$. Thus, we can express the process of long division as follows: There exist polynomials q and r such that $f = qg + r$, and $\deg(r) < \deg(g)$. It makes sense to ask whether such a process of division can be carried out in a polynomial ring. Note that in the above example, the coefficients of f and g lie in $\mathbf{Z}$, but nevertheless, the coefficients of q and r involve rational numbers. This suggests that it would be wise to confine our investigations to polynomial rings over a field. With this restriction, the process of long division can be carried over to polynomial rings. The resulting process in $F[X]$ is called the *division algorithm* in $F[X]$.

THEOREM 6: Let F be a field, X an indeterminate over F, and let $f, g \in F[X]$, $g \neq \mathbf{0}$. Then there exist polynomials q and r belonging to $F[X]$ such that

$$\text{(a) } f = qg + r,$$

and

$$\text{(b) } \deg(r) < \deg(g).$$

Proof: If $f = \mathbf{0}$, then we may set $q = r = \mathbf{0}$. Then $\deg(g) \geq 0$, since $g \neq \mathbf{0}$, so that $\deg(r) < \deg(g)$. Thus, we may assume that $\deg(f) \geq 0$. Let us proceed by induction on the degree of f. We leave it to the reader to verify the theorem if $\deg(f) = 0$, in which case f is a nonzero constant. Let us assume the theorem for polynomials of degree $\leq n$, and let us assume that f has degree $n + 1$. Suppose that

$$f = a_{n+1}X^{n+1} + a_nX^n + \cdots + a_0, \qquad a_i \in F, \quad a_{n+1} \neq 0,$$
$$g = b_mX^m + \cdots + b_0, \qquad b_i \in F, \quad b_m \neq 0.$$

If $m > n + 1$, then we may set $f = r$, $\mathbf{0} = q$. Therefore, we may assume that $n + 1 \geq m$.
Set

$$f' = f - a_{n+1}b_m^{-1}X^{n+1-m}g.$$

Then a quick computation shows that $\deg(f')$ is at most n. Therefore, by induction there exist polynomials q' and r such that $f' = q'g + r$, $\deg(r) < \deg(g)$. But, then, if we set

$$q = a_{n+1}b_m^{-1}X^{n+1-m} + q',$$

we see that $f = qg + r$, $\deg(r) < \deg(g)$. Thus, the induction is complete.

Remark: The polynomials q and r satisfying (a) and (b) are unique. We leave the proof of this fact as an exercise for the reader.

Thus far, we have considered polynomial rings in only one indeterminate X. However, if $R[X]$ is a polynomial ring and Y is an indeterminate over

$R[X]$, then we may form the polynomial ring $R[X][Y]$. A typical element of $R[X][Y]$ is of the form

$$a_0(X) + a_1(X)Y + \cdots + a_n(X)Y^n, \qquad a_i(X) \in R[X],$$

or

$$b_0 + b_{01}X + b_{10}Y + b_{20}X^2 + b_{11}XY + b_{02}Y^2 + \cdots.$$

Let us make the convention that $XY = YX$ in $R[X][Y]$. Then, in particular, we see that $R[X][Y] = R[Y][X]$. We will denote $R[X][Y]$ simply by $R[X, Y]$. Then by what we have said, $R[X, Y] = R[Y, X]$. This ring is said to be a polynomial ring in two variables (indeterminates) over R. Similarly, we can construct rings $R[X_1, \ldots, X_n]$ in n variables over R. Such rings are of critical importance to geometric investigations, since geometrical objects (curves, surfaces, etc.) are described by equations in several variables.

4.2 Exercises

1. Prove that multiplication of polynomials is associative, and that addition and multiplication of polynomials satisfy the left distributive law $(g + h)\cdot f = g\cdot f + h\cdot f$.
2. Assume that R is a noncommutative ring. Show that $R[X]$ is noncommutative.
3. Prove Proposition 5.
4. Let $F = \mathbf{Z}_5$, $R = F[X]$. Calculate the following:
   (a) $(\bar{1}X^2 + \bar{1}X + \bar{1})^2$.
   (b) $(\bar{1}X^2 + \bar{2})^3$.
   (c) $(\bar{1}X + \bar{3})(\bar{4}X + \bar{2})(\bar{2}X + \bar{1})$.
5. Let R be a commutative ring, X an indeterminate over R, $f = a_0 + a_1X + \cdots + a_nX^n \in R[X]$, $\alpha \in R$. Define the *value of f* at α, denoted $f(\alpha)$, by

   $$f(\alpha) = a_0 + a_1\alpha + \cdots + a_n\alpha^n.$$

   Show that
   (a) $(f + g)(\alpha) = f(\alpha) + g(\alpha)$.
   (b) $(fg)(\alpha) = f(\alpha)g(\alpha)$.
6. Let $f = \bar{1}X^5 - \bar{1}X \in \mathbf{Z}_5[X]$. Find all $\alpha \in \mathbf{Z}_5$ such that $f(\alpha) = \bar{0}$.
7. (Lagrange's Interpolation Formula) Let F be a field, $\alpha_1, \ldots, \alpha_{n+1} \in F$, $x_1, \ldots, x_{n+1} \in F$, x_i distinct. Show that there exists a polynomial $f \in F[X]$ of degree at most n such that $f(x_i) = \alpha_i$ $(1 \leq i \leq n + 1)$. [*Hint:* Let

   $$f = \sum_{i=1}^{n+1} \alpha_i f_i,$$

   where

   $$f_i = [(x_i - x_1)\cdots(x_i - x_{i-1})(x_i - x_{i+1})\cdots(x_i - x_{n+1})]^{-1}(X - x_1)\cdots(X - x_{i-1})\cdot(X - x_{i+1})\cdots(X - x_{n+1}).]$$
8. Let F be a field and let X be an indeterminate over F, and let $f \in F[X]$ be a

nonconstant polynomial, a an element of F. Show that the value of f at a is zero if and only if $f = (X - a)g$ for some $g \in F[X]$. (See Exercise 5.)

9. Refer to Exercise 8.
   (a) Prove that if f has degree n, then f has at most n zeros in F [that is, there are at most n elements a of F such that $f(a) = 0$].
   (b) Let **H** denote the division ring of Hamilton's quaternions. (See Exercise 15, Section 4.1.) Prove that there exist infinitely many elements of **H** whose square is $(-1)\mathbf{1}$.
   (c) Does an analogue of the division algorithm hold in **H**?

4.3 Subrings and Quotient Rings

In Chapter 3 we started out by introducing the notion of a group. Then we showed how to manufacture new groups from a given group by forming subgroups and quotient groups. In this section we will describe the analogous process of manufacture for rings—we will study the subrings and quotient rings of a given ring R. Let us begin our investigation with subrings.

DEFINITION 1: Let R be a ring. A subset S of R is called a *subring* of R if S is a ring with respect to the operations of addition and multiplication in R.

EXAMPLE 1: $\mathbf{Z}$ is a subring of $\mathbf{Q}$.

EXAMPLE 2: $\mathbf{Q}$ is a subring of $\mathbf{R}$, the field of real numbers.

EXAMPLE 3: $2\mathbf{Z}$ is a subring of $\mathbf{Z}$.

EXAMPLE 4: $\mathbf{Z}[\sqrt{d}\,]$ is a subring of $\mathbf{Q}(\sqrt{d}\,)$.

EXAMPLE 5: Let $M_2(\mathbf{Q})$ denote the ring of 2×2 matrices over $\mathbf{Q}$, and let

$$T_2(\mathbf{Q}) = \left\{ \begin{pmatrix} a & b \\ 0 & d \end{pmatrix} \middle| \; a, b, d \in \mathbf{Q} \right\},$$

$$U_2(\mathbf{Q}) = \left\{ \begin{pmatrix} a & b \\ 0 & 0 \end{pmatrix} \middle| \; a, b \in \mathbf{Q} \right\}.$$

We assert that $T_2(\mathbf{Q})$ and $U_2(\mathbf{Q})$ are subrings of $M_2(\mathbf{Q})$. It is not necessary to check the associativity laws or the commutativity of addition, since these already hold in $M_2(\mathbf{Q})$. Thus, in order to check that $T_2(\mathbf{Q})$ and $U_2(\mathbf{Q})$ are subrings, it suffices to check that they are subgroups of $M_2(\mathbf{Q})$ under addition and that they are closed with respect to multiplication. But this is easy, since

$$\begin{pmatrix} a & b \\ 0 & d \end{pmatrix} - \begin{pmatrix} a' & b' \\ 0 & d' \end{pmatrix} = \begin{pmatrix} a - a' & b - b' \\ 0 & d - d' \end{pmatrix},$$

$$\begin{pmatrix} a & b \\ 0 & d \end{pmatrix} \cdot \begin{pmatrix} a' & b' \\ 0 & d' \end{pmatrix} = \begin{pmatrix} a \cdot a' & a \cdot b' + b \cdot d' \\ 0 & d \cdot d' \end{pmatrix},$$

$$\begin{pmatrix} a & b \\ 0 & 0 \end{pmatrix} - \begin{pmatrix} a' & b' \\ 0 & 0' \end{pmatrix} = \begin{pmatrix} a - a' & b - b' \\ 0 & 0 \end{pmatrix},$$

$$\begin{pmatrix} a & b \\ 0 & 0 \end{pmatrix} \cdot \begin{pmatrix} a' & b' \\ 0 & 0' \end{pmatrix} = \begin{pmatrix} a \cdot a' & b \cdot b' \\ 0 & 0 \end{pmatrix}.$$

[Note that we need Proposition 2 of Section 3.3 to draw the conclusion that $T_2(\mathbf{Q})$ and $U_2(\mathbf{Q})$ are subgroups of $M_2(\mathbf{Q})$ under addition.]

EXAMPLE 6: Let R be any ring. Then R and $\{0\}$ are subrings of R, called the *trivial subrings*.

The observations made in Example 5 lead to

THEOREM 2: Let S be a nonempty subset of the ring R. Then S is a subring of R if and only if for $a, b \in S$, we have $a - b \in S$, $a \cdot b \in S$.

Proof: $\Rightarrow$ If S is a subring of R and $a, b \in S$, we clearly must have $a - b \in S$, $a \cdot b \in S$.
$\Leftarrow$ Suppose that whenever $a, b \in S$, we have $a - b \in S$, $a \cdot b \in S$. Then by Proposition 2 of Section 3·3, S is a subgroup of R with respect to addition. Moreover, S is closed under multiplication. And the associativity of multiplication and distributivity laws hold because they hold in R.

As an easy consequence of Theorem 2, we can prove the following proposition.

PROPOSITION 3: Let R be a ring and let $\mathcal{C}$ be a collection of subrings of R. Then

$$\bigcap_{S \in \mathcal{C}} S$$

is a subring of R.

Proof: Let

$$S^* = \bigcap_{S \in \mathcal{C}} S,$$

and let $a, b \in S^*$. It suffices to show that $a - b \in S^*$ and $a \cdot b \in S^*$. But, since $a, b \in S^*$, we have $a, b \in S$ for all $S \in \mathcal{C}$. And since $S \in \mathcal{C}$ implies that S is a subring of R, we see that $a - b, a \cdot b \in S$ for all $S \in \mathfrak{C}$ by Theorem 2. Thus,

$$a - b, a \cdot b \in \bigcap_{S \in \mathfrak{C}} S = S^*.$$

Therefore, by Theorem 2, S^* is a subring of R.

Suppose that R is a ring and that $C \subseteq R$ is any subset. Let us ask the question: What subrings of R contain C? There clearly are such subrings. For example, R is one. Let $\mathfrak{C}$ be the collection of all subrings of R which contain C, and let us consider

$$[C] = \bigcap_{S \in \mathfrak{C}} S.$$

By Proposition 3, $[C]$ is a subring of R. Moreover, $[C]$ certainly contains C, since every $S \in \mathfrak{C}$ contains C. Further, if T is a subring of R which contains C, then $T \in \mathfrak{C}$, and therefore $T \supseteq [C]$. Thus, we see that $[C]$ is a subring of R which contains C and which is contained in every subring of R which contains $[C]$. We have therefore proved

PROPOSITION 4: Let C be a subset of the ring R. Then $[C]$ is the smallest subring of R which contains C.

The subring $[C]$ is called the *subring generated by the set* C. Let us give some examples. If $R = \mathbf{Z}$ and $C = \{2\}$, then $[C] = \{\ldots, -4, -2, 0, 2, 4, \ldots\}$. That is, $[C]$ is the subring $2\mathbf{Z}$. Indeed, $2\mathbf{Z}$ is a subring of R which contains $\{2\}$, and every element of $2\mathbf{Z}$ must automatically be contained in any subring containing $\{2\}$. As a second example, let F be a field and let $R = F[X]$, $C = F \cup \{X\}$. Then $[C] = F[X]$, since the smallest subring of R which contains every element of F and X is R itself. Similarly, if $C = F \cup \{X^2\}$,

$$[C] = \{a_0 + a_2X^2 + a_4X^4 + \cdots + a_{2t}X^{2t} \mid a_{2i} \in F\}.$$

After our very brief study of subrings of a ring, let us now turn our attention to quotient rings. Let R be a ring, S a subring of R. In particular, S is a subgroup of the additive group of R. And since the additive group of R is commutative, S is a normal subgroup of the additive group of R. Thus, by our group-theoretical results, the set of cosets R/S becomes a group with respect to the law of addition:

$$(r + S) + (r' + S) = (r + r') + S,$$

where $r + S$ and $r' + S$ are typical cosets belonging to R/S. It is clear that with respect to this law of addition, R/S is an abelian group. Our problem is: How can we define a law of multiplication on R/S so that R/S becomes a ring? There may be many ways to define such a multiplication, but there is one which has a compelling simplicity:

$$(r + S) \cdot (r' + S) = r \cdot r' + S. \tag{1}$$

The principal difficulty with this definition is that the product of $r + S$ and $r' + S$ depends, ostensibly, on the choice of the coset representatives r and r', and not just on the cosets $r + S, r' + S$. In order for our law of multiplication to be well defined, it is both necessary and sufficient that

$$(r + s + S)\cdot(r' + s' + S) = (r + S)\cdot(r' + S), \tag{2}$$

for all $s, s' \in S$. But the left-hand side equals $(r + s)\cdot(r' + s') + S$, while the right-hand side equals $r\cdot r' + S$. Therefore, (2) holds if and only if

$$r\cdot r' + s\cdot r' + r\cdot s' + s\cdot s' + S = r\cdot r' + S.$$

Since $s\cdot s' \in S$, this is equivalent to saying that

$$s\cdot r' + r\cdot s' \in S, \tag{3}$$

where (3) must hold for all $s, s' \in S$ and $r, r' \in R$. Putting $s' = 0$, we see that we must have $s\cdot r' \in S$ for $s \in S, r' \in R$. Similarly, putting $s = 0$, we see that we must have $r\cdot s' \in S$ for $r \in R, s' \in S$. Conversely, if $s\cdot r' \in S$ and $r\cdot s' \in S$ for all $r, r' \in R, s, s' \in S$, then (3) holds. Thus, to sum up, we have shown that the consistency condition (2) is equivalent to

$$r\cdot s' \in S, \qquad s\cdot r' \in S \tag{4}$$

for $r, r' \in R, s, s' \in S$. The condition (4) may or may not be satisfied for a given subring S. But whenever it is satisfied, we have shown that it is possible to use (1) to define a law of multiplication on R/S.

Let $\sim$ denote the equivalence relation on R defined by: $a \sim b \leftrightarrow a - b \in S$. Then R/S is the set of equivalence classes with respect to $\sim$. The condition (2) merely states that the binary operation $\cdot$ on R is compatible with $\sim$. The law of multiplication (1) on R/S is then the binary operation introduced in Section 2.6.

DEFINITION 5: Let R be a ring. An *ideal* of R is a subring I of R such that if $a \in I, r \in R$, then $a\cdot r \in I$ and $r\cdot a \in I$.

Note that if R is commutative, then the two conditions $a\cdot r \in I, r\cdot a \in I$ can be replaced by the single condition $r\cdot a \in I$. Note also that an ideal of R is a subring, but there may be subrings which are not ideals. For example, $\mathbf{Z}$ is a subring of $\mathbf{Q}$, but $\mathbf{Z}$ is not an ideal of $\mathbf{Q}$ since $\frac{1}{2} \in \mathbf{Q}, 1 \in \mathbf{Z}$, but $\frac{1}{2}\cdot 1 \notin \mathbf{Z}$. Let us give some examples of ideals.

EXAMPLE 7: Let R be any ring. Then $\{0\}$ and R are subrings of R and it is clear that these subrings are ideals. These ideals are called the *trivial ideals.*

EXAMPLE 8: Let $R = \mathbf{Z}$. We have seen that $n\mathbf{Z} = \{n\cdot r \mid r \in \mathbf{Z}\}$ is a subring of $\mathbf{Z}$. If $s \in \mathbf{Z}$ and $n\cdot r \in n\mathbf{Z}$, then

$$(n\cdot r)\cdot s = n\cdot(r\cdot s) \in n\mathbf{Z}.$$

Therefore, $n\mathbf{Z}$ is an ideal of $\mathbf{Z}$. In particular, we see that every subgroup of $\mathbf{Z}$ is a subring.

EXAMPLE 9: Let F be a field, X an indeterminate over $F, f \in F[X]$. In analogy with Example 8, set

$$fF[X] = \{f \cdot g \mid g \in F[X]\}.$$

Then $fF[X]$ is an ideal of $F[X]$.

EXAMPLE 10: Let F be a field, $F[X, Y]$ the ring of polynomials in two indeterminates X and Y over F. Set

$$(X, Y) = \{X \cdot f + Y \cdot g \mid f, g \in F[X, Y]\}.$$

Then (X, Y) consists of all polynomials with zero constant term and is an ideal of $F[X, Y]$.

EXAMPLE 11: Let R be any commutative ring, $a \in R$. Then set

$$aR = \{a \cdot r \mid r \in R\}.$$

Then aR is an ideal of R. If $R = \mathbf{Z}$, then this example reduces to Example 8. If $R = F[X]$, then this example reduces to Example 9.

Before delving any further into the theory of ideals, let us complete our construction of quotient rings. Let R be a ring and let us recall how we were led to the notion of an ideal. We started with a subring S of R and asked whether the multiplication (1) actually makes sense. We found that this is the case if and only if S is an ideal of R.

PROPOSITION 6: Let R be a ring, I an ideal of R, $R/I =$ the set of cosets of the form $a + I$ $(a \in R)$. Define addition of cosets by

$$(a + I) + (b + I) = (a + b) + I \qquad (a, b \in R), \tag{5}$$

and define multiplication of cosets by

$$(a + I) \cdot (b + I) = ab + I. \tag{6}$$

Then with respect to these operations, R/I becomes a ring, called the quotient ring of R with respect to I.

Proof: From our discussion above, R/I is an abelian group with respect to the law of addition (5). Moreover, we showed that the law of multiplication (6) makes sense, since I is an ideal. Multiplication is associative: For if $a, b, c \in R$,

$$\begin{aligned}(a + I) \cdot [(b + I) \cdot (c + I)] &= (a + I) \cdot (b \cdot c + I)\\ &= a \cdot (b \cdot c) + I\\ &= (a \cdot b) \cdot c + I \quad \text{(by the associativity of multiplication in } R)\\ &= (a \cdot b + I) \cdot (c + I)\\ &= [(a + I) \cdot (b + I)] \cdot (c + I).\end{aligned}$$

Similarly, the distributive laws in R imply the corresponding laws in R/I. Thus, R/I is a ring.

PROPOSITION 7: Let R be a ring, I an ideal of R.

(1) If R is commutative, then R/I is commutative.
(2) If R is a ring with identity 1, then R/I is a ring with identity $1 + I$.

Proof: Exercise.

Let us return to the examples of ideals which we gave above and let us describe the corresponding quotient rings.

EXAMPLE 12: $R = \mathbf{Z}, I = n\mathbf{Z}, n > 0$. In this case, R/I consists of the residue classes modulo n. Moreover, it is immediate that the operations defined by (5) and (6) are just addition and multiplication of residue classes, as defined in Chapter 2. Therefore, in this case, $R/I = \mathbf{Z}_n$.

EXAMPLE 13: Let R be any ring, I a trivial ideal of R—that is, $I = \{0\}$ or $I = R$. In case $I = \{0\}$, then R/I consists of the cosets

$$a + \{0\} \qquad (a \in R).$$

Two such cosets $a + \{0\}$, $a' + \{0\}$ are equal if and only if $a - a' \in \{0\}$; that is, $a = a'$. Thus, the elements in $R/\{0\}$ are in one-to-one correspondence with the elements of R, via the mapping

$$a \longrightarrow a + \{0\}.$$

Addition and multiplication in $R/\{0\}$ corresponds to addition and multiplication in R under this correspondence. Thus, $R/\{0\}$ is "essentially" R. (Strictly speaking, $R/\{0\}$ is isomorphic to R, but we won't introduce isomorphisms between rings until the next section.) Assume now that $I = R$. Then the elements of R/I are the cosets $a + I$ $(a \in R)$ and two cosets $a + I$, $a' + I$ are equal if and only if $a - a' \in I$. Therefore, since $I = R$, any two cosets of R/I are the same and R/I is the trivial ring.

EXAMPLE 14: Let F be a field, $R = F[X], f \in F[X]$ a polynomial such that $n = \deg(f) \geq 1, I = fF[X]$. The elements of R/I are the cosets $g + I$ $(g \in F[X])$. Note, however, that many different g can represent the same coset. In fact,

$$g + I = g + f \cdot h + I, \qquad h \in F[X].$$

From all these different representations for the coset $g + I$, let us pick out a "natural" one. By the division algorithm in $F[X]$ (Theorem 6 of Section 4.2), there exist polynomials $q, r \in F[X]$ such that $g = f \cdot q + r$ and either $r = \mathbf{0}$ or $0 \leq \deg(r) < \deg(f) = n$. However, since $f \cdot q \in I$,

$$g + I = f \cdot q + r + I = r + I.$$

Therefore, each coset $g + I$ of R/I can be written in the form $r + I$, where

$$r = a_0 + a_1 X + \cdots + a_{n-1} X^{n-1} \qquad (a_i \in F). \tag{7}$$

Moreover, if $r + I = r' + I$, then $r - r'$ is a multiple of f. But since r and r' have degree less than $\deg(f)$, this implies that $r = r'$. We have therefore proved that every coset $g + I$ can be written uniquely in the form $r + I$, where r is given by (7).

Let us consider a special case. Let $R = \mathbf{Q}[X], f = X^2 - 2$. Then the elements of R/I are the cosets

$$a_0 + a_1 X + I, \qquad a_0, a_1 \in \mathbf{Q}.$$

The addition of cosets is given by

$$(a_0 + a_1 X + I) + (b_0 + b_1 X + I) = (a_0 + b_0) + (a_1 + b_1)X + I.$$

The multiplication of cosets is defined by

$$\begin{aligned}(a_0 + a_1 X + I)\cdot(b_0 + b_1 X + I) &= a_0 b_0 + (a_1 b_0 + a_0 b_1)X + a_1 b_1 X^2 + I\\ &= (a_0 b_0 + 2a_1 b_1) + (a_1 b_0 + a_0 b_1)X\\ &\quad + a_1 b_1 (X^2 - 2) + I\\ &= (a_0 b_0 + 2a_1 b_1) + (a_1 b_0 + a_0 b_1)X + I.\end{aligned}$$

Note that the above addition and multiplication are very similar to the corresponding operations in the ring $\mathbf{Q}(\sqrt{2})$. In fact, if we let correspond to the coset $a_0 + a_1 X + I$ the element $a_0 + a_1\sqrt{2}$ of $\mathbf{Q}(\sqrt{2})$, then addition and multiplication in the two rings $\mathbf{Q}[X]/(X^2 - 2)\mathbf{Q}[X]$ and $\mathbf{Q}(\sqrt{2})$ correspond to one another. Thus, essentially, the ring $\mathbf{Q}[X]/(X^2 - 2)\mathbf{Q}[X]$ is the "same" as the ring $\mathbf{Q}(\sqrt{2})$. In Section 4.4 we will see that $\mathbf{Q}[X]/(X^2 - 2)\mathbf{Q}[X]$ and $\mathbf{Q}(\sqrt{2})$ are *isomorphic* to one another.

Let us close this section by determining all subrings of a quotient ring R/I.

PROPOSITION 8: Let R be a ring, I an ideal of R. If S is a subring of R containing I, then S/I is a subring of R/I. Conversely, every subring of R/I is of the form S/I, where S is a subring of R which contains I.

Proof: $S/I = \{s + I \mid s \in S\}$, and therefore $S/I \subseteq R/I$. Let $s + I, t + I \in S/I$. Then $s - t, s\cdot t \in S$ since S is a subring of R. Therefore,

$$(s + I) - (t + I) = (s - t) + I \in S/I,$$

$$(s + I)\cdot(t + I) = s\cdot t + I \in S/I.$$

Thus, by Proposition 3, S/I is a subring of R/I. Conversely, let $U \subseteq R/I$ be a subring. Let $S = \{s \in R \mid s + I \in U\}$. Then S is a subring of R containing I. (Exercise.) And it is clear that $S/I = U$.

4.3 Exercises

1. Let F be a field, $a \in F$.
   (a) Show that
   $$F[X]_a = \{a_1(X - a) + a_2(X - a)^2 + \cdots + a_n(X - a)^n \mid a_i \in F\}$$
   is a subring of $F[X]$.
   (b) Show that $F[X]_a =$ the smallest ideal of $F[X]$ which contains $X - a$.
2. Let n be a positive integer and let
   $$\mathbf{Q}_n = \left\{x \in \mathbf{Q} \mid x = a_0 + \frac{a_1}{n} + \frac{a_2}{n^2} + \cdots + \frac{a_r}{n^r} \quad \text{for } a_0, \ldots, a_r \in \mathbf{Z}\right\}.$$
   (a) Show that $\mathbf{Q}_n$ is a subring of $\mathbf{Q}$.
   (b) Show that $\mathbf{Q}_n = [1/n]$.
3. Let F be a field.
   (a) Describe the quotient ring $F[X]/(X^2 + 1)F[X]$.
   *(b) When is the quotient ring an integral domain?
4. Let $X^2 + X + \bar{1} \in \mathbf{Z}_2[X]$.
   (a) Describe $\mathbf{Z}_2[X]/[X^2 + X + \bar{1}]$ (it has four elements). Write down tables describing addition and multiplication in $\mathbf{Z}_2[X]/[X^2 + X + \bar{1}]$.
   (b) Show that $\mathbf{Z}_2[X]/(X^2 + X + \bar{1})\mathbf{Z}_2[X]$ is a field with four elements.
5. Find all ideals of $\mathbf{Q}$.
6. Give an example of a subring which is not an ideal. (*Hint:* Use Exercises 2 and 5.)

4.4 Isomorphisms and Homomorphisms

In this section we will imitate our approach in the theory of groups, whereby we introduced isomorphisms and homomorphisms between groups.

DEFINITION 1: Let R and S be rings. A *ring homomorphism from R to S* is a function $f: R \longrightarrow S$ such that for $a, b \in R$, we have

(a) $f(a + b) = f(a) + f(b)$,

(b) $f(a \cdot b) = f(a) \cdot f(b)$.

Note that in the condition (a) the sum on the left refers to addition in R, while the sum on the right refers to addition in S. Similar comments apply to the multiplication in (b).

Condition (a) implies that a ring homomorphism is a homomorphism from the additive group of R to the additive group of S. Therefore, we may apply our results on group homomorphisms to get that

$$f(-a) = -f(a) \qquad (a \in R), \tag{1}$$

$$f(0) = 0. \tag{2}$$

[Note that there is a slight ambiguity in (2). The zero on the left refers to the zero element of R, while the zero on the right refers to the zero element of S.]

DEFINITION 2: Let R and S be rings. A ring homomorphism $f: R \longrightarrow S$ is said to be an *isomorphism* if f is injective. We say that R is *isomorphic* to S if there exists a surjective isomorphism from R to S, and write $R \approx S$.

EXAMPLE 1: Let R be any commutative ring, X an indeterminate over R, $a \in R$. If $f = a_0 + a_1X + \cdots + a_nX^n \in R[X]$, let us define the *value of f at a*, denoted $f(a)$, by

$$f(a) = a_0 + a_1a + \cdots + a_na^n .$$

Let ϕ_a: $R[X] \longrightarrow R$ be the mapping defined by

$$\phi_a(f) = f(a) \qquad (f \in R[X]).$$

Then ϕ_a is a homomorphism. Indeed, if $g = b_0 + b_1X + \cdots + b_nX^n + \cdots \in R[X]$.

$$\begin{aligned}\phi_a(f + g) &= \phi_a((a_0 + b_0) + (a_1 + b_1)X + \cdots + (a_n + b_n)X^n + \cdots) \\ &= (a_0 + b_0) + (a_1 + b_1)a + \cdots + (a_n + b_n)a^n + \cdots \\ &= (a_0 + a_1a + \cdots + a_na^n + \cdots) \\ &\quad + (b_0 + b_1a + \cdots + b_na^n + \cdots) \\ &= \phi_a(f) + \phi_a(g).\end{aligned}$$

Similarly, since

$$f \cdot g = c_0 + c_1X + \cdots + c_nX^n + \cdots ,$$

where

$$c_i = a_0b_i + a_1b_{i-1} + \cdots + a_ib_0,$$

we see that

$$\begin{aligned}\phi_a(f \cdot g) &= c_0 + c_1a + c_2a^2 + \cdots + c_na^n + \cdots \\ &= (a_0 + a_1a + \cdots)\cdot(b_0 + b_1a + \cdots) \quad \text{(since } R \text{ is commutative)} \\ &= \phi_a(f)\cdot\phi_a(g).\end{aligned}$$

Therefore, ϕ_a is a ring homomorphism. We usually refer to ϕ_a as "evaluation at a."

EXAMPLE 2: Let $R = \mathbf{Z}$, $S = \mathbf{Z}_n$. For $x \in \mathbf{Z}$, define $f(x) = \bar{x}$, where $\bar{x} \in \mathbf{Z}_n$ denotes the residue class modulo n which contains x. Then

$$f(x + y) = \overline{x + y} = \bar{x} + \bar{y} = f(x) + f(y),$$
$$f(x \cdot y) = \overline{x \cdot y} = \bar{x} \cdot \bar{y} = f(x) \cdot f(y).$$

Therefore f: $R \longrightarrow S$ is a homomorphism.

EXAMPLE 3: Let $R = S = \mathbf{Z}[\sqrt{2}]$ and define $f : R \longrightarrow S$ by

$$f(a + b\sqrt{2}) = a - b\sqrt{2} \; (a + b\sqrt{2} \in \mathbf{Z}[\sqrt{2}])$$

Then f is a ring homomorphism. It is trivial to see that f is injective, so that f is an isomorphism.

EXAMPLE 4: Let $R = \mathbf{Q}(\sqrt{2})$, $S = \mathbf{Q}[X]/(X^2 - 2)\mathbf{Q}[X]$, and let $f: S \to R$ be defined by

$$f(a_0 + a_1X + I) = a_0 + a_1\sqrt{2} \qquad (a_0, a_1 \in \mathbf{Q}, I = (X^2 - 2)\mathbf{Q}[X]).$$

Then the mapping f was discussed in Section 4.3. It is a consequence of that discussion that f is a surjective isomorphism. Therefore,

$$\mathbf{Q}(\sqrt{2}) \approx \mathbf{Q}[X]/(X^2 - 2)\mathbf{Q}[X].$$

EXAMPLE 5: Let R be any ring and let S be any ring. The mapping

$$f: R \longrightarrow S,$$
$$f(x) = 0 \qquad (x \in R)$$

is called the *zero homomorphism*.

EXAMPLE 6: Let R be any ring, I an ideal of R. Let us define the mapping

$$\phi_I: R \longrightarrow R/I,$$
$$\phi_I(x) = x + I \qquad (x \in R).$$

Then by the definitions of addition and multiplication in R/I,

$$\begin{aligned}\phi_I(x + y) &= (x + y) + I = (x + I) + (y + I)\\ &= \phi_I(x) + \phi_I(y),\\ \phi_I(x \cdot y) &= x \cdot y + I\\ &= (x + I) \cdot (y + I)\\ &= \phi_I(x) \cdot \phi_I(y).\end{aligned}$$

Therefore, ϕ_I is a ring homomorphism, called the *canonical ring homomorphism* from R to R/I. Note that if we consider ϕ_I only as a homomorphism from the additive group of R to the additive group of R/I, then ϕ_I is the canonical homomorphism which we considered in our development of the first isomorphism theorem for groups. This naturally prompts us to ask whether the latter theorem can be extended to ring homomorphisms. This will be taken up below.

LEMMA 3: Let R and S be rings and let $f: R \to S$ be a surjective isomorphism. Then $f^{-1}: S \to R$ is a surjective isomorphism.

Proof: First note that it makes sense to speak of f^{-1} since f is bijective. Let $x, y \in S$. Then we must show that

$$f^{-1}(x + y) = f^{-1}(x) + f^{-1}(y), \qquad f^{-1}(x \cdot y) = f^{-1}(x) \cdot f^{-1}(y). \qquad (3)$$

Since f is a homomorphism,

$$f(f^{-1}(x) + f^{-1}(y)) = f(f^{-1}(x)) + f(f^{-1}(y)) \\ = x + y, \tag{4}$$

$$f(f^{-1}(x) \cdot f^{-1}(y)) = f(f^{-1}(x)) \cdot f(f^{-1}(y)) \\ = x \cdot y. \tag{5}$$

By applying f^{-1} to both sides of (4) and (5), we get (3). Therefore, f^{-1} is a homomorphism. But it is clear that f^{-1} is bijective. Thus, f^{-1} is a surjective isomorphism.

LEMMA 4: Let $f: R \longrightarrow S$ and $g: S \longrightarrow T$ be ring homomorphisms. Then $gf: R \longrightarrow T$ is a ring homomorphism.

Proof: Let $a, b \in R$. Then $gf(a + b) = g(f(a + b)) = g(f(a) + f(b)) = g(f(a)) + g(f(b)) = gf(a) + gf(b)$. Similarly, $gf(ab) = gf(a) \cdot gf(b)$.

PROPOSITION 5: Ring isomorphism is an equivalence relation. That is, the following properties hold: Let R, S, T be rings.

(1) $R \approx R$.

(2) If $R \approx S$, then $S \approx R$.

(3) If $R \approx S$ and $S \approx T$, then $R \approx T$.

Proof: (1) If $i: R \longrightarrow R$ is the identity function, then i is a surjective isomorphism.

(2) If $f: R \longrightarrow S$ is a surjective isomorphism, then $f^{-1}: S \longrightarrow R$ is a surjective isomorphism by Lemma 3, so that $S \approx R$.

(3) If $f: R \longrightarrow S$, $g: S \longrightarrow T$ are surjective isomorphisms, then $gf: R \longrightarrow T$ is bijective and is a homomorphism by Lemma 4. Therefore, gf is a surjective isomorphism and $R \approx T$.

Just as was the case in the theory of groups, it is easy to see that isomorphic rings have identical properties differing only in the naming of the elements. The surjective ring isomorphisms preserve completely the structure of a ring. Although homomorphisms do not preserve all the properties of a ring, they nevertheless preserve a number of very significant properties. The next proposition gives some examples of properties preserved by homomorphisms.

PROPOSITION 6: Let R and S be rings, $f: R \longrightarrow S$ a homomorphism.

(1) If U is a subring of R, then $f(U)$ is a subring of S. Moreover, if f is Moreover, if f is surjective, then the following are true:

(2) If R is commutative, then S is commutative.

(3) If R has an identity 1, then S has an identity $f(1)$.

Proof: (1) Let $x, y \in f(U)$, and let $x = f(r)$, $y = f(t)$, where $r, t \in U$. Then

$$x - y = f(r) - f(t) = f(r - t),$$
$$xy = f(r)f(t) = f(rt)$$

Therefore, $x - y$ and xy belong to $f(U)$, so that $f(U)$ is a subring of S.

(2) Let $x, y \in S$, and let $x = f(r)$, $y = f(t)$, where $r, t \in U$. Then, since R is commutative, we see that $xy = f(r)f(t) = f(rt) = f(tr) = f(t)f(r) = yx$. Therefore, S is commutative.

(3) Let $x \in S$, and let $x = f(r)$. Then $f(1)x = f(1)f(r) = f(1 \cdot r) = f(r) = x$. Similarly, $xf(1) = x$. Therefore, $f(1)$ is an identity for S.

The reader should note that without the surjectivity of f, assertions (2) and (3) of Proposition 6 are false. (See the Exercises.)

In analogy with the situation in group theory, let us define the *kernel* of the ring homomorphism $f: R \longrightarrow S$ as

$$\{r \in R \mid f(r) = 0\}.$$

Let us denote the kernel of f by $\ker(f)$. Let us compute the kernels of the homomorphisms considered in Examples 1–6:

1′. If $f = \phi_a$, where $\phi_a: R[X] \longrightarrow R$ is evaluation at a, then $\ker(f) = \{f \in R[X] \mid f(a) = 0\}$.

2′. If $f: \mathbf{Z} \longrightarrow \mathbf{Z}_n$, $f(x) = \bar{x}$, then $\ker(f) = \{n \cdot r \mid r \in \mathbf{Z}\}$.

3′. If $f: \mathbf{Z}[\sqrt{2}] \longrightarrow \mathbf{Z}[\sqrt{2}]$, $f(a + b\sqrt{2}) = a - b\sqrt{2}$, then $\ker(f) = \{0\}$.

4. If $f: \mathbf{Q}[\sqrt{2}] \longrightarrow \mathbf{Q}[X]/(X^2 - 2)\mathbf{Q}[X]$, $f(a_0 + a_1\sqrt{2}) = a_0 + a_1 X + I$, then $\ker(f) = \{0\}$.

5. If R is any ring and $f: R \longrightarrow \{0\}$, $f(x) = 0$ for all $x \in R$, then $\ker(f) = R$.

6. If R is any ring, I an ideal in R, $\phi_I: R \longrightarrow R/I$, $\phi_I =$ the canonical homomorphism, then $\ker(\phi_I) = I$.

In group theory, the kernel of a homomorphism is always a normal subgroup, whereas the kernel of an isomorphism consists of just the identity element. These facts generalize to ring homomorphisms:

PROPOSITION 7: Let R and S be rings, $f: R \longrightarrow S$ a ring homomorphism.

(1) $\ker(f)$ is an ideal of R.

(2) f is an isomorphism if and only if $\ker(f) = \{0\}$.

Proof: Since f is a homomorphism from the additive group of R to the additive group of S, our results on group homomorphisms imply (a) that

$\ker(f)$ is a subgroup of the additive group of R and (b) that f is injective if and only if $\ker(f) = \{0\}$. Let $x \in \ker(f)$, $r \in R$. Then

$$\begin{aligned} f(x \cdot r) &= f(x) \cdot f(r) && \text{(since } f \text{ is a ring homomorphism)} \\ &= 0 \cdot f(r) && [\text{since } x \in \ker(f)] \\ &= 0. \end{aligned}$$

Therefore, $x \cdot r \in \ker(f)$. Similarly, $r \cdot x \in \ker(f)$. Thus by (*a*), we conclude that part (1) holds. Furthermore, (b) implies part (2).

Let us now take up the ring analogue of the first isomorphism theorem of group theory. Let R and S be rings and let $f: R \longrightarrow S$ be a ring homomorphism. Then, in particular, R and S are groups under addition and f is a homomorphism of the additive group of R into the additive group of S. Moreover, the kernel of f as a homomorphism of rings is just the same as the kernel of f as a homomorphism of additive groups. Therefore, the first isomorphism theorem of group theory implies that there exists a unique group isomorphism $\bar{f}: R/\ker(f) \longrightarrow S$ such that the diagram

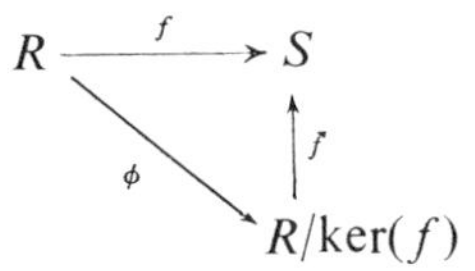

is commutative—that is, $\bar{f}\phi = f$. Here ϕ denotes the canonical homomorphism $\phi: R \longrightarrow R/\ker(f)$. Moreover, the proof of the first isomorphism theorem yields an explicit formula for $\bar{f}$, namely $\bar{f}(r + \ker(f)) = f(r)$. However, from the way in which we defined multiplication in $R/\ker(f)$, we see that $\bar{f}((r + \ker(f))(r' + \ker(f))) = \bar{f}(rr' + \ker(f)) = f(rr') = f(r)f(r') = \bar{f}(r + \ker(f))\bar{f}(r' + \ker(f))$. Therefore, the unique group isomorphism which is provided by the first isomorphism theorem for group theory is also a ring isomorphism. Thus, we may summarize our findings in the following theorem.

THEOREM 8 (FIRST ISOMORPHISM THEOREM OF RING THEORY): Let R and S be rings, and let $f: R \longrightarrow S$ be a homomorphism, $\phi: R \longrightarrow R/\ker(f)$ the canonical homomorphism associated to $\ker(f)$. Then there exists a unique ring isomorphism $\bar{f}: R/\ker(f) \longrightarrow S$ such that the diagram

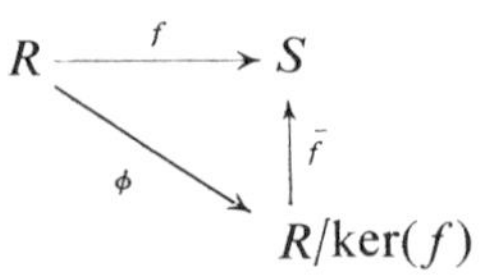

is commutative. In particular, $f(R) \approx R/\ker(f)$.

The interpretation of Theorem 8 is analogous to the interpretation of the corresponding theorem for groups. On the one hand, the theorem asserts that up to isomorphism, all homomorphisms of rings can be obtained as canonical homomorphisms. On the other hand, the first isomorphism theorem gives us a way of guaranteeing the existence of isomorphisms without having to actually write them down.

4.4 *Exercises*

1. Determine which of the following mappings are ring homomorphisms. For those which are ring homomorphisms, determine their respective kernels and images. Determine which are isomorphisms.
   (a) $f\colon \mathbf{Z} \longrightarrow \mathbf{Z}, f(x) = 9x$.
   (b) $f\colon \mathbf{Z} \longrightarrow \mathbf{Z}, f(x) = x^3$.
   (c) $f\colon \mathbf{Z} \longrightarrow \mathbf{Z}_n, f(x) = \bar{x}$.
   (d) $f\colon \mathbf{Z}_3 \longrightarrow \mathbf{Z}_3, f(x) = x^3$.
   (e) $f\colon M_2(\mathbf{R}) \longrightarrow M_2(\mathbf{R}), f\left(\begin{pmatrix} a & b \\ c & d \end{pmatrix}\right) = \begin{pmatrix} 2a + 3b & a + 4b \\ 2c + 3d & c + 4d \end{pmatrix}$.
   (f) $f\colon \mathbf{Z}_{3n} \to \mathbf{Z}_n, f(x + 3n\mathbf{Z}) = x + n\mathbf{Z}$.
2. Give an example of two rings R and S with identities 1_R and 1_S, respectively, and a ring homomorphism $f\colon R \longrightarrow S$ such that $f(1_R) \neq 1_S$.
3. Let F be a field, R a ring, $f\colon F \longrightarrow R$ a ring homomorphism. Show that either f is the zero homomorphism or f is an isomorphism.
4. Let $f\colon \mathbf{Z}_{kn} \longrightarrow \mathbf{Z}_n$ be the mapping
$$f(x + kn\mathbf{Z}) = x + n\mathbf{Z} \quad (x \in \mathbf{Z}).$$
Show that f is a homomorphism and that $\ker(f) = n\mathbf{Z}_{kn} = \{n\bar{x} \mid \bar{x} \in \mathbf{Z}_{kn}\}$. Thus, show that
$$\mathbf{Z}_{kn}/n\mathbf{Z}_{kn} \approx \mathbf{Z}_n.$$
5. Show that $\mathbf{Q}(\sqrt{d}) \approx \mathbf{Q}[X]/(X^2 - d)\mathbf{Q}[X]$, provided that d is not a perfect square of an element of $\mathbf{Q}$.

*6. Let F be a field, X an indeterminate over F, $f \in F[X]$. A *zero of f* is an element $x \in F$ such that $\phi_x(f) = 0$.
   (a) Show that if x is a zero of f, then there exists a polynomial $g \in F[X]$ such that $f = (X - x)g$. (*Hint*: Division algorithm.)
   (b) Show that if $f \neq \mathbf{0}$, f has only a finite number of zeros and the number of zeros is at most deg (f).
   (c) Let $f\colon \mathbf{R} \longrightarrow \mathbf{R}$ be a polynomial function—that is, there exist $a_0, \ldots, a_n \in \mathbf{R}$ such that $f(x) = a_0 + a_1x + \cdots + a_nx^n$ $(x \in \mathbf{R})$. Show that if $f(x) = 0$ for all $x \in \mathbf{R}$, then $a_0 = a_1 = \cdots = a_n = 0$.

*7. A *polynomial function $f\colon F \longrightarrow F$* is a function of the form $f(x) = a_0 + a_1x + \cdots + a_nx^n$ $(x \in F, a_i \in F)$. Let P_F denote the set of all polynomial functions $f\colon F \longrightarrow F$.
   (a) Define the sum and product of two polynomials by
$$(f + g)(x) = f(x) + g(x) \qquad (x \in F),$$
$$(fg)(x) = f(x)g(x) \qquad (x \in F),$$
respectively. Prove that with respect to these operations P_F becomes a ring.

(b) Show that the mapping $\psi: F[X] \longrightarrow P_F$ defined by

$$\psi(a_0 + a_1 X + \cdots + a_n X^n)(x) = a_0 + a_1 x + \cdots + a_n x^n$$

is a surjective homomorphism.

(c) Show that if F is infinite, then ψ is an isomorphism.

(d) Is ψ necessarily an isomorphism if F is finite?

8. (Universal Mapping Property for Quotient Rings) Let R be a ring, S an ideal of R. Suppose that there exists a ring T and a ring homomorphism $\alpha: R \longrightarrow T$ with the following property: (Q) Whenever there is given a ring U and a ring homomorphism $\lambda: R \longrightarrow U$ such that $\lambda(S) = \{0_U\}$, then there exists a unique ring isomorphism $\theta: T \longrightarrow U$ such that the diagram

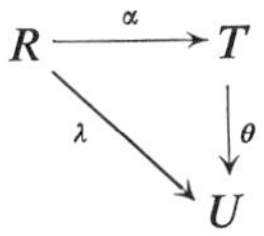

commutes.

(a) Show that if $T = R/S$ and $\alpha =$ the canonical homomorphism $R \longrightarrow R/S$, then Q holds.

(b) If T is given satisfying the property Q, then $T \approx R/S$. Thus, the quotient ring R/S can be defined entirely in terms of mapping property Q.

9. Give an example of two rings R and S and a homomorphism $f: R \longrightarrow S$ such that R is commutative whereas S is not.

10. Give an example of two rings R and S and a homomorphism $f: R \longrightarrow S$ such that R is an integral domain but S is not.

11. Give an example of two rings R and S and a homomorphism $f: R \longrightarrow S$ such that R is a field but S is not.

12. Prove that $\mathbf{R}[X]/(X^2 - 1)\mathbf{R}[X] \approx \mathbf{R} \oplus \mathbf{R}$; $\mathbf{R}[X]/(X^2 - 1)^2\mathbf{R}[X] \not\approx \mathbf{R} \oplus \mathbf{R}$.

13. Let F be a field, X an indeterminate over F, and let $f,g \in F[X]$ be polynomials such that there exist $a,b \in F[X]$ such that $af + bg = 1$. (We will see later that this condition is equivalent to the fact that f and g have no nonconstant factors in common.) Show that $F[X]/fgF[X] \approx F[X]/fF[X] \oplus F[X]/gF[X]$.

14. Let $F[X]$ be a polynomial ring over the field F, $a \in F$. Prove that $F[X]/(X - a)F[X] \approx F$.

4.5 *Two Useful Theorems*

In this section we prove two extremely useful results concerning rings, both of which make use of the notion of an embedding.

DEFINITION 1: Let R and S be rings. We say that R is *embedded* in S if there exists an isomorphism $f: R \longrightarrow S$. The isomorphism f is said to be an *embedding* of R in S.

If R is embedded in S via the embedding f, then $f(R)$ is a subring of S (Proposition 2 of Section 4.4) and f is a surjective isomorphism of R onto $f(R)$. Therefore, $R \approx f(R)$. We may identify R with $f(R)$ and therefore view R as a subring of S.

In Section 4.1 we saw that $2\mathbf{Z}$ is a ring without an identity element. Note, however, that $2\mathbf{Z}$ can be embedded in $\mathbf{Z}$, which is a ring with identity. Our first result shows that this is a quite general phenomenon.

THEOREM 2: Let R be a ring. Then R can be embedded in a ring with identity.

Before proceeding with the proof of Theorem 2, let us define a general notion which will come up rather often throughout the book. Let R be a ring and let $r \in R, n \in \mathbf{Z}$. We seek to define the product $n \cdot r$. Set

$$\begin{aligned} n \cdot r &= \underbrace{r + r + \cdots + r}_{n \text{ times}} && \text{if } n > 0 \\ &= 0 && \text{if } n = 0 \\ &= \underbrace{-(r + r + \cdots + r)}_{-n \text{ times}} && \text{if } n < 0. \end{aligned}$$

Note that $n \cdot r$ is just the "nth power of r" in the additive group of R. Therefore, the following properties of $n \cdot r$ are just restatements of the laws of exponents in an abelian group. We leave the translation from multiplicative notation to additive notation to the reader.

$$n \cdot (r_1 + r_2) = n \cdot r_1 + n \cdot r_2, \tag{1}$$

$$n(-r) = -(n \cdot r), \tag{2}$$

$$(n_1 + n_2) \cdot r = n_1 \cdot r + n_2 \cdot r, \tag{3}$$

$$(n_1 \cdot n_2) \cdot r = n_1 \cdot (n_2 \cdot r), \tag{4}$$

$$1 \cdot r = r. \tag{5}$$

Proof of Theorem 2: Let us consider the Cartesian product

$$R \times \mathbf{Z} = \{(r, n) \mid r \in R, n \in \mathbf{Z}\}.$$

Let us define addition and multiplication on $R \times \mathbf{Z}$ by

$$(r_1, n_1) + (r_2, n_2) = (r_1 + r_2, n_1 + n_2),$$

$$(r_1, n_1) \cdot (r_2, n_2) = (r_1 r_2 + n_1 \cdot r_2 + n_2 \cdot r_1, n_1 \cdot n_2).$$

Then, with respect to these operations, $R \times \mathbf{Z}$ is a ring. (Exercise.) Moreover, $R \times \mathbf{Z}$ has an identity $(0, 1)$, since

$$(r_1, n) \cdot (0, 1) = (r_1 \cdot 0 + n \cdot 0 + 1 \cdot r_1, 1 \cdot n) = (r_1, n),$$

$$(0, 1) \cdot (r_1, n) = (0 \cdot r_1 + 1 \cdot r_1 + n \cdot 0, 1 \cdot n) = (r_1, n).$$

Define the mapping

$$f: R \longrightarrow R \times \mathbf{Z},$$
$$f(r) = (r, 0).$$

Then

$$f(r_1 + r_2) = (r_1 + r_2, 0) = f(r_1) + f(r_2),$$
$$f(r_1 \cdot r_2) = (r_1 \cdot r_2, 0) = f(r_1) \cdot f(r_2).$$

Therefore, f is a ring homomorphism. Moreover, $\ker(f) = \{0\}$, so that f is an isomorphism (Proposition 7 of Section 4.4). Therefore, R is embedded in the ring with identity $R \times \mathbf{Z}$ via the embedding f.

Let us now pose the following problem: Can an integral domain D always be embedded in a field? In order to get a feel for the problem as well as its solution, let us consider the special case $D = \mathbf{Z}$. In this special case, we know that the answer is yes, since $\mathbf{Z} \subseteq \mathbf{Q}$, the field of rational numbers, and the mapping $i: \mathbf{Z} \longrightarrow \mathbf{Q}$ defined by $i(x) = x$ is an embedding of $\mathbf{Z}$ into $\mathbf{Q}$. In order to understand the relationship between $\mathbf{Z}$ and $\mathbf{Q}$, let us examine how $\mathbf{Q}$ can be gotten from $\mathbf{Z}$. Roughly speaking, $\mathbf{Q}$ is gotten by forming "quotients" a/b of elements $a, b \in \mathbf{Z}, b \neq 0$. However, several different quotients can represent the same element of $\mathbf{Q}$. For example, $\frac{2}{3} = \frac{4}{6} = \frac{6}{9}$, etc. The element $a \in \mathbf{Z}$ is viewed as the quotient $a/1$, and the embedding i is actually the mapping $i(a) = a/1$. Note that every element of $\mathbf{Q}$ is of the form $i(a)i(b)^{-1}$ $(a, b \in \mathbf{Z}, b \neq 0)$, since

$$a/b = (a/1)(b/1)^{-1}.$$

Our goal in the second theorem is to generalize these facts to the case of an arbitrary integral domain in place of $\mathbf{Z}$. Our main result is the following theorem.

THEOREM 3: Let R be an integral domain. There exists a field F_R and an embedding $f: R \longrightarrow F_R$ such that every element of F_R is of the form $f(a) \cdot f(b)^{-1}$ $(a, b \in R, b \neq 0)$.

Proof: Let us mimic the situation in the case of the rational numbers. Let

$$S = \{(a, b) \in R \times R \mid b \neq 0\},$$

and let $\sim$ denote the relation on S defined by

$$(a, b) \sim (c, d) \longleftrightarrow ad - bc = 0.$$

Then $\sim$ is an equivalence relation on S. (Exercise.) Let F_R denote the set of equivalence classes of S with respect to $\sim$, and let us denote the equivalence class containing (a, b) by a/b. (The above construction is necessary to guarantee that the fractions a/b and ac/bc are equal.) We define addition and multiplication on F_R by

$$a/b + c/d = (ad + bc)/(bd),$$
$$(a/b)\cdot(c/d) = (a\cdot c)/(b\cdot d).$$

(Note that $bd \neq 0$ since $b \neq 0$, $d \neq 0$, and R is an integral domain.)

Let us prove that these operations are consistent. By way of example, let us supply the details only in the case of addition. Suppose that $a/b = a'/b'$, $c/d = c'/d'$. Let us show that

$$a/b + c/d = a'/b' + c'/d'. \tag{6}$$

Indeed, $a/b + c/d = (ad + bc)/bd$, $a'/b' + c'/d' = (a'd' + b'c')/b'd'$. Moreover, since $a/b = a'/b'$, we see that $ab' = a'b$. Similarly, $cd' = c'd$. The statement (6) is equivalent to

$$(ad + bc)b'd' = (a'd' + b'c')bd. \tag{7}$$

But, $(ad + bc)b'd' = ab'dd' + b'bcd' = a'bdd' + b'bc'd = (a'd' + b'c')bd$, which proves (7). Thus addition is consistent. We leave the proof that multiplication is consistent as an exercise. We also leave as an exercise the fact that F_R becomes a field with respect to these operations.

The mapping

$$f: R \longrightarrow F_R,$$
$$f(r) = r/1$$

is an embedding. Moreover, if $a/b \in F_R$, then

$$\begin{aligned} a/b = (a/1)\cdot(1/b) &= (a/1)\cdot(b/1)^{-1} \\ &= f(a)\cdot f(b)^{-1}. \end{aligned}$$

Thus, F_R satisfies the conditions of the theorem.

DEFINITION 4: F_R is called the *quotient field* of the integral domain R.

Usually, we will dispense with the embedding f and will merely identify $r \in R$ with the element $r/1 \in F_R$. Then R becomes a subring of F_R.

EXAMPLE 1: If $R = \mathbf{Z}$, then $F_R = \mathbf{Q}$.

EXAMPLE 2: Let $R = F[X]$. Then

$$F_R = \{f/g \mid f \in F[X], g \in F[X], g \neq \mathbf{0}\}.$$

Two elements, f/g and f'/g', of F_R are equal if and only if

$$fg' = f'g.$$

The elements of F_R are called *rational functions* in X.

4.5 Exercises

1. (a) Show that $\{a_0 + a_1X + \cdots + a_nX^n \mid a_2 \in 2\mathbf{Z}\}$ is a subring of $\mathbf{Z}[X]$ which does not contain an identity.
   (b) Embed the ring of (a) in a ring with identity.
2. Let $F[X, Y]$ be the polynomial ring in two variables over the field F. Describe its quotient field.

*3. Let D be an integral domain which contains only finitely many elements.
   (a) Show that D is a field.
   (b) What is the quotient field of D?

4. Prove that $R \times \mathbf{Z}$ used in the proof of Theorem 2 is a ring with respect to the operations defined.
5. Let F denote the quotient field of $\mathbf{Z}[\sqrt{2}]$. Show that $F \approx \mathbf{Q}(\sqrt{2})$.
6. (Universal Mapping Property for Quotient Fields) Let D be an integral domain, F its quotient field and $\eta\colon D \longrightarrow F$ the embedding of D in F. Let there be given a field G and an embedding $\lambda\colon D \longrightarrow G$. Show that there exists a unique isomorphism $\theta\colon F \longrightarrow G$ such that the diagram

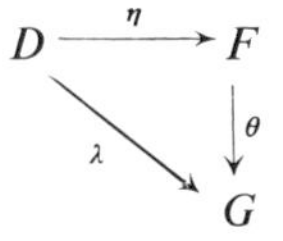

   commutes.
7. Let D be an integral domain, F its quotient field. Let G be any field containing D and let G_0 denote the smallest subfield of G which contains D. Show that $G_0 \approx F$. (*Hint*: Apply Exercise 6.)
8. Prove formulas (1)–(5).
9. (a) Prove that multiplication in F_R is consistent.
   (b) Prove that with respect to the defined addition and multiplication, F_R is a ring.

4.6 The Real Numbers

In Section 4.5 we introduced the field $\mathbf{Q}$ of rational numbers and showed that every element of $\mathbf{Q}$ is of the form a/b $(a, b \in \mathbf{Z}, b \neq 0)$,that is, every rational number is the ratio of two integers. The Greek geometers were the first to notice that very elementary geometric constructions give rise to quantities which are not ratios of integers. For example, a 45°–45°–90° triangle with two sides of length 1 has a hypotenuse of length $\sqrt{2}$.
And we have

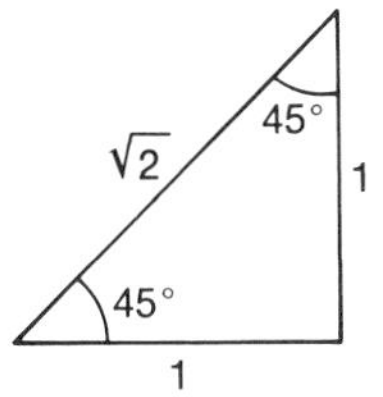

FIGURE 4-1: 45°—45°—90° Triangle.

THEOREM 1 (PYTHAGORAS): There is no rational number x such that $x^2 = 2$.

Proof: Suppose on the contrary that $x = a/b$ $(a, b \in \mathbf{Z}, b \neq 0)$ has the property $x^2 = 2$. Then

$$a^2 = 2b^2. \tag{1}$$

Without loss of generality, we may assume that $(a, b) = 1$. [For if $(a, b) = k$, then $a = ka^*, b = kb^*$ for some $a^*, b^* \in \mathbf{Z}$. Therefore, $(a^*, b^*) = 1$ and $x = a^*/b^*$.] Note that $2 \mid 2b^2$, so that $2 \mid a$ by (1). Therefore,

$$\begin{aligned} 4 \mid a^2 &\Longrightarrow 4 \mid 2b^2 \qquad \text{by (1)} \\ &\Longrightarrow 2 \mid b^2 \\ &\Longrightarrow 2 \mid b. \end{aligned}$$

Therefore, $2 \mid a$ and $2 \mid b$, which contradicts the assumption that $(a, b) = 1$.

Note that we could have phrased Theorem 1 as follows: $\sqrt{2}$ is not a rational number. However, we refrained from doing so, because at this point, it is not clear what sort of number $\sqrt{2}$ is. The reader is probably accustomed to thinking of $\sqrt{2}$ as the infinite decimal, whose expansion begins

$$\sqrt{2} = 1.414 \cdots.$$

More generally, it is possible to think of any real number (for example, $0, \frac{1}{3}, \sqrt{2}, \pi, e, \ldots$) as an infinite decimal. And this is the definition of real number which prevailed for many centuries. It was not until the nineteenth century that mathematicians attempted to clarify this definition by precisely specifying the notion of an infinite decimal. The result of the efforts of Cauchy, Dedekind, and Cantor is to provide us with several ways of defining the real numbers in terms of the rational numbers. In each case, the ordering of $\mathbf{Q}$ is the essential ingredient in the definition.

Cauchy's contribution to the construction of the real numbers was the precise definition of convergence of a sequence. Both Dedekind and Cantor gave constructions of the real numbers. Dedekind's construction uses what are now known as *Dedekind cuts*. Cantor, who founded the modern theory of sets, relies on *Cauchy sequences* for his construction. We will follow Cantor, since his construction allows us to give a nice application of the theory of quotient rings. After we construct the real numbers, we will verify all the axioms which are usually stated in a calculus course.

In order to construct the real numbers, we need some further properties of the rational numbers. If $x \in \mathbf{Q}$, let us define the absolute value of x, denoted $|x|$, by

$$|x| = \begin{cases} x, & x > 0, \\ 0, & x = 0, \\ -x, & x < 0. \end{cases}$$

PROPOSITION 2: Let $x, y \in \mathbf{Q}$.

(1) $|x| \geq 0$.
(2) $|x| = 0$ if and only if $x = 0$.
(3) $|x \cdot y| = |x||y|$.
(4) $|x + y| \leq |x| + |y|$ (triangle inequality).

Proof: Parts (1) and (2) are obvious.

(3) Follows by considering separately the cases x or $y = 0$; $x > 0$, $y > 0$; $x < 0, y < 0$; $x > 0, y < 0$; $x < 0, y > 0$.

(4) Let us consider the cases of part (3). If x or $y = 0$, then part (4) is obvious. If $x > 0$, $y > 0$, then $x + y > 0$ and $|x + y| = x + y$, $|x| = x$, $|y| = y$, so (4) follows. If $x < 0, y < 0$, then $x + y < 0$ and $|x + y| = -(x + y)$, $|x| = -x, |y| = -y$, so again (4) follows. If $x < 0, y > 0$, then $|x| = -x, |y| = y$. Moreover, $|x + y| = x + y$ or $-(x + y)$. In the first case, (4) follows since $x + y \leq -x + y$. In the second case, (4) follows since $-(x + y) = -x - y \leq -x + y$. If $x > 0, y < 0$, then (4) follows similarly.

Let S be a set and let $\mathbf{N}_+$ denote the set of positive natural numbers. A *sequence of elements of* S is a function $f: \mathbf{N}_+ \longrightarrow S$. If $n \in \mathbf{N}_+$, we will write f_n instead of $f(n)$. Then the sequence f is uniquely determined by the set

$$\{f_1, f_2, f_3, \ldots\} \tag{2}$$

of elements of S, and every such set determines a sequence of elements of S. If $S = \mathbf{Q}$, then typical sequences of elements in S are given by

(a) $\{0, 1, 0, 1, 0, 1, \ldots\}$,
(b) $\{1, \frac{1}{2}, \frac{1}{3}, \frac{1}{4}, \ldots\}$,
(c) $\{1, \frac{3}{2}, \frac{7}{8}, \frac{15}{16}, \ldots\}$,
(d) $\{3, 3.14, 3.141, 3.1415, \ldots\}$,
(e) $\{1, 1, 1, 1, 1, \ldots\}$.

We denote the sequence (2) by $\{f_n\}_{1 \leq n < \infty}$ (or just $\{f_n\}$). The element f_n is called the nth *term of the sequence*.

Let $\mathcal{S}$ denote the set of all sequences of rational numbers. Let us define addition and multiplication in $\mathcal{S}$ by

$$\{a_n\}_{1\le n<\infty} + \{b_n\}_{1\le n<\infty} = \{a_n + b_n\}_{1\le n<\infty}, \tag{3}$$

$$\{a_n\}_{1\le n<\infty}\cdot\{b_n\}_{1\le n<\infty} = \{a_n\cdot b_n\}_{1\le n<\infty}. \tag{4}$$

Thus, for example,

$$\{0, 1, 0, 1, \ldots\} + \{-1, 0, -1, 0, \ldots\} = \{-1, 1, -1, 1, \ldots\},$$

$$\{0, 1, 0, 1, \ldots\}\cdot\{-1, 0, -1, 0, \ldots\} = \{0, 0, 0, 0, \ldots\}.$$

Note that (3) and (4) are just the usual definitions of addition and multiplication of real-valued functions. It is easy to verify that $\mathcal{S}$ is a commutative ring with identity with respect to the operations (3) and (4). Our idea is to let certain elements of $\mathcal{S}$ play the role of real numbers. For example, the real number 1 will be the sequence

$$\{1, 1, 1, \ldots\};$$

the real number π will be the sequence

$$\{3, 3.1, 3.14, 3.141, 3.1415, \ldots\},$$

and so forth. Let us characterize the sequences which we have in mind.

DEFINITION 3: A sequence $\{a_n\}_{1\le n<\infty}$ of rational numbers is a *Cauchy sequence* if for every $\epsilon > 0$, $\epsilon \in \mathbf{Q}$ there exists a positive integer N, depending on ϵ, such that $|a_n - a_m| < \epsilon$ whenever $n, m \geq N$.

Thus, for example the sequences (b), (c), (d), and (e) are Cauchy sequences, which (a) is not. By way of illustration, let us prove that (b) is a Cauchy sequence. Suppose that $\epsilon > 0$ is given. By Corollary 3 of Section 2.3 there exists a positive integer n_0 such that $\epsilon^{-1} < n_0$. Then $n_0^{-1} < \epsilon$. Set $N = 2n_0$. Then, if $n, m \geq N$, we have

$$\begin{aligned} |1/n - 1/m| &\leq 1/n + 1/m && \text{(by the triangle inequality)} \\ &\leq 1/N + 1/N && \text{(since } n, m \geq N\text{)} \\ &= 2\cdot(1/2n_0) && \text{(since } N = 2n_0\text{)} \\ &= 1/n_0 < \epsilon. \end{aligned}$$

The reader should think of a Cauchy sequence of rational numbers as approximating a real number, in the sense that it "tends" to the real number.

LEMMA 4: Let $\{a_n\}_{1\le n<\infty} \in \mathcal{S}$ be a Cauchy sequence. Then there exists $M \in \mathbf{Q}$, $M > 0$ such that $|a_n| \leq M$ for all n.

Proof: Since $\{a_n\}_{1\le n<\infty}$ is a Cauchy sequence, there exists a positive integer N such that, whenever $n, m \geq N$, we have

$$|a_n - a_m| < 1. \tag{5}$$

Set $A =$ the largest of $|a_1|, |a_2|, \ldots, |a_N|$, and set $M = A + 1$. It is clear that $|a_n| \leq M$ for $n \leq N$. If $n \geq N + 1$, then by (5),

$$\begin{aligned} |a_n| &= |(a_n - a_N) + a_N| \\ &\leq |a_n - a_N| + |a_N| \qquad \text{(by the triangle inequality)} \\ &< 1 + A = M. \end{aligned}$$

Let $\mathfrak{R}$ denote the set of all Cauchy sequences in $\mathfrak{S}$.

LEMMA 5: $\mathfrak{R}$ is a subring of $\mathfrak{S}$.

Proof: Let $\{a_n\}, \{b_n\} \in \mathfrak{R}$. It suffices to show that

$$\{a_n\} - \{b_n\} = \{a_n - b_n\}, \tag{6}$$

$$\{a_n\} \cdot \{b_n\} = \{a_n \cdot b_n\} \tag{7}$$

are Cauchy sequences. Let $\epsilon > 0, \epsilon \in \mathbf{Q}$ be given. Since $\{a_n\}$ and $\{b_n\}$ are Cauchy sequences, there exist positive integers N_1 and N_2 such that

$$|a_n - a_m| < \epsilon/2 \qquad \text{if } n, m \geq N_1,$$

$$|b_n - b_m| < \epsilon/2 \qquad \text{if } n, m \geq N_2.$$

Set $N_3 =$ the larger of N_1 and N_2. Then, if $n, m \geq N_3$, we have

$$\begin{aligned} |(a_n - b_n) - (a_m - b_m)| &= |(a_n - a_m) + (b_m - b_n)| \\ &\leq |a_n - a_m| + |b_n - b_m| \\ &\leq \epsilon/2 + \epsilon/2 = \epsilon. \end{aligned}$$

Therefore, (6) is a Cauchy sequence. By Lemma 4, there exists a positive rational number M such that $|a_n| \leq M, |b_n| \leq M$ for all n. Then, since $\{a_n\}$ and $\{b_n\}$ are Cauchy sequences, there exist positive integers N_4 and N_5 such that

$$|a_n - a_m| < \epsilon/2M \qquad \text{if } n, m \geq N_4,$$

$$|b_n - b_m| < \epsilon/2M \qquad \text{if } n, m \geq N_5.$$

If $N_6 =$ the larger of N_4 and N_5, then for $n, m \geq N_6$, we have

$$\begin{aligned} |a_n b_n - a_m b_m| &= |a_n(b_n - b_m) + b_m(a_n - a_m)| \\ &\leq |a_n||b_n - b_m| + |b_m||a_n - a_m| \\ &\leq M \cdot \frac{\epsilon}{2M} + M \cdot \frac{\epsilon}{2M} = \epsilon. \end{aligned}$$

Therefore, (7) is a Cauchy sequence.

The basic idea of our construction of the real numbers is to let the Cauchy sequences play the role of real numbers. However, things cannot be quite so simple, since our experience leads us to believe that a number of Cauchy sequences should represent the same real number. For example, each of the sequences

$$\{2, 2, 2, 2, \ldots\},$$
$$\{1, 1.9, 1.99, 1.999, \ldots\},$$
$$\{1, 1.99, 1.9999, 1.999999, \ldots\},$$
$$\{0, 0, 0, 2, 2, 2, 2, 2, \ldots\}$$

should represent the real number 2. Thus, let us isolate all the Cauchy sequences which represent 0 and consider two Cauchy sequences to represent the same real number if they differ by such a sequence. The Cauchy sequences which represent 0 are called null sequences.

DEFINITION 6: A *null sequence* is a Cauchy sequence $\{a_n\}$ of rational numbers such that if $\epsilon > 0$, $\epsilon \in \mathbf{Q}$ is given, there exists a positive integer N, depending on ϵ, such that $|a_n| < \epsilon$ whenever $n \geq N$. The set of all null Cauchy sequences will be denoted $\mathcal{I}$.

For example,

$$\{1, \tfrac{1}{2}, \tfrac{1}{3}, \tfrac{1}{4}, \ldots\},$$
$$\{0, 0, 0, 0, \ldots\}$$

are null sequences.

LEMMA 7: $\mathcal{I}$ is an ideal of $\mathcal{R}$.

Proof: Let $\{a_n\}, \{b_n\} \in \mathcal{I}, \{c_n\} \in \mathcal{R}$. We must show that

$$\{a_n\} - \{b_n\} = \{a_n - b_n\} \in \mathcal{I}, \tag{8}$$
$$\{c_n\} \cdot \{a_n\} = \{c_n a_n\} \in \mathcal{I}. \tag{9}$$

Since $\{a_n\}$ and $\{b_n\}$ are null Cauchy sequences, if $\epsilon > 0$ is given there exist positive integers N_1 and N_2, depending on ϵ, such that

$$|a_n| < \epsilon/2 \qquad \text{if } n \geq N_1,$$
$$|b_n| < \epsilon/2 \qquad \text{if } n \geq N_2.$$

Let N_3 be the larger of N_1 and N_2. Then, if $n \geq N_3$,

$$\begin{aligned} |a_n - b_n| &\leq |a_n| + |b_n| \\ &< \epsilon/2 + \epsilon/2 = \epsilon. \end{aligned}$$

Thus, (8) holds. By Lemma 4, there exists $M \in \mathbf{Q}$, $M > 0$ such that $|c_n| \leq M$ for all n. Since $\{a_n\}$ is a null Cauchy sequence, given $\epsilon \in \mathbf{Q}, \epsilon > 0$, there exists a positive integer N_4 such that $|a_n| < \epsilon/M$ whenever $n \geq N_4$. Therefore, if $n \geq N_4$,

$$\begin{aligned} |c_n a_n| &= |c_n||a_n| \\ &< M \cdot \epsilon/M = \epsilon, \end{aligned}$$

which proves (9).

Let us set $\mathbf{R} = \mathcal{R}/\mathcal{I}$. We will show below that $\mathbf{R}$ is a field.

DEFINITION 8: $\mathbf{R}$ is called the field of *real numbers.* An element of $\mathbf{R}$ is called a real number.

A typical real number is a coset $\{a_n\} + \mathcal{I}$, where $\{a_n\}$ is a Cauchy sequence of rational numbers. Let us now derive some of the properties of the real numbers.

PROPOSITION 9: $\mathbf{R}$ is a field.

Proof: Let $\{a_n\} + \mathcal{I} \in \mathbf{R}$ be nonzero. We must show that $\{a_n\} + \mathcal{I}$ has an inverse with respect to multiplication in $\mathbf{R}$. Since $\{a_n\} + \mathcal{I}$ is nonzero, we see that $\{a_n\} \notin \mathcal{I}$, so that $\{a_n\}$ is not a null Cauchy sequence. Therefore, there exists a positive rational number ϵ_0 such that

$$|a_n| \geq \epsilon_0 \tag{10}$$

for infinitely many values of n. But since $\{a_n\}$ is a Cauchy sequence, there exists a positive rational number N, depending on ϵ_0, such that

$$|a_n - a_m| < \epsilon_0/2 \tag{11}$$

whenever $n,m \geq N$. Let us choose $n \geq N$ so that (10) holds. Then, for $m \geq N$, we have, by (10) and (11),

$$\begin{aligned}\epsilon_0 \leq |a_n| &= |(a_n - a_m) + a_m| \\ &\leq |a_n - a_m| + |a_m| \\ &< \epsilon_0/2 + |a_m|.\end{aligned}$$

Therefore,

$$|a_m| \geq \epsilon_0 - \epsilon_0/2 = \epsilon_0/2 = A \qquad (m \geq N). \tag{12}$$

Set

$$\{\delta_n\} = \{A - a_1, A - a_2, \ldots, A - a_N, 0, 0, 0, \ldots\}. \tag{13}$$

It is clear that $\{\delta_n\}$ is a null Cauchy sequence, so that

$$\{a_n\} + \mathcal{I} = \{a_n\} + \{\delta_n\} + \mathcal{I}. \tag{14}$$

Moreover, by (12) and (13), we see that

$$|a_n + \delta_n| \geq A \qquad \text{for all } n. \tag{15}$$

Now (14) and (15) imply that there exists a Cauchy sequence $\{a'_n\}$ such that $\{a_n\} + \mathcal{I} = \{a'_n\} + \mathcal{I}$ and $|a'_n| \geq A$ for all n. Let us choose one such sequence and hold it fixed throughout the remainder of the argument. Since $A > 0$, we see that $a'_n \neq 0$ for any n. Therefore, it makes sense to speak of the sequence

$$\{a_1'^{-1}, a_2'^{-1}, \ldots\} = \{a_n'^{-1}\}.$$

We assert that $\{a_n'^{-1}\}$ is a Cauchy sequence. Let $\epsilon > 0, \epsilon \in \mathbf{Q}$ be given. Since $\{a'_n\}$ is a Cauchy sequence, there exists a positive integer M such that $|a'_n -$

$a'_m| < \epsilon A^2$, whenever $m, n \geq M$. Then, for $m, n \geq M$, we have

$$\begin{aligned}|a'^{-1}_n - a'^{-1}_m| &= |1/a'_n - 1/a'_m| \\ &= |(a'_m - a'_n)/a'_m a'_n| \\ &= |a'_m - a'_n|/|a'_m a'_n| \\ &\leq \epsilon A^2/A^2 \qquad (\text{since } |a'_m| \geq A, |a'_n| \geq A) \\ &= \epsilon.\end{aligned}$$

Therefore, $\{a'^{-1}_n\}$ is a Cauchy sequence and $\{a'^{-1}_n\} + \mathscr{I} \in \mathbf{R}$. But

$$[\{a'_n\} + \mathscr{I}]\cdot[\{a'^{-1}_n\} + \mathscr{I}] = e + \mathscr{I},$$

where $e = \{1, 1, 1, \ldots\}$. And since $e + \mathscr{I}$ is the identity element of $\mathbf{R}$, this implies that $\{a'^{-1}_n\} + \mathscr{I}$ is the inverse of $\{a'_n\} + \mathscr{I} = \{a_n\} + \mathscr{I}$. This completes the proof.

Now that we have established that the real numbers form a field, let us introduce in $\mathbf{R}$ the usual order relation. We say that a Cauchy sequence $\{a_n\}$ is *positive* if there exists $C \in \mathbf{Q}$, $C > 0$ and a positive integer N such that

$$a_n \geq C \qquad \text{for all } n \geq N. \tag{16}$$

LEMMA 10: Let $\{a_n\}$ and $\{b_n\}$ be positive Cauchy sequences and let $\{c_n\}$ be a null Cauchy sequence.

(1) $\{a_n + b_n\}$ and $\{a_n \cdot b_n\}$ are positive Cauchy sequences.
(2) $\{a_n + c_n\}$ is a positive Cauchy sequence.

Proof: (1) Exercise.

(2) Let C and N be determined so that (16) holds. Since $\{c_n\}$ is a Cauchy null sequence, there exists $N' \geq N$ such that

$$|c_n| < C/2 \qquad \text{for all } n \geq N'.$$

But then, for $n \geq N'$, we have

$$\begin{aligned}C \leq |a_n| &= |(a_n + c_n) - c_n| \\ &\leq |a_n + c_n| + |c_n| \\ &\leq |a_n + c_n| + C/2 \\ \Longrightarrow |a_n + c_n| &\geq C/2 \qquad \text{for all } n \geq N'.\end{aligned}$$

Therefore, part (2) is proved.

Let $x = \{a_n\} + \mathscr{I}$ be a real number. We say that x is *positive* if $\{a_n\}$ is a positive Cauchy sequence. By Lemma 10, part (2), we see that if x is positive, then every Cauchy sequence in the coset $\{a_n\} + \mathscr{I}$ is positive. Let P denote the set of positive real numbers.

PROPOSITION 11: (1) Let $x, y \in P$. Then

$$x + y \in P, \qquad x \cdot y \in P.$$

(2) Let $x \in \mathbf{R}$. Exactly one of the following holds:

$$\text{(a)} \quad x \in P, \qquad \text{(b)} \quad x = 0, \qquad \text{(c)} \quad -x \in P.$$

Proof: (1) Immediate from Lemma 10, part (1).

(2) Note that $0 \notin P$. Therefore, (a) and (b) or (b) and (c) cannot hold simultaneously. If (a) and (c) both hold, then by (1), we have $0 = x + (-x) \in P$. But $0 \notin P$, so this is a contradiction and (a) and (c) cannot both hold. Thus at most one of the possibilities (a), (b), or (c) can hold. Let us now show that at least one of them holds. Suppose that $x \notin P$ and $-x \notin P$. We must prove that $x = 0$. Thus, if $x = \{a_n\} + \mathfrak{I}$, we must show that $\{a_n\} \in \mathfrak{I}$. Since $x \notin P$ and $-x \notin P$, the Cauchy sequences $\{a_n\}$, $\{-a_n\}$ are not positive. Thus, given $\epsilon > 0$, $\epsilon \in \mathbf{Q}$, and a positive integer N,

$$a_n < \epsilon/2 \qquad \text{for some } n \geq N, \tag{17}$$

$$-a_m < \epsilon/2 \qquad \text{for some } m \geq N. \tag{18}$$

Since $\{a_n\}$ is a Cauchy sequence, there exists a positive integer $N' \geq N$ such that

$$|a_m - a_n| < \epsilon/2 \qquad \text{for } m, n \geq N'. \tag{19}$$

Let us choose $n \geq N'$ so that (17) holds. Then, by (19),

$$\begin{aligned} a_m - a_n \leq |a_m - a_n| < \epsilon/2 \qquad & (m \geq N'), \\ \Longrightarrow a_m < \epsilon/2 + a_n < \epsilon \qquad & (m \geq N'). \end{aligned} \tag{20}$$

Let us now choose $m \geq N'$ so that (18) holds. Then, by (19),

$$\begin{aligned} a_m - a_p \leq |a_m - a_p| < \epsilon/2 \qquad & (p \geq N'), \\ \Longrightarrow a_p - a_m > -\epsilon/2 \qquad & (p \geq N'), \\ \Longrightarrow a_p + \epsilon/2 > -\epsilon/2 \qquad & (p \geq N'), \\ \Longrightarrow a_p > -\epsilon \qquad & (p \geq N'). \end{aligned} \tag{21}$$

Combining (20) and (21), we get

$$\begin{aligned} -\epsilon < a_n < \epsilon \qquad & (n \geq N'), \\ \Longrightarrow |a_n| < \epsilon \qquad & (n \geq N'). \end{aligned}$$

Thus, $\{a_n\}$ is a null Cauchy sequence and part (2) is proved.

Now that we have defined the positive real numbers, it is an easy matter to put an order structure on $\mathbf{R}$. Let $x, y \in \mathbf{R}$. We say that x is *greater than* y, denoted $x > y$, if $x - y$ is positive. If x is greater than y, we also say that y is *less than* x, denoted $y < x$. We say that x is *greater than or equal to* y, denoted $x \geq y$, if either $x > y$ or $x = y$. Similarly, we say that y is *less than or equal to* x, denoted $y \leq x$, if either $y < x$ or $y = x$. The properties of the order relation $>$ are summarized in the following result.

PROPOSITION 12: Let $x, y, z, w \in \mathbf{R}$.

(1) $x \in P$ if and only if $x > 0$.
(2) If $x > y$ and $y > z$, then $x > z$.
(3) If $x > y$ and $z > w$, then $x + z > y + w$.
(4) If $x > y$ and $z > 0$, then $xz > yz$; if $x > y$ and $z < 0$, then $yz > xz$.
(5) Exactly one of the following holds: $x > 0$, $x = 0$, $x < 0$.

Proof: (1) $x > 0 \leftrightarrow x - 0 \in P$.

(2) If $x > y$ and $y > z$, then $x - y \in P$ and $y - z \in P$. Therefore, by Proposition 11, part (1), we have $x - z = (x - y) + (y - z) \in P$. Thus $x > z$.

(3) If $x > y$ and $z > w$, then $x - y \in P$, $z - w \in P$. Therefore, by Proposition 11, part (1), $(x + z) - (y + w) \in P$, so that $x + z > y + w$.

(4) If $x > y$ and $z > 0$, then $x - y \in P$ and $z \in P$. Therefore, by Proposition 11, part (1), $z \cdot (x - y) = z \cdot x - z \cdot y \in P$. Thus, $z \cdot x > z \cdot y$. If $x > y$ and $z < 0$, then $x - y \in P$ and $-z = 0 - z \in P$. Therefore, $y \cdot z - x \cdot z = (-z) \cdot (x - y) \in P$, and thus $yz > xz$.

(5) This statement is just a translation of Proposition 11, part (2), in terms of the order relation.

If $x \in \mathbf{R}$, let us define the *absolute value* of x, denoted $|x|$, by

$$|x| = \begin{cases} x & \text{if } x > 0, \\ 0 & \text{if } x = 0, \\ -x & \text{if } x < 0. \end{cases}$$

The same arguments as we used to prove Proposition 2 can be used to prove

PROPOSITION 13: Let $x, y \in \mathbf{R}$.
(1) $|x| \geq 0$.
(2) $|x| = 0$ if and only if $x = 0$.
(3) $|x \cdot y| = |x| \cdot |y|$.
(4) $|x + y| \leq |x| + |y|$ (triangle inequality).

The mapping

$$\psi : \mathbf{Q} \longrightarrow \mathbf{R}$$

$$\psi(a) = \{a, a, a, a, \ldots\} + \mathcal{I}$$

is an isomorphism and is therefore an embedding of $\mathbf{Q}$ in $\mathbf{R}$. Moreover, if

$a \in \mathbf{Q}$ is positive, then $\{a, a, a, \dots\}$ is a positive Cauchy sequence, so that $\psi(a)$ is positive. Thus, ψ preserves the notion of positivity. Also, ψ preserves order. That is, if $a, b \in \mathbf{Q}$ are such that $a < b$, then $\psi(a) < \psi(b)$. Indeed, if $a < b$, then $b - a \in P$, which implies that $\psi(b) - \psi(a) = \psi(b - a) \in P$. Thus, $\psi(a) < \psi(b)$. Similarly, ψ preserves absolute values:

$$\psi(|a|) = |\psi(a)| \quad (a \in \mathbf{Q}).$$

We leave the proof of this last fact to the reader. Thus, since ψ preserves all arithmetic operations, positivity, order, and absolute values, every result which we have proved about $\mathbf{Q}$ is true for $\psi(\mathbf{Q})$. Thus, we may identify $\mathbf{Q}$ with the subring $\psi(\mathbf{Q})$ of $\mathbf{R}$. In what follows, we will always assume that this identification has been made. Thus, for example, if we speak of the "real number $\frac{2}{3}$" we will really mean the real number $\psi(\frac{2}{3})$—that is,

$$(\tfrac{2}{3}, \tfrac{2}{3}, \tfrac{2}{3}, \dots) + \mathcal{J}.$$

The following property of the ordering of $\mathbf{R}$ is very significant.

PROPOSITION 14: Let $x \in \mathbf{R}$. Then there exists a positive integer n such that $n > x$.

Proof: If $x \leq 0$, then we may take $n = 1$. Thus, assume $x > 0$. Let

$$x = \{a_m\} + \mathcal{J}.$$

By Lemma 4, there exists a positive rational number M such that $|a_m| \leq M$ for all m. Assume that $M = a/b$, where $a,b \in Z$ and $a \geq 1$, $b \geq 1$. Then we may set $n = a + 1$, since

$$n - a_m \geq a + 1 - M = \frac{a(b - 1) + b}{b} \qquad \text{for all } m,$$

which implies that $n - x$ is positive—that is, $n > x$.

An interesting consequence of Proposition 14 is given by

COROLLARY 15: Let $x \in \mathbf{R}$, $x > 0$.

(1) There exists a positive integer n such that $x < 2^n$.
(2) There exists a positive integer m such that $x > 1/2^m$.

Proof: (1) By induction, we can see that $2^n \geq n$ for all positive integers n. By Proposition 14, there exists a positive integer n such that $x < n$. Then $x < 2^n$.

(2) Since $x > 0$, we have $x^{-1} > 0$. Therefore, by part (1), there exists a positive integer m such that $x^{-1} < 2^m$. Multiplying both sides of this last inequality by $x/2^m$, we see that $x > 1/2^m$.

The form in which Corollary 15 will be used most is given by

COROLLARY 16: Let x and a be real numbers.

(1) If $x \leq a + 1/2^n$ for all positive integers n, then $x \leq a$.
(2) If $x \geq a - 1/2^n$ for all positive integers n, then $x \geq a$.

Proof: (1) If $x > a$, then $x - a > 0$, so that by Corollary 15, part (2), we can find a positive integer m so that $x - a > 1/2^m$. This is a contradiction, so that $x \leq a$.

(2) The proof of part (2) is similar to the proof of part (1).

The properties of the real numbers which we have established up to this point do not reflect any marked difference between **Q** and **R**. Let us now establish a property of **R** which shows that **R** is very different from **Q**.

Let S be a subset of **R**. An *upper bound* for S is a real number u such that $s \leq u$ for all $s \in S$. An upper bound for S is called a *least upper bound* if, whenever v is an upper bound for S, we have $u \leq v$. For example, let $S = \{x \in \mathbf{R}, x \leq 1\}$. Then 1, 2, 5, and $\frac{17}{3}$ are all upper bounds for S. Only 1 is a least upper bound for S. The essential difference between **Q** and **R** is contained in the following result, which is called the *least upper bound property*.

THEOREM 17: Let S be a nonempty subset of **R**. If S has an upper bound, then S has a least upper bound.

Before proving this result, let us study an example. Let us define a sequence of rational numbers $\{x_n\}$ as follows:

$$x_1 = 1, \quad x_{n+1} = \frac{x_n + 2x_n^{-1}}{2} \qquad (n = 1, 2, 3, \ldots).$$

The first few terms of the sequence are

$$\begin{aligned} x_1 &= 1, \\ x_2 &= \tfrac{3}{2} = 1.5000, \\ x_3 &= \tfrac{17}{12} = 1.41666\cdots, \\ x_4 &= \tfrac{577}{408} = 1.41421\cdots. \end{aligned}$$

Note that the decimal expansions suggest that x_n provides a rational approximation to the square root of 2. Indeed, this is so. For $n = 1, 2, 3, \ldots$, we have

$$\begin{aligned} x_{n+1}^2 - 2 &= \left(\frac{x_n + 2x_n^{-1}}{2}\right)^2 - 2 \\ &= \frac{1}{4}\left(x_n - \frac{2}{x_n}\right)^2. \end{aligned} \tag{22}$$

Therefore, since $(x_n - 2/x_n)^2 \geq 0$, we see that $x_{n+1}^2 - 2 \geq 0$, so that

$$x_n^2 \geq 2 \qquad \text{for all } n > 1. \tag{23}$$

Moreover, we assert that

$$x_n^2 - 2 \leq \frac{1}{2^{2^{n-1}}} \qquad \text{for all } n. \tag{24}$$

Let us prove (24) by induction on n. It is clearly true for $n = 1, 2$. If (24) holds for some n, then

$$\begin{aligned} x_{n+1}^2 - 2 &= \frac{1}{4}\left(x_n - \frac{2}{x_n}\right)^2 \\ &= \frac{1}{4}\frac{(x_n^2 - 2)^2}{x_n^2} \\ &\leq \frac{1}{2^2}(x_n^2 - 2)^2 \qquad \text{by (23)} \\ &\leq \frac{1}{2^2}\left(\frac{1}{2^{2^{n-1}}}\right)^2 \leq \frac{1}{2^{2^n}} \qquad \text{(by the induction assumption).} \end{aligned}$$

Thus (24) holds for $n + 1$, so that the induction is complete. From (24), it follows that

$$x_n^2 - 2 \leq \frac{1}{2^n}, \tag{25}$$

since $2^{n-1} \geq n$ for all positive integers n (induction).

Let $S = \{x \in \mathbf{Q} \mid x > 0, x^2 \leq 2\}$. By Theorem 17, S has a least upper bound r. Let us prove that $r^2 = 2$. If $x \in S$, then $x^2 \leq 2 \leq x_n^2$ for all n, by (23). Therefore, since $x > 0$, we see that $x < x_n$, so that x_n is an upper bound for S. Since r is a least upper bound for S, this implies that $r \leq x_n$ for all n, so that $r^2 \leq x_n^2$. Thus, by (25), $r^2 \leq 2 + \frac{1}{2^n}$ for all n, and thus by Corollary 16, part (1) we derive that

$$r^2 \leq 2.$$

By (23), $x_n^2 \geq 2$, so that $x_n \geq 1$ for all n. Now

$$\begin{aligned} \left(x_n - \frac{1}{2^{2^n}}\right)^2 &= x_n^2 - \frac{x_n}{2^{n-1}} + \frac{1}{2^n} \\ &\leq 2 + \frac{1}{2^n} - \frac{1}{2^{n-1}} + \frac{1}{2^{2^n}} \quad \text{[by (24) and the fact that } x_n \geq 1\text{]} \\ &= 2. \end{aligned} \tag{26}$$

Therefore, $x_n - 1/2^n \in S$, and $r \geq x_n - 1/2^n$ since r is an upper bound for S. Finally, by (23), $r^2 \geq (x_n - 1/2^n)^2 \geq 2 - 1/2^{n-2}$, so that $r^2 \geq 2$. Combining this last inequality with (26), we see that $r^2 = 2$, as we asserted.

At the beginning of this chapter we showed that $\mathbf{Q}$ contained no element whose square is 2. What we have just shown is that **R** contains such an element. In fact, we actually exhibited the element as the least upper bound of a set of rational numbers. The real number we constructed is called the square root of 2 and is denoted $\sqrt{2}$. The procedure which we used to construct $\sqrt{2}$ gives us a great deal of information on how we can approximate $\sqrt{2}$ by rational numbers. Indeed, we showed that

$$x_n - \frac{1}{2^n} \leq \sqrt{2} \leq x_n \qquad \text{for all } n. \tag{27}$$

Thus, for example, for $n = 4$, we see that

$$\frac{1103}{816} \leq \sqrt{2} \leq \frac{577}{408}.$$

Translating this into decimals, we get

$$1.351 \leq \sqrt{2} \leq 1.415.$$

We will pursue these ideas further in the exercises. Now that we have demonstrated the utility of the least upper bound property, let us prove it.

Proof of Theorem 17: Let a be an upper bound for S and let A be a positive integer such that $A \geq a$. Such an A exists by Proposition 14. Let $b \in S$ and let B be a positive integer such that $B \geq -b$. Then, $-B \leq b$. Moreover, since A is an upper bound for S, we have

$$-B \leq b \leq A. \tag{28}$$

For each positive integer n, let us consider the set S_n of all rational numbers of the form $c/2^n$ ($c \in \mathbf{Z}$) such that

$$-B \leq \frac{c}{2^n} \leq A.$$

Clearly, $A, -B \in S_n$, since $A = A \cdot 2^n/2^n$, $-B = (-B \cdot 2^n)/2^n$. And S_n contains only finitely many elements, since $-2^nB \leq c \leq 2^nA$. Since $A \in S_n$, S_n contains upper bounds for S. Let a_n be the smallest upper bound contained in S_n. In particular, $a_n - 1/2^n$ is not an upper bound for S. Note that $S_1 \subseteq S_2 \subseteq S_3 \subseteq \cdots$, and therefore if $m \geq n$, we have

$$a_n - \frac{1}{2^n} < a_m \leq a_n. \tag{29}$$

Suppose that $\epsilon > 0$, $\epsilon \in \mathbf{Q}$ is given. By Corollary 15, part (2), there exists a positive integer N such that $1/2^N < \epsilon$. Therefore, if $m \geq n \geq N$, (29) implies that

$$0 \leq a_n - a_m < \tfrac{1}{2^n} \leq \tfrac{1}{2^N},$$
$$\Longrightarrow |a_n - a_m| < \epsilon.$$

Therefore, $\{a_m\}$ is a Cauchy sequence of rational numbers. Let $x = \{a_m\} + \mathcal{I}$ be the real number corresponding to $\{a_m\}$. From (29), we see that if n is a positive integer, then

$$a_n - a_m \geq 0, \quad a_m - \left(a_n - \frac{1}{2^n}\right) \geq 0 \qquad \text{for } m \geq n,$$

$$\Longrightarrow a_n - x \geq 0, \quad x - \left(a_n - \frac{1}{2^n}\right) \geq 0 \qquad \text{(Why?)} \tag{30}$$

$$\Longrightarrow a_n - \frac{1}{2^n} \leq x \leq a_n.$$

We assert that x is an upper bound for S. Suppose not. Then there exists $y \in S$ such that $x < y$. By Corollary 15, part (2), there is a positive integer N such that $\frac{1}{2^N} < y - x$. But (30) implies that $a_N - \frac{1}{2^N} \leq x$. Therefore,

$$y > x + \tfrac{1}{2^N}, \qquad x \geq a_N - \tfrac{1}{2^N} \Longrightarrow y > a_N,$$

which contradicts the fact that a_N is an upper bound for S. Thus, x is an upper bound for S.

Now let us show that x is a least upper bound for S. Suppose not. Then there is an upper bound z of S such that $z < x$. By Corollary 15, part (2), there is a positive integer N such that $\frac{1}{2^N} < x - z$. Since $a_N - \frac{1}{2^N}$ is not an upper bound for S, there exists $w \in S$ such that $w > a_N - \frac{1}{2^N}$. However, since z is an upper bound for S, we have $z \geq w$. Therefore, $z > a_N - \frac{1}{2^N}$. But $x > z + \frac{1}{2^N}$, so that $x > a_N$, which contradicts (30). Thus, x is a least upper bound for S.

In our example above, we observed that $\sqrt{2}$ can be approximated with arbitrary accuracy by rational numbers. This is not an isolated phenomenon. Indeed, we have the following theorem.

THEOREM 18: Let $x \in \mathbf{R}$ and let $\epsilon > 0$ be a given real number. Then there exists a rational number a/b such that

$$|x - a/b| < \epsilon.$$

Proof: By Corollary 15, part (2), there exists a positive integer N such that $\frac{1}{2^N} < \epsilon$. It suffices to prove the result with ϵ replaced by $\frac{1}{2^N}$. Thus, without loss of generality, we may assume that $\epsilon \in \mathbf{Q}$. Let $x = \{a_n\} + \mathcal{I}$. Since $\{a_n\}$ is a Cauchy sequence of rational numbers, there exists a positive integer M such that $|a_n - a_m| < \epsilon$ whenever $n, m \geq M$. Set $a/b = a_M$. Then

$$\begin{aligned} |x - a/b| &= |\{a_n - a_M\} + \mathcal{I}| \\ &< \epsilon. \qquad \text{(Why?)} \end{aligned}$$

Thus, the theorem is proved.

We have now reached the point at which most calculus books start. We have constructed a number system which verifies the set of axioms usually cited in calculus books. But what of our original notion of a real number as an infinite decimal? Let us now define precisely what is meant by an infinite decimal. Let

$$b_r b_{r-1} \dots b_1 b_0 \cdot a_1 a_2 a_3 \dots \tag{31}$$

be a sequence of integers such that $0 \leq a_i, b_j \leq 9$. Such a sequence is called an *infinite decimal*. Let us associate with each infinite decimal a real number as follows: Let x_n be the rational number

$$b_r \cdot 10^r + b_{r-1} \cdot 10^{r-1} + \cdots + b_1 \cdot 10 + b_0 + \frac{a_1}{10} + \cdots + \frac{a_n}{10^n}.$$

If $n > m$, then

$$x_n - x_m = \frac{a_{m+1}}{10^{m+1}} + \cdots + \frac{a_n}{10^n},$$

and therefore

$$\begin{aligned} |x_n - x_m| &\leq \frac{9}{10^{m+1}} + \frac{9}{10^{m+2}} + \cdots + \frac{9}{10^n} \\ &= \frac{9}{10^{m+1}}\left(1 + \frac{1}{10} + \cdots + \frac{1}{10^{n-m-1}}\right) \\ &= \frac{9}{10^{m+1}}\frac{10}{9}\left(1 - \frac{1}{10^{n-m}}\right) \\ &\leq \frac{1}{10^m}. \end{aligned}$$

Thus, $\{x_n\}$ is a Cauchy sequence of rational numbers and we associate to the infinite decimal (31) the real number $\{x_n\} + \mathcal{I}$. Actually, every real number is associated to some infinite decimal via this correspondence. We will leave this result for the exercises.

We have shown that **R** contains a positive element x whose square is 2—that is, **R** contains a square root of 2. In Exercise 5 we will show that a positive real number x has an nth root in **R**, where n is any positive integer.

4.6 *Exercises*

1. (a) Show that there does not exist a rational number x such that $x^2 = 3$.
   (b) Show that there exists a real number x such that $x^2 = 3$. [*Hint*: Define x via the Cauchy sequence $\{x_n\}_{1 \leq n < \infty}$, where $x_1 = 1$, $x_{n+1} = (x_n + 3x_n^{-1})/2$ $(n = 1, 2, \dots)$.]

2. Let $n \in \mathbf{Z}$ be such that $n \neq m^2$ for any $m \in \mathbf{Z}$. Show that there does not exist $x \in \mathbf{Q}$ such that $x^2 = n$.

*3. (a) Let n be a nonnegative rational number. Show that there exists $x \in \mathbf{R}$ such that $x^2 = n$.
(b) Let n be a negative rational number. Show that there does not exist $x \in \mathbf{R}$ such that $x^2 = n$.

*4. Show that every real number can be written as an infinite decimal.

*5. (a) Let $x \in \mathbf{R}$, $x \geq 0$. Show that there exists $y \in \mathbf{R}$ such that $y^n = x$. [*Hint:* Let $y_1 \in \mathbf{N}$ be such that $y_1^n \leq x < (y_1 + 1)^n$. Define y_{m+1} $(1 \leq m < \infty)$ by

$$y_{m+1} = y_m - \frac{y_m^n - x}{ny_m^{n-1}}.$$

Show that $\{y_m\}$ is a Cauchy sequence of rationals equal to x.]
(b) Show that y is unique. It is called the *real* nth *root of* x and is usually denoted $\sqrt[n]{x}$.

6. Use Exercise 5 to compute $\sqrt[3]{2}$, $\sqrt[4]{3}$, and $\sqrt[5]{2}$ to three decimal places.

4.7 The Complex Numbers

The complex numbers $\mathbf{C}$ are introduced in high school as the set of all quantities of the form $a + bi$, where a and b are real numbers and $i = \sqrt{-1}$. This last condition means simply that $i^2 = -1$. We say that two complex numbers $a + bi$ and $c + di$ are *equal* if $a = c$ and $b = d$. Further, we define addition and multiplication of complex numbers by

$$(a + bi) + (c + di) = (a + c) + (b + d)i, \tag{1}$$

$$(a + bi)\cdot(c + di) = (ac - bd) + (ad + bc)i \tag{2}$$

With respect to these operations, $\mathbf{C}$ becomes a field. The additive identity is $0 + 0i$ and the multiplicative identity is $1 + 0i$. The additive inverse of $a + bi$ is $(-a) + (-b)i$, while if $a + bi$ is nonzero, then either $a \neq 0$ or $b \neq 0$ and the multiplicative inverse of $a + bi$ is

$$\frac{a}{a^2 + b^2} + \frac{-b}{a^2 + b^2}\, i.$$

We leave the verification of these facts, as well as the associativity, commutativity, and distributivity laws for the reader to check. Note that we may regard the real numbers $\mathbf{R}$ to be a subfield of $\mathbf{C}$ by identifying $a \in \mathbf{R}$ with the complex number $a + 0i$. More precisely, the mapping

$$\mathbf{R} \longrightarrow \mathbf{C}$$
$$a \longrightarrow a + 0i$$

is an isomorphism. In what follows, we will always regard $\mathbf{R}$ as a subfield of $\mathbf{C}$.

All of the above is probably more or less familiar from high school algebra. However, there is a basic defect with the above definition of $\mathbf{C}$. In our above explanation, we have introduced $i = \sqrt{-1}$. And it is not at all clear what i is or where it comes from. This gives the complex numbers a somewhat mysterious quality and accounts for the terminology which calls i an "imaginary number." The way out of this logical morass was discovered by Gauss in his dissertation. Instead of considering quantities of the form $a + bi$, let us consider ordered pairs (a, b). We say that two ordered pairs (a, b) and (c, d) are equal if $a = c$ and $b = d$. We define addition and multiplication of the ordered pairs via

$$(a, b) + (c, d) = (a + c, b + d), \tag{1'}$$

$$(a, b) \cdot (c, d) = (ac - bd, ad + bc). \tag{2'}$$

We define **C** to be the set of all these ordered pairs. With respect to the addition (1′) and multiplication (2′), **C** becomes a field. Moreover, **R** may be identified with the subfield $\{(a, 0) \mid a \in \mathbf{R}\}$ of **C**. Thus, we have explicitly constructed a field **C** containing **R**. Consider the element (0, 1) of **C**. From the definition of multiplication in **C**,

$$(0, 1) \cdot (0, 1) = (-1, 0).$$

That is, $(0, 1)^2$ equals -1. Thus, our field **C** contains a square root of -1! Let us set $i = (0, 1)$. Then every complex number can be uniquely written in the form $a + b \cdot i$, where $a,b \in \mathbf{R}$. [Strictly speaking, we should write $(a, 0) + (b, 0) \cdot i$ instead of $a + bi$.] And the field **C** which we have constructed is precisely what we are accustomed to think of as the field of complex numbers.

Let us proceed to study the properties of the complex numbers in somewhat more detail than we have done up to this point. If $\alpha = a + ib \in \mathbf{C}$, we define the *complex conjugate* $\bar{\alpha}$ of α by

$$\bar{\alpha} = a - ib.$$

The reader should have no trouble in verifying the following properties of complex conjugation: Let $\alpha, \beta \in \mathbf{C}$.

I. $\overline{\alpha + \beta} = \bar{\alpha} + \bar{\beta}$.

II. $\overline{\alpha \cdot \beta} = \bar{\alpha} \cdot \bar{\beta}$.

III. $\overline{(\bar{\alpha})} = \alpha$.

IV. $\bar{\alpha} = \alpha$ if and only if α is real.

V. The mapping $\alpha \longrightarrow \bar{\alpha}$ of **C** onto **C** is a surjective isomorphism which is the identity isomorphism on **R**. (This follows from I, II, and IV.)

VI. $\alpha \cdot \bar{\alpha}$ is a real number and $\alpha \cdot \bar{\alpha} \geq 0$. (In fact, if $x = a + ib$, then $\alpha \cdot \bar{\alpha} = a^2 + b^2 \geq 0$.)

Let us define the *absolute value of* α, denoted $|\alpha|$, by

$$|\alpha| = \sqrt{\alpha \cdot \bar{\alpha}} = \sqrt{a^2 + b^2} \qquad (\alpha = a + ib).$$

By VI, $\alpha \cdot \bar{\alpha}$ is a nonnegative real number, so $|\alpha|$ makes sense and is a nonnegative real number. It is possible to give the absolute value of a complex number a geometric interpretation as follows: Let us associate to the complex number $\alpha = a + ib$ the point of the Cartesian plane with coordinates (a, b). Then this sets up a one-to-one correspondence between complex numbers and points of the Cartesian plane. Moreover, it is clear from Figure 4.2 that $|a + ib|$ is just the distance of the point (a, b) from the origin.

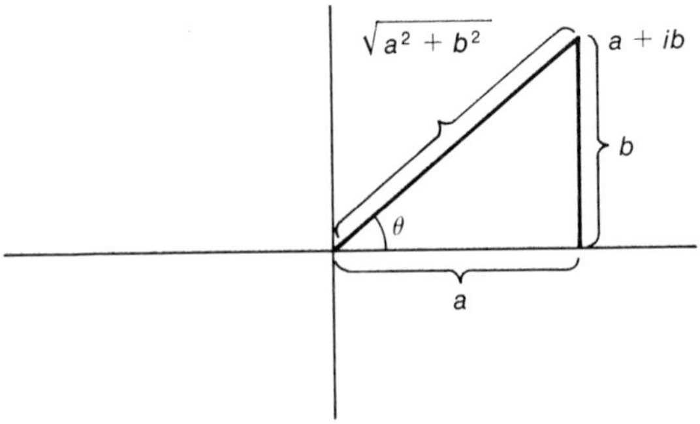

FIGURE 4-2: One-to-One Correspondence Between $\mathbf{R}^2$ and $\mathbf{C}$.

Let $\rho = |a + ib| = \sqrt{a^2 + b^2}$ and let θ be the angle which the line connecting the origin and (a, b) makes with the positive half of the x-axis (see Figure 6.1). By the definitions of $\sin \theta$ and $\cos \theta$, we know that

$$a = \rho \cos \theta, \qquad b = \rho \sin \theta. \tag{3}$$

Therefore, we may represent the complex number $a + ib$ in the form

$$a + ib = \rho(\cos \theta + i \sin \theta). \tag{4}$$

This representation is called the *polar form* of $a + ib$ and θ is called the *argument* of $a + ib$.

Let us illustrate geometrically the meaning of addition and multiplication of complex numbers. Let $a + ib, c + id \in \mathbf{C}$. Then

$$(a + ib) + (c + id) = (a + c) + i(b + d)$$

corresponds geometrically to the point $(a + c, b + d)$. Thus, complex numbers add according to the usual "parallelogram law" of elementary physics (see Figure 4.3).

In order to describe multiplication, it is best to use the polar form. Let

$$a + ib = \rho_1(\cos \theta_1 + i \sin \theta_1),$$
$$c + id = \rho_2(\cos \theta_2 + i \sin \theta_2).$$

Then

$$\begin{aligned}(a + ib)\cdot(c + id) &= \rho_1\rho_2([\cos \theta_1 \cos \theta_2 - \sin \theta_1 \sin \theta_2] \\ &\qquad + i[\cos \theta_1 \sin \theta_2 + \cos \theta_2 \sin \theta_1]) \\ &= \rho_1\rho_2[\cos(\theta_1 + \theta_2) + i \sin(\theta_1 + \theta_2)].\end{aligned} \tag{5}$$

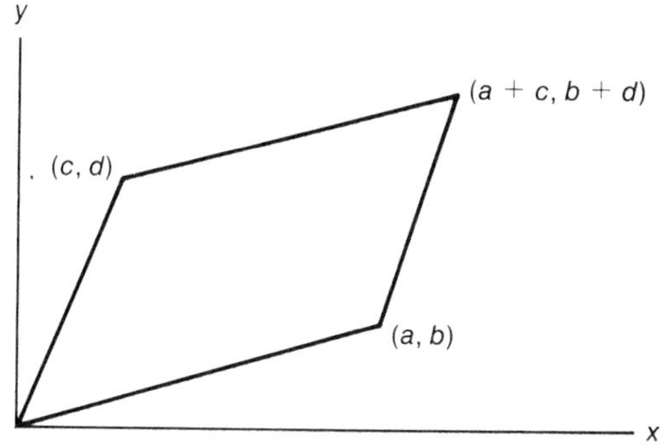

FIGURE 4-3: Addition in **C**.

Thus, $(a + ib)\cdot(c + id)$ corresponds to the point at distance $\rho_1\rho_2$ from the origin and has argument $\theta_1 + \theta_2$. Note that we have used the addition formulas for sin and cos, which should be familiar from trigonometry. From (5), an induction argument immediately shows that for $n \geq 0$,

$$(a + ib)^n = \rho_1^n[\cos(n\theta_1) + i\sin(n\theta_1)]. \tag{6}$$

This last formula is usually known as de Moivre's theorem. Let us use de Moivre's theorem to get some information about nth roots of complex numbers.

Let $\alpha = \rho(\cos\theta + i\sin\theta)$ be a complex number. Let us inquire as to whether α has any nth roots in **C**, where n is a positive integer. (We say that β is an nth root of α if $\beta^n = \alpha$.) If $\alpha = 0$, then $\beta = 0$ is an nth root, and is, in fact, the only nth root. Thus, let us assume that $\alpha \neq 0$. Then we have $\rho > 0$. First observe that α has at most n nth roots in **C**, since any nth root of α is a zero of the polynomial $X^n - \alpha$, and this polynomial has at most n zeros (Exercise 6b, Section 4-4). Next, consider the following n distinct complex numbers:

$$\beta_k = \rho^{1/n}\left[\cos\left(\frac{\theta}{n} + \frac{360}{n}\cdot k\right) + i\sin\left(\frac{\theta}{n} + \frac{360}{n}\cdot k\right)\right]$$
$$(k = 0, 1, \ldots, n - 1), \tag{7}$$

where $\rho^{1/n}$ is the positive nth root of ρ. By De Moivre's theorem,

$$\begin{aligned}\beta_k^n &= (\rho^{1/n})^n[\cos(\theta + 360\cdot k) + i\sin(\theta + 360\cdot k)]\\ &= \rho[\cos\theta + i\sin\theta]\\ &= \alpha.\end{aligned}$$

Therefore, β_k is an nth root of α and we have proved

THEOREM 1: Every nonzero complex number has n nth roots in **C**.

The formula (7) can be used to compute nth roots. For example, let us find the nth roots of 1. Since

$$1 = 1\cdot(\cos 0 + i\sin 0),$$

we see that the nth roots of 1 are given by

$$\cos\left(\frac{360k}{n}\right) + i\sin\left(\frac{360k}{n}\right) \qquad (k = 0, 1, \ldots, n-1). \tag{8}$$

Let us examine this formula for some low values of n:

$$n = 1:\ \cos\left(\frac{360\cdot 0}{1}\right) + i\sin\left(\frac{360\cdot 0}{1}\right) = 1 + 0i = 1,$$

$$n = 2:\ \cos\left(\frac{360\cdot 0}{2}\right) + i\sin\left(\frac{360\cdot 0}{2}\right) = 1 + 0i = 1,$$

$$\cos\left(\frac{360\cdot 1}{2}\right) + i\sin\left(\frac{360\cdot 1}{2}\right) = \cos(180) + i\sin(180) = -1 + 0\cdot i = -1.$$

$$n = 3:\ \cos\left(\frac{360\cdot 0}{3}\right) + i\sin\left(\frac{360\cdot 0}{3}\right) = 1 + 0i = 1,$$

$$\cos\left(\frac{360\cdot 1}{3}\right) + i\sin\left(\frac{360\cdot 1}{3}\right) = \cos(120) + i\sin(120) = \frac{-1 + \sqrt{3}\,i}{2},$$

$$\cos\left(\frac{360\cdot 2}{3}\right) + i\sin\left(\frac{360\cdot 2}{3}\right) = \frac{-1 - \sqrt{3}\,i}{2},$$

Let

$$\zeta_n = \cos\left(\frac{360}{n}\right) + i\sin\left(\frac{360}{n}\right).$$

Then, from (8), ζ_n is an nth root of 1, and, in fact, corresponds to $k = 1$. Moreover, from de Moivre's theorem, we see that

$$\zeta_n^k = \cos\left(\frac{360k}{n}\right) + i\sin\left(\frac{360k}{n}\right).$$

Therefore, from (8), we see that all nth roots of unity are powers of ζ_n. Thus, we have

THEOREM 2: The nth roots of 1 in **C** form a cyclic group of order n. A generator of this group is ζ_n.

We will denote the group of nth roots of unity by X_n. A generator of X_n is called a *primitive* nth *root of unity*. Since the order of ζ_n^k is $n/(k, n)$, we see that the order of ζ_n^k is n if and only if $(k, n) = 1$. Therefore, the primitive nth roots of unity are given by

$$\zeta_n^k \qquad [0 \le k \le n-1, \quad (k, n) = 1].$$

There are exactly $\phi(n)$ such roots of unity.

We found above that the n nth roots of a nonzero complex number

$\alpha = \rho(\cos\theta + i\sin\theta)$ are given by β_k $(k = 0, \ldots, n-1)$, where β_k is defined by (7). Note, however, that by (5), we have

$$\beta_k = \left\{\rho^{1/n}\left[\cos\left(\frac{\theta}{n}\right) + i\sin\left(\frac{\theta}{n}\right)\right]\right\}\left[\cos\left(\frac{360k}{n}\right) + i\sin\left(\frac{360k}{n}\right)\right]$$
$$= \beta_0\zeta_n^k.$$

Thus, we obtain the n nth roots of α by multiplying the fixed nth root β_0 by all possible nth roots of unity. Therefore, we have

THEOREM 3: Let α be a nonzero complex number. Let β be an nth root of α. Then the nth roots of α are all of the form $\beta \cdot \zeta$, where ζ is an nth root of unity. Conversely, every such quantity is an nth root of α.

4.7 Exercises

1. Show that with respect to the operation of addition (1′) and the operation of multiplication (2′), $\mathbf{C}$ is a field.
2. Show that every complex number can be written uniquely in the form $a + bi$ $(a, b \in \mathbf{R})$.
3. Verify properties I–VI.
4. Let $a, b, c \in \mathbf{C}$, $a \neq 0$. Show that there exists a complex number x such that $ax^2 + bx + c = 0$. Show that if $b^2 - 4ac \neq 0$, then the number of such x is 2, whereas if $b^2 - 4ac = 0$, the number of such x is 1.
5. Find the polar forms of the following complex numbers:
   (a) i (b) $1 + i$ (c) $2 + 2i$
   (d) $3 + 2i$ (e) $\frac{\sqrt{2}}{2} + \frac{\sqrt{2}}{2}i$ (f) $-\frac{1}{2} + \frac{\sqrt{3}}{2}i$
   (g) -1 (h) $-\frac{1}{2} + \frac{\sqrt{3}}{2}i$
6. Find all cube roots of $1 + i$.
7. Find all primitive nth roots of 1 for $n = 1, 2, 3, 4$.
8. Find all solutions in $\mathbf{C}$ of the equation $X^5 - 2 = 0$.
9. Prove that $\mathbf{C} \approx \mathbf{R}[X]/(X^2 + 1)\mathbf{R}[X]$.

5 UNIQUE FACTORIZATION

5.1 Unique Factorization and Fermat's Last Theorem

In Euclidean geometry one often meets right triangles whose sides are integers. For example, there is the famous 3, 4, 5 triangle. If the sides of a right triangle are x and y and the hypotenuse is z, then the Pythagorean theorem asserts that

$$x^2 + y^2 = z^2. \tag{1}$$

And determining all right triangles whose sides are integers is equivalent to determining all integral solutions of (1)—that is, solutions for which x, y, and z are integers. It can be shown that all solutions of (1) are of the form

$$x = c(a^2 - b^2), \quad y = 2abc, \quad z = c(a^2 + b^2), \qquad a, b, c \in \mathbf{Z}. \tag{2}$$

It is easy to check that (2) provides a solution to (1) for every a and b. The problem of solving polynomial equations in integers goes back to the ancient Greeks, and such equations are called Diophantine equations after the Greek mathematician Diophantus of Alexandria (250 AD?), who first made an extensive study of them. In Chapter 2 we used the theory of congruences to solve the linear Diophantine equation

$$ax + by = c, \qquad a, b, c \in \mathbf{Z}. \tag{3}$$

This was not an especially difficult task. However, this is not a very representative example. For Diophantine equations are usually very difficult to solve. A given Diophantine equation may have no solutions, a finite number of solutions, or an infinite number of solutions. Equation (1) is of the latter type. The equation $x^2 + y^2 = 1$ has only four solutions: $(x, y) = (1, 0)$, $(0, -1)$, $(-1, 0)$, $(0, 1)$; the equation $x^2 + y^2 = -1$ has no solutions.

Because of the challenges which Diophantine equations present, they have been the fascination of mathematical amateurs for many centuries. One of the great amateurs was Pierre Fermat, a seventeenth-century French jurist who took up mathematics as a hobby. In Chapter 3 we met one of Fermat's more famous contributions, Fermat's Little Theorem. Fermat was very interested in the solution of Diophantine equations and carefully studied the works of Diophantus. Fermat often made marginal notes in his copy of Diophantus's works. On the page on which (1) was discussed, Fermat asserted that if $n > 2$, then

$$x^n + y^n = z^n \tag{4}$$

has no solution in integers x, y, z, none 0. Furthermore, he claimed to have discovered "a truly marvellous demonstration which this margin is too narrow to contain." It is highly doubtful that Fermat had a correct proof of his assertion, which has become known as *Fermat's last theorem*. Indeed, in spite of the tremendous progress which mathematics has made in the three centuries since Fermat, his last theorem is still unproved and is one of the most celebrated unsolved problems in mathematics. The attempts to prove it have led to the creation of much beautiful and important mathematics. In fact, the modern theory of rings is the intellectual descendant of these attempts.

Although Fermat's last theorem is not known to be true or false, it has been proved correct for special values of n. For example, Fermat had a proof for $n = 4$, as did Leibniz and Euler. The case $n = 3$ was settled by Euler, while the case $n = 5$ was disposed of by Dirichlet and Legendre.

PROPOSITION 1: It suffices to prove Fermat's last theorem for $n = 4$ and for $n = p$, where p is an odd prime.

Proof: Let us assume Fermat's theorem for $n = 4$ and $n = p$, and let us prove it for general $n > 2$. Assume that $n > 2$ is not prime. If n is divisible by an odd prime p, then $n = pk$ for some $k \in \mathbf{Z}$. If $x^n + y^n = z^n$ has a solution (x_0, y_0, z_0), x_0, y_0, z_0 not 0, then (x_0^k, y_0^k, z_0^k) is a nonzero solution of $x^p + y^p = z^p$, which is a contradiction. If n is not divisible by an odd prime, then $n = 2^r$ $(r \geq 2)$. If $x^n + y^n = z^n$ has a solution (x_0, y_0, z_0), x_0, y_0, z_0 not 0, then $(x_0^{2^{r-2}}, y_0^{2^{r-2}}, z_0^{2^{r-2}})$ is a nonzero solution of $x^4 + y^4 = z^4$, which is a contradiction.

Let us assume Fermat's result for $n = 4$ and let us concentrate on the Diophantine equation

$$x^p + y^p = z^p, \qquad p \text{ an odd prime.} \tag{5}$$

The first organized attack on (5) for all p was made by Kummer in 1835. Kummer's idea was as follows: Let ζ be a primitive pth root of 1, and let

$$\mathbf{Z}[\zeta] = \{a_0 + a_1\zeta + \cdots + a_r\zeta^r \mid a_i \in \mathbf{Z}, r \geq 0\}.$$

It is easy to see that $\mathbf{Z}[\zeta]$ is a subring of $\mathbf{C}$ containing 1, so that $\mathbf{Z}[\zeta]$ is an integral domain. The ring $\mathbf{Z}[\zeta]$ is a generalization of the ring $\mathbf{Z}$ and is called the ring of *p-cyclotomic integers.* Kummer supposed that it was possible to write every nonzero element of $\mathbf{Z}[\zeta]$ uniquely as a product of "prime elements." In order to undertand the precise meaning of this statement, we need a few definitions.

DEFINITION 2: Let R be a ring, $x \in R^\times$. We say that x is *irreducible* if

(1) Whenever $x = ab$, $a,b \in \mathbf{R}$, either a or b is a unit of R.
(2) x is not a unit of R.

In the ring $\mathbf{Z}$, the units are ± 1, so that if $x \in \mathbf{Z}$, then x is irreducible if and only if $x = \pm p$, where p is a prime.

DEFINITION 3: Let R be a ring, $x, y \in R$. We say that x and y are *associates* if $x = \epsilon y$ for some unit ϵ of R.

Thus, if $x, y \in \mathbf{Z}$, then x and y are associates if and only if $x = \pm y$. Using Definitions 2 and 3, we may state the fundamental theorem of arithmetic (Theorem 10 of Section 2.4) as follows:

THEOREM 4: Let $x \in \mathbf{Z}^\times$, $x \neq \pm 1$.

(1) x can be written as a product of irreducible elements of $\mathbf{Z}$.

(2) If $x = p_1 \ldots p_s = q_1 \ldots q_t$ are two expressions of x as a product of irreducible elements, then $s = t$ and it is possible to renumber the p_i so that p_i and q_i are associates $(1 \leq i \leq s)$.

The fundamental theorem of arithmetic in the above form prompts us to formulate the following definition.

DEFINITION 5: Let R be an integral domain. We say that R is a unique factorization domain (UFD) if the following two conditions are satisfied:

(1) If $x \in R^\times$ is not a unit of R, then x can be written as a product of irreducible elements of R.

(2) If $x \in R^\times$ is not a unit of R, and if

$$x = \pi_1 \ldots \pi_s = \lambda_1 \ldots \lambda_t$$

are two expressions of x as a product of irreducible elements, then $s = t$ and it is possible to renumber $\pi_1, \ldots, \pi_s$ so that π_i and λ_i are associates $(1 \leq i \leq s)$.

Kummer assumed that $\mathbf{Z}[\zeta]$ is a UFD and on the basis of this assumption proved that Fermat's last theorem is true! However, Dirichlet pointed out to Kummer that his assumption was false! For example, Dedekind showed that $\mathbf{Z}[\sqrt{-5}]$ is not a UFD. A sketch of the proof of this fact was given in Exercises 8–10 of Section 2.4. Kummer investigated the rings $\mathbf{Z}[\zeta]$ more closely and found that usually they are not UFDs. Therefore, Kummer's proof of Fermat's last theorem was invalid, owing to his mistaken assumption. But what a fortunate mistake for mathematics!

In attempting to recoup the loss of unique factorization in $\mathbf{Z}[\zeta]$, Kummer was led to invent the notion of an ideal. And Kummer's deep results on the arithmetic of ideals of $\mathbf{Z}[\zeta]$ are the beginning of modern ring theory. In Chapter 5 we will study some elementary ideal theory and will indicate the connection between the theory of ideals and unique factorization. At the end of the chapter, we will return to Fermat's last theorem and will use the theoretical apparatus of the chapter to describe the achievements of Kummer somewhat more explicitly than we could in this modest introduction.

5.1 *Exercises*

1. The following exercises will outline a proof of the fact that the solutions to the Diophantine equation (1) are given by (2).
   (a) Show that it is enough to prove that the only solutions to $x^2 + y^2 = z^2$ such that x, y, z have no common factor are given by
   $$x = a^2 - b^2, \quad y = 2ab, \quad z = a^2 + b^2, \qquad a, b \in \mathbf{Z}.$$
   (b) Show that either x or y is even. Henceforth, assume that y is even, say $y = 2q$. Then $4q^2 = z^2 - x^2 = (z + x)(z - x)$. Show that x and z are odd, so that $|z - x| = 2c$, $|z + x| = 2d$.
   (c) Show that c and d are relatively prime and that $q^2 = cd$.
   (d) Show that c and d are perfect squares, say $c = a^2$, $d = b^2$.
   (e) Show that $x = a^2 - b^2$, $y = 2ab$, $z = a^2 + b^2$.
2. Let $d < -1$. Show that the Diophantine equation $x^2 - dy^2 = 1$ has the solutions $(x, y) = (\pm 1, 0)$ and only these.
3. Let $d < 0$. Show that the Diophantine equation $x^2 - dy^2 = 4$ has the solutions $(x, y) = (\pm 2, 0)$, $(0, \pm 2)$ for $d = -1$; $(x, y) = (\pm 2, 0)$ for $d = -2$; $(x, y) = (\pm 2, 0)$, $(\pm 1, \pm 1)$ for $d = -3$; $(x, y) = (\pm 2, 0)$, $(0, \pm 1)$ if $d = -4$; $(x, y) = (\pm 2, 0)$ for $d < -4$.
4. Let $\epsilon = 1 + \sqrt{2}$, and let $n \in \mathbf{Z}$.
   (a) Show that $\epsilon^n = a + b\sqrt{2}$ $(a, b \in \mathbf{Z})$.
   (b) Show that $a^2 - 2b^2 = \pm 1$.
   *(c) Show that all solutions of the Diophantine equation $x^2 - 2y^2 = -1$ are of the form $(\pm a, \pm b)$, where a and b are as constructed in (a).

5.2 The Arithmetic of Ideals

Let R be a commutative ring with identity. In this section we will begin to study the arithmetic of ideals of R.

LEMMA 1: Let $\mathfrak{C}$ be a nonempty collection of ideals of R. Then

$$A = \bigcap_{I \in \mathfrak{C}} I$$

is an ideal of R.

Proof: From Theorem 3 of Section 3.3 we know that A is a subgroup of the additive group of R. Let $r \in R$, $a \in A$. Then $a \in I$ for all $I \in \mathfrak{C}$, so that $r \cdot a \in I$ for all $I \in \mathfrak{C}$, since I is an ideal of R. Therefore, $r \cdot a \in A$ and A is an ideal of R.

Let S be a subset of R. Then there exists a smallest ideal of R which contains S. For let $\mathfrak{C}$ be the set of all ideals of R which contain S. Then

$$A = \bigcap_{I \in \mathfrak{C}} I$$

is an ideal of R by Lemma 1 and it is clear that A contains S. Furthermore, if A_0 is an ideal of R which contains S, then $A_0 \in \mathfrak{C}$ and $A \subseteq A_0$. Therefore A is the smallest ideal of R which contains S and A is called the *ideal generated by* S and is denoted (S). If I is an ideal and S is a subset of R, then we say that S is a *set of generators for* I if $I = (S)$.

If $S = \{a_1, \ldots, a_n\}$, we will write $(a_1, \ldots, a_n)$ instead of (S). An ideal of the form $(a_1, \ldots, a_n)$ is said to be *finitely generated*. An ideal of the form (a) is called a *principal ideal*.

PROPOSITION 2: Let $I = (S)$. Then I is the set of sums of the form

$$r_1 s_1 + r_2 s_2 + \ldots + r_t s_t \qquad (r_i \in R, s_i \in S). \tag{1}$$

Proof: It is clear that every sum of the form (1) belongs to I, since I is an ideal containing S. Therefore, it suffices to show that the set of sums (1) form an ideal containing S, since I is the smallest such ideal. But this is straightforward and is left as an exercise.

COROLLARY 3: Let $I = (a_1, \ldots, a_n)$. Then I consists of all elements of the form

$$r_1 a_1 + \cdots + r_n a_n \qquad (r_i \in R).$$

EXAMPLE 1: Let R be arbitrary. Then (0) is the ideal consisting of only 0, and (1) contains $r \cdot 1$ for all $r \in R$, so that $(1) = R$.

EXAMPLE 2: Let $R = \mathbf{Z}$, and let I be an ideal. Either $I = (0)$ or I contains a nonzero element a. In the latter case, I also contains $-a$, and thus I contains a positive element. Let a be the smallest positive element in I. We assert that $I = (a)$. Let x be a nonzero element of I. By the division algorithm in $\mathbf{Z}$, there exist integers q and r such that

$$x = q \cdot a + r, \qquad 0 \leq r < a.$$

Since $x \in I$, $a \in I$, we have $x - q \cdot a = r \in I$. Therefore, since a is the smallest positive element in I, and since $0 \leq r < a$, we see that $r = 0$. Therefore, $x = q \cdot a \in (a)$. Thus, $I \subseteq (a)$. But it is clear that $(a) \subseteq I$, since $a \in I$. Therefore, $I = (a)$. We have just proved that every ideal of $\mathbf{Z}$ is a principal ideal.

EXAMPLE 3: Let F be a field. Then (0) and (1) are ideals of F. Let $I \neq (0)$ be an ideal of F. Then I contains an element $x \neq 0$. But since F is a field and I is an ideal, $x^{-1} \cdot x = 1 \in I$. But if $y \in F$, then $y \cdot 1 = y \in I$. Thus, $I = (1)$, and the only ideals of F are (0) and (1).

EXAMPLE 4: Let F be a field, X an indeterminate over F, $R = F[X]$. Let $I \neq (0)$ be an ideal of R. Then I contains a nonzero polynomial f. Let g be the nonzero polynomial of smallest degree in I. We assert that $I = (g)$. Since $g \in I$, we see that $(g) \subseteq I$. Let $h \in I$. By the division algorithm in $F[X]$, there exist polynomials $q, r \in F[X]$ such that

$$h = q \cdot g + r, \qquad \deg(r) < \deg(g).$$

Since $\deg(r) < \deg(g)$ and since $r = h - q \cdot g \in I$, we see that $r = 0$ by the choice of g. Therefore, $h = q \cdot g$ and $h \in (g)$. Thus, we have proved that $I \subseteq (g)$, so that $I = (g)$. Thus, every ideal of $F[X]$ is principal.

EXAMPLE 5: Let us give an example of a ring in which not every ideal is principal. Let F be a field and let X and Y be indeterminates over F. Let us consider the ideal (X, Y). This ideal consists of all polynomials in $F[X, Y]$ having zero constant term. If $(X, Y) = (f)$ for some $f \in F[X, Y]$, then $X = k_1 f$, $Y = k_2 f$ for some $k_1, k_2 \in F[X, Y]$. But this is clearly impossible. Thus, (X, Y) is not a principal ideal.

A ring R having the property that every ideal is principal is called a *principal ideal ring* (PIR). If, in addition, R is an integral domain, then R is called a *principal ideal domain* (PID). For example, $\mathbf{Z}$, a field F, a polynomial ring $F[X]$ over a field F are all examples of principal ideal domains. However, $F[X, Y]$ is not a principal ideal domain, as we have seen in Example 5.

Let A and B be ideals of R. The *sum* $A + B$ of A and B is the ideal generated by $A \cup B$. By Lemma 2, $A + B$ consists of all sums of the form

$$r_1 a_1 + \cdots + r_n a_n + s_1 b_1 + \cdots + s_m b_m \qquad (r_i, s_j \in R, a_i \in A, b_j \in B).$$

But since A and B are ideals, $r_i a_i \in A$ $(1 \leq i \leq n)$, $r_j b_j \in B$ $(1 \leq j \leq m)$, and thus $r_1 a_1 + \cdots + r_n a_n \in A$, $s_1 b_1 + \cdots + s_m b_m \in B$. Therefore, every element of $A + B$ is of the form $a + b$ for $a \in A$, $b \in B$. Conversely, every such element belongs to $A + B$, so that

$$A + B = \{a + b \mid a \in A, b \in B\}.$$

For example, if $R = \mathbf{Z}$, then

$$\begin{aligned}(2) + (4) &= \{2m + 4n \mid m, n \in \mathbf{Z}\} \\ &= \{2k \mid k \in \mathbf{Z}\} \\ &= (2); \\ (5) + (7) &= \{5m + 7n \mid m, n \in \mathbf{Z}\} \\ &= (1),\end{aligned}$$

since $1 = 5 \cdot 3 + 7 \cdot (-2) \in (5) + (7)$. More generally, if $a, b \in \mathbf{Z}$, then

$$(a) + (b) = \{am + bn \mid m, n \in \mathbf{Z}\}.$$

If $a = b = 0$, then $(a) + (b) = (0)$. Thus, we may assume that a and b are not both 0. Then $(a) + (b)$ contains $a = a \cdot 1 + b \cdot 0$ and $b = a \cdot 0 + b \cdot 1$, and is therefore not (0). We have seen above that

$$(a) + (b) = (c),$$

where c is the smallest positive integer contained in $(a) + (b)$. However, in Chapter 2 we showed that the smallest positive integer of the form $am + bn$ $(m, n \in \mathbf{Z})$ equals $(a, b) =$ the *g. c. d.* of a and b. Therefore, we have proved

LEMMA 4: Let $a, b \in \mathbf{Z}$, a, b not both 0. Then the sum of the ideals (a) and (b) is the ideal generated by the greatest common divisor (g.c.d.) of a and b:

$$(a) + (b) = ((a, b)).$$

By Lemma 4 it is reasonable to regard the sum of two ideals (a), (b) of a general ring R as some sort of greatest common divisor of a and b. This is Kummer's great idea. Define the g.c.d. of two elements $a, b \in \mathbf{Z}[\zeta]$ to be the ideal $(a) + (b)$ and replace the usual arithmetic properties of the elements $\mathbf{Z}[\zeta]$ (for example, divisibility, primes, factorization) by corresponding properties of ideals of $\mathbf{Z}[\zeta]$. As a first step in this direction, let us define the product of two ideals.

Let A and B be ideals of R. The *product* $A \cdot B$ of A and B is the ideal generated by all products of the form $a \cdot b$ $(a \in A, b \in B)$. Thus, by Lemma 2, $A \cdot B$ consists of all elements of the form

$$a_1 \cdot b_1 + \cdots + a_n \cdot b_n, \qquad a_i \in A, \quad b_j \in B.$$

It is easy to see that

$$(a) \cdot (b) = (a \cdot b), \tag{2}$$

$$(a_1, \ldots, a_m) \cdot (b_1, \ldots, b_n) = (a_1 \cdot b_1, a_1 \cdot b_2, \ldots, a_1 \cdot b_n, a_2 \cdot b_1, \ldots, a_m \cdot b_n). \tag{3}$$

Thus, for example, in $F[X, Y]$, we have

$$(X, Y)\cdot(X^2, Y) = (X^3, XY, X^2Y, Y^2).$$

The following properties of sums and products are left as exercises: Let A, B, and C be ideals of R. Then

$$A + B = B + A, \tag{4}$$

$$A + (B + C) = (A + B) + C, \tag{5}$$

$$A\cdot B = B\cdot A, \tag{6}$$

$$A\cdot(B\cdot C) = (A\cdot B)\cdot C, \tag{7}$$

$$A\cdot(B + C) = A\cdot B + A\cdot C. \tag{8}$$

One surprising property of the product is

$$A\cdot B \subseteq A \cap B. \tag{9}$$

5.2 *Exercises*

1. Prove formulas (4)–(9).
2. We proved that $\mathbf{Z}$ is a PID. Find a generator for each of the following ideals:
   (a) $(3, 1)$ (b) $(6, 2)$ (c) $(15, 27)$
   (d) $(8, 12)$ (e) $(2, 4, 5)$. (f) $(3, 6, 24)$.
3. We proved that $F[X]$ is a PID. Find a generator for each of the following ideals.
   (a) (a, X) $(a \in F)$. (b) (X^2, X^3). (c) $(X^2 + 1, X^4)$.
4. Let R be a commutative ring with identity $\neq 0$ and assume that R has only the ideals (0) and (1). Show that R is a field.
5. Is $\mathbf{Z}[X]$ a PID?
6. Let a and b be integers. Prove that
   $$[(a) + (b)]\cdot[(a) \cap (b)] = (a)\cdot(b).$$
   (*Hint:* Reduce to the case $a \neq 0$, $b \neq 0$ and write a, b as products of powers of primes.)
7. Describe the following ideals of $\mathbf{R}[X]$:
   (a) $(X + 1, X^2 + 8)\cdot(X + 2)$.
   (b) $(X^2 + 1) \cap [(X - 1)\cdot(X + 1)]$.
   (c) $(X^2 + 1) + (X + 1)$.
8. Let F be a field, X an indeterminate over F. If f and g belong to $F[X]$, let us say that h in $F[X]$ is a *greatest common divisor* of f and g if $(f) + (g) = (h)$. Throughout this exercise, assume that at least one of f and g is nonzero.
   (a) Show that if h and h' are two g.c.d.s of f and g, then $h = ah'$ for some $a \in F$.
   (b) Show that f and g have a g. c. d.
   (c) Let h be a g.c.d. of f and g. Show that there exist $k, k' \in F[X]$ such that $f = kh$, $g = k'h$. Moreover, show that if h_0 is any element of $F[X]$ such

that there exist $k, k' \in F[X]$ for which $f = kh_0$, $g = k'h_0$, then h is of the form rh_0 for some $r \in F[X]$.

(d) By using the Euclidean algorithm as a model, describe a procedure for explicitly determining the g.c.d. of two elements of $F[X]$.

5.3 Arithmetic in a Principal Ideal Domain

We have seen that $\mathbf{Z}$ is a unique factorization domain and also a principal ideal domain. On the other hand, $\mathbf{Z}[\sqrt{-5}]$ is not a unique factorization domain and we will show in the exercises for this Section that $\mathbf{Z}[\sqrt{-5}]$ is not a principal ideal domain. These facts suggest a connection between the validity of unique factorization and the existence of nonprincipal ideals. And such a connection really exists! Our main result in this section will be the following theorem.

THEOREM 1: Let R be a principal ideal domain. Then R is a unique factorization domain.

Let us begin by developing the theory of divisibility in an integral domain. Throughout this section, let R be an integral domain. If $a, b \in R$, then we say that *a divides b* (denoted $a \mid b$) if $b = xa$ for some $x \in R$. If a does not divide b, we write $a \nmid b$. The elementary facts about divisibility are summarized in the following result:

LEMMA 2: Let $a, b, c \in R$.

(1) $a \mid b,\ b \mid c \Rightarrow a \mid c$.
(2) $a \mid b \Rightarrow a \mid bc$ *for all* $c \in R$.
(3) $a \mid b,\ a \mid c \Rightarrow a \mid b + c$.
(4) $a \mid b,\ a \mid c \Rightarrow a \mid bx + cy$ *for all* $x, y \in R$.
(5) Let $b \neq 0$. Then $a \mid b,\ b \mid a \Rightarrow a$ and b are associates.

Proof:

(1) $b = xa,\ c = yb \Rightarrow c = (xy)a \Rightarrow a \mid c$.
(2) $b = xa \Rightarrow bc = (xc)a \Rightarrow a \mid bc$.
(3) $b = xa,\ c = ya \Rightarrow b + c = (x + y)a \Rightarrow a \mid b + c$.
(4) By (ii) and (iii), $a \mid bx,\ a \mid cy \Rightarrow a \mid bx + cy$.
(5) $b = xa,\ a = yb \Rightarrow b = (xy)b \Rightarrow b(1 - xy) = 0 \Rightarrow 1 - xy = 0$ since $b \neq 0$ and since R is an integral domain $\Rightarrow xy = 1 \Rightarrow x$ is a unit of $R \Rightarrow a$ and b are associates since $b = xa$.

It is possible to formulate the notion of divisibility in terms of ideals. For if $a \mid b$, then $b = xa \Rightarrow b \in (a) \Rightarrow by \in (a)$ for all $y \in R \Rightarrow (b) \subseteq (a)$. Conversely, if $(b) \subseteq (a)$, then $b \in (a)$, so that $b = xa$ and $a \mid b$. Thus, we have

$$a \mid b \text{ if and only if } (b) \subseteq (a). \tag{1}$$

LEMMA 3: Let $a, b \in R$. Then a and b are associates if and only if $(a) = (b)$.

Proof: $\Rightarrow$ If a and b are associates, then $a = \epsilon b$, where ϵ is a unit of R. Therefore, $b = \epsilon^{-1}a$, so that $b \mid a$ and $a \mid b$. Thus, by (1),

$$(a) \subseteq (b), \qquad (b) \subseteq (a).$$
$$\Longrightarrow (a) = (b).$$

$\Leftarrow$ Suppose that $(a) = (b)$. Then, by (1), we have $a \mid b$ and $b \mid a$, so that a and b are associates by Lemma 2, part 5.

DEFINITION 4: Let $a, b \in R$. A *greatest common divisor* (g.c.d) of a and b is an element c of R having the following properties:

(1) $c \mid a$, $c \mid b$.
(2) If $d \in R$ is such that $d \mid a$ and $d \mid b$, then $d \mid c$.

In case $R = \mathbf{Z}$, Definition 4 is essentially the same as the definition of g.c.d. which we gave in Chapter 2. There is one small difference, however. In Chapter 2 we required that a g.c.d. be positive. In a general ring, the notion of positivity makes no sense, and therefore we cannot include the condition of positivity in Definition 4. Therefore, if $a = 3$, $b = 6$, then $(a, b) = 3$ using the definition of Chapter 2, whereas both 3 and -3 are g.c.d.s in the sense of Definition 4. Note that 3 and -3 are associates. This is not an accident.

LEMMA 5: Let $a, b \in R$, and let c and c' be two g.c.d.s of a and b. Then c and c' are associates.

Proof: Since $c \mid a$ and $c \mid b$, and since c' is a g.c.d. of a and b, condition (2) of Definition 4 implies that $c \mid c'$. Reversing the roles of c and c', we see that $c' \mid c$. If $c = 0$, then $c' = 0$ and the lemma holds. If $c \neq 0$, then Lemma 2, part (5) implies that c and c' are associates.

Thus, we see that if c_0 is a g.c.d. of a and b, then all g.c.d.s of a and b are of the form $c_0\epsilon$, where ϵ is a unit of R.

DEFINITION 6: Let $a, b \in R$. We say that a and b are *relatively prime*, denoted $(a, b) = 1$, if 1 is a g.c.d. of a and b.

In an arbitrary integral domain R, it is not usually true that every pair of elements $a, b \in R$ have a g.c.d. We will provide an example of this phenomenon in the exercises. However, if R is a PID, then no pathologies occur and g.c.d.s exist.

PROPOSITION 7: Let R be a principal ideal domain and let $a, b \in R$, a and b not both 0. Then a and b have a g.c.d. in R.

Proof: Let $I = (a, b)$. Since R is a PID, $I = (c)$ for some $c \in R$. Moreover, since a and b are not both 0, $I \neq (0)$, so that $c \neq 0$. Let us show that c is a g.c.d. of a and b. Since $a, b \in I$, we see that $a = cx$, $b = cy \Rightarrow c \mid a$, $c \mid b \Rightarrow$ (1). Suppose that $d \in R$ is such that $d \mid a$ and $d \mid b$. Since $I = (a, b) = \{ax + by \mid x, y \in R\}$, we have $c = ax + by$ for some $x, y \in R$. But then by Lemma 2, part (4), we have $d \mid ax + by \Rightarrow d \mid c \Rightarrow$ (2).

COROLLARY 8: Let R be a PID, and let $a, b \in R$, a, b not both 0. Then every g.c.d. of a and b can be written in the form $ax + by$ for some $x,y \in R$. In particular, if a and b are relatively prime, then there exist $x, y \in R$ such that $ax + by = 1$.

In Chapter 2 we proved that if p is prime and a and b are integers such that $p \mid ab$, then either $p \mid a$ or $p \mid b$. Let us generalize this result to a PID.

THEOREM 9: Let R be a principal ideal domain and let $a, b \in R$. Let π be an irreducible element of R such that $\pi \mid ab$. Then either $\pi \mid a$ or $\pi \mid b$.

Proof: Let us assume that π does not divide a. We must then show that $\pi \mid b$. Since π is irreducible, $\pi \neq 0$ and thus π and a have a g.c.d. λ by Proposition 7. But $\lambda \mid \pi$ so that λ is either an associate of π or a unit of R. In the former case, $\pi \mid a$ since $\lambda \mid a$. And this contradicts our assumption. Therefore, λ is a unit of R and $1 = \lambda\lambda^{-1}$ is a g.c.d. of π and a, since λ^{-1} is a unit of R. But then, by Corollary 8, there exist $x, y \in R$ such that

$$\pi x + ay = 1.$$

Thus, we see that $\pi(bx) + (ab)y = b$. However, $\pi \mid \pi$ and $\pi \mid ab$, so that by Lemma 2, part (4), we see that $\pi \mid \pi bx + aby$, so that $\pi \mid b$.

COROLLARY 10: Let R be a PID, $a_1, \ldots, a_n \in R$, π an irreducible element of R. If $\pi \mid a_1 \ldots a_n$, then $\pi \mid a_i$ for some i.

Proof: Exercise.

We now come to the proof of Theorem 1. Our proof will be divided into two parts. First we will show that if $x \in R^\times$, $x \notin U_R$, then x can be written as a product of irreducible elements. Second, we will show that the expression of x as a product of irreducible elements is unique up to order and multiplication of the irreducible elements by units.

THEOREM 11: Let $x \in R^\times$, $x \notin U_R$. Then x can be written as a product of irreducible elements.

Proof: Let us reason by contradiction. Assume that x cannot be written as a product of irreducible elements. In particular, x is not irreducible, so that we may write

$$x = a_1 b_1, \qquad a_1, b_1 \text{ not units.}$$

Either a_1 or b_1 cannot be written as a product of irreducible elements. For otherwise x could be so written, Thus, suppose that a_1 is not a product of irreducible elements. Since $x = a_1 b_1$, we see that $a_1 \mid x$ and thus $(a_1) \supseteq (x)$ by (1). Moreover, since b_1 is not a unit, $(a_1) \neq (x)$ by Lemma 3. Thus, (x) is properly contained in (a_1). Let us denote this fact by $(a_1) \supset (x)$. Note that $a_1 \in R^\times$, $a_1 \notin U_R$, and a_1 cannot be written as a product of irreducible elements. Thus, we can apply the same reasoning to a_1 as we applied to x to find $a_2 \in R^\times$, $a_2 \neq U_R$, $a_2 \neq$ a product of irreducible elements, and $(a_2) \supset (a_1)$. Let us proceed in this way and construct $a_3, a_4, a_5, \ldots$. We eventually arrive at the following chain of ideals:

$$(a_0) \subset (a_1) \subset (a_2) \subset \ldots, \tag{2}$$

where $a_0 = x$. Set

$$I = \bigcup_{i \geq 0} (a_i). \tag{3}$$

Let us prove that I is an ideal of R. Let $a, b \in I$. Then $a \in (a_i)$, $b \in (a_j)$ for some i and j. Let $j \geq i$. Then $(a_i) \subset (a_j)$ and $a, b \in (a_j)$. But since (a_j) is an ideal $a \pm b \in (a_j)$, so that $a \pm b \in I$. Thus, I is an additive subgroup of R. A similar argument shows that I is closed with respect to multiplication by elements of R. This shows that I is an ideal of R. Since R is a PID, $I = (a)$ for some $a \in R$. Now $a \in I$, so that $a \in (a_k)$ for some k. Therefore, $(a) \subseteq (a_k)$. However, $(a_k) \subseteq I = (a)$, so that $(a) = (a_k)$. In particular, $(a_{k+1}) \subseteq (a_k)$ by (3). But this is a contradiction to (2), which asserts that $(a_k) \subset (a_{k+1})$.

THEOREM 12: Let $x \in R^\times$, $x \notin U_R$ and let

$$x = \pi_1 \ldots \pi_s = \lambda_1 \ldots \lambda_t$$

be two expressions of x as a product of irreducible elements. Then $s = t$ and by renumbering $\lambda_1, \ldots, \lambda_s$, we can guarantee that π_i and λ_i are associates $(1 \leq i \leq s)$.

Proof: If $s = 1$, then x is irreducible and the assertion is clear. Assume that $s > 1$ and let us proceed by induction on s. Since $\pi_1 \mid x$, we see that

$\pi_1 \mid \lambda_1 \ldots \lambda_t$. Therefore, by Corollary 10, $\pi_1 \mid \lambda_i$ for some i $(1 \leq i \leq t)$. By renumbering $\lambda_1, \ldots, \lambda_t$, we may assume that $i = 1$, so that $\pi_1 \mid \lambda_1$. But since λ_1 is irreducible and π_1 is not a unit we see that π_1 and λ_1 are associates, say $\epsilon\pi_1 = \lambda_1$, $\epsilon \in U_R$. Therefore,

$$\pi_1 \ldots \pi_s = \epsilon\pi_1\lambda_2 \ldots \lambda_t,$$
$$\pi_1(\pi_2 \ldots \pi_s - \epsilon\lambda_2 \ldots \lambda_t) = 0,$$
$$\Longrightarrow \pi_2 \ldots \pi_s = \epsilon\lambda_2 \ldots \lambda_t \qquad \text{since } R \text{ is an integral domain.}$$

Set $\lambda_2' = \epsilon\lambda_2$, $\lambda_3' = \lambda_3, \ldots, \lambda_t' = \lambda_t$. Then λ_i' and λ_i are associates and $\pi_2 \ldots \pi_s = \lambda_2' \ldots \lambda_t'$ are two decompositions of $\pi_2 \ldots \pi_s$ into irreducible factors. Therefore, by the induction hypothesis, $s - 1 = t - 1$ and, after renumbering $\lambda_2', \ldots, \lambda_t'$, we can guarantee that π_i and λ_i' are associates $(2 \leq i \leq s)$. But then $s = t$ and π_i and λ_i are associates $(1 \leq i \leq s)$. Thus, the induction step is established.

We have shown in Section 5.2 that if F is a field and X is an indeterminate over F, then $F[X]$ is a PID. Therefore, by Theorem 1, we see that $F[X]$ is a UFD. Let us see exactly what this says: The units of $F[X]$ are just the nonzero constant polynomials. Therefore, if $f \in F[X]$ is a nonzero. nonconstant polynomial, then f can be written in the form

$$f = f_1 \ldots f_t, \tag{4}$$

where $f_i \in F[X]$ is an irreducible polynomial. Moreover, the decomposition (3) is unique up to rearrangement and multiplication of the f_i by constants. Let $f = a_nX^n + a_{n-1}X^{n-1} + \cdots + a_0$, $a_n \neq 0$. Then a_n is called the *leading coefficient* of f. If f has leading coefficient 1, then we say that f is *monic*. It is clear that a nonzero polynomial $g = b_mX^m + \cdots + b_0$, $b_m \neq 0$ can be written as the product of a constant and a monic polynomial:

$$g = b_m(X^m + b_m^{-1}b_{m-1}X^{m-1} + \cdots + b_m^{-1}b_0).$$

Therefore, by (3), we can write

$$f = af_1^* \ldots f_t^*, \tag{5}$$

where a is a constant and f_i^* is monic and irreducible. By comparing leading coefficients on both sides of (5), we see that $a = a_n$.

PROPOSITION 13: Let F be a field, X an indeterminate over F, f a nonzero, nonconstant polynomial in $F[X]$, a the leading coefficient of f. Then f can be written in the form

$$f = af_1^* \ldots f_t^*,$$

where f_i^* is an irreducible, monic polynomial. Moreover, this decomposition is unique up to the order of the f_i^*.

Proof: All that must be proved is the uniqueness. Let

$$f = af_1^* \ldots f_t^* = ag_1^* \ldots g_s^*$$

be two decompositions of f, where f_i^* and g_i^* are irreducible and monic. Since $F[X]$ is a UFD, we have $s = t$ and we may renumber $f_1^*, \ldots, f_t^*$ so that f_i^* and g_i^* are associates $(1 \leq i \leq s)$. But since f_i^* and g_i^* are both monic, $f_i^* = g_i^*$.

5.3 *Exercises*

1. Let R be a PID, $\pi \in R^\times$ a nonunit. Show that π is irreducible if and only if whenever $a, b \in R$ are such that $\pi \mid ab$, then either $\pi \mid a$ or $\pi \mid b$.
2. Let R be a UFD. Prove that if $a, b \in R$ are not both 0, then a and b have a g.c.d. in R.
3. Let R be a UFD and let $a, b \in R^\times$. Further, let us fix a set P of irreducible elements of R such that no two elements of P are associates and every irreducible element is an associate of some element of P.
   (a) Show that a and b can be uniquely written in the form
   $$a = \epsilon_1 \cdot \prod_{\pi \in P} \pi^{\nu_\pi(a)} \qquad [\nu_\pi(a) \geq 0],$$
   $$b = \epsilon_2 \cdot \prod_{\pi \in P} \pi^{\nu_\pi(b)} \qquad [\nu_\pi(b) \geq 0],$$
   where ϵ_i $(i = 1, 2)$ are units of R and $\nu_\pi(a)$ [respectively, $\nu_\pi(b)$] is zero for all but a finite number of π.
   (b) Show that a g.c.d. of a and b is given by
   $$\prod_{\pi \in P} \pi^{\nu_\pi},$$
   where $\nu_\pi =$ the smaller of $\nu_\pi(a)$ and $\nu_\pi(b)$.
4. Let R be a UFD, $a, b \in R$. A *least common multiple* (l.c.m.) of a and b is an element c of R such that (1) $a \mid c$, $b \mid c$ and (2) if d is such that $a \mid d$ and $b \mid d$, then $c \mid d$.
   (a) Show that if both a and b are nonzero, then a and b have a l.c.m.
   (b) Show that any two l.c.m.s are associates.
   (c) Let a and b be nonzero, let c be a g.c.d. of a and b, and let d be a l.c.m. of a and b. Show that ab and cd are associates.
5. There are examples of UFDs which are not PIDs. Does this contradict Theorem 1?
6. (a) Show that **9** and $\mathbf{6} + \mathbf{3}\sqrt{-\mathbf{5}}$ do not have a g.c.d in $\mathbf{Z}[\sqrt{-5}]$. [*Hint*: Determine all common divisors of **9** and $\mathbf{6} + \mathbf{3}\sqrt{-\mathbf{5}}$.]
   (b) Show that $\mathbf{Z}[\sqrt{-5}]$ is not a PID. [Hint: Is $(9, 6 + 3\sqrt{-5})$ principal?]

5.4 *Euclidean Domains*

In Section 5.3 we proved that a principal ideal domain possesses a theory of unique factorization. Now we are faced with the problem of determining whether or not a given integral domain is a principal ideal domain. This is generally a very difficult question. However, a simple partial solution is available. Notice that we were able to prove that **Z** and $F[X]$ are PIDs by using the division algorithm in **Z** and $F[X]$, respectively. In this section let us introduce a class of integral domains in which an analogue of the division algorithm holds. We will show that every such ring is a PID.

DEFINITION 1: Let R be an integral domain. Then R is called a *Euclidean domain* if there exists a function $\phi: R \longrightarrow \mathbf{Z}$ such that

(1) $\phi(x) \geq 0$ for all $x \in R$.
(2) $\phi(x) = 0$ if and only if $x = 0$.
(3) $\phi(xy) = \phi(x)\phi(y)$ for all $x, y \in R$.
(4) Let $x, y \in R$, $y \neq 0$. Then there exist $q, r \in R$ such that $x = qy + r$, $0 \leq \phi(r) < \phi(y)$. (This condition is a generalization of the division algorithm.)

EXAMPLE 1: Let $R = \mathbf{Z}$, $\phi(x) = |x|$. Then conditions (1)–(3) are elementary properties of absolute values. Condition (4) is just the division algorithm in **Z**.

EXAMPLE 2: Let $R = F[X]$. Set $\phi(x) = 2^{\deg(x)}$ $(x \neq 0)$, $= 0$ $(x = 0)$. Then (1)–(3) are easy to check. Moreover, by the division algorithm in $F[X]$, given $x, y \in R, y \neq 0$, there exist $q, r \in R$ such that $x = qy + r$, $\deg(r) < \deg(y)$. But then $0 \leq \phi(r) < \phi(y)$ and condition (4) holds.

THEOREM 2: Let R be a Euclidean domain. Then R is a principal ideal domain.

Proof: Let I be a nonzero ideal of R. It suffices to show that I is principal. Let $a \in I$ be chosen so that $a \neq 0$ and $\phi(a)$ is as small as possible. We assert that $I = (a)$. It is clear that $(a) \subseteq I$. Let $x \in I$. Since R is a Euclidean domain, there exist $q,r \in R$ so that

$$x = qa + r, \qquad 0 \leq \phi(r) < \phi(a).$$

Note that $r = x - qa \in I$. If $r \neq 0$, then $\phi(r) < \phi(a)$, $r \in I$, which contradicts the choice of a. Therefore, $r = 0$ and $x = qa \in (a)$. Thus, $I \subseteq (a)$.

Let us now give an example of a Euclidean domain which is connected with Fermat's last theorem.

THEOREM 3: Let $\zeta = (-1 + \sqrt{3}\,i)/2$. Then ζ is a primitive cube root of 1 and $\mathbf{Z}[\zeta]$ is a Euclidean domain.

Before proving Theorem 3, it is necessary to describe $\mathbf{Z}[\zeta]$ more explicitly. Note that $\zeta^2 + \zeta + 1 = 0$, so that

$$\zeta^2 \in \{a + b\zeta \mid a, b \in \mathbf{Z}\}.$$

Also, $\zeta^3 = -\zeta^2 - \zeta$, so that

$$\zeta^3 \in \{a + b\zeta \mid a, b \in \mathbf{Z}\}.$$

Proceeding by induction, we see that

$$\zeta^n \in \{a + b\zeta \mid a, b \in \mathbf{Z}\}$$

for all $n \geq 1$. Therefore,

$$\mathbf{Z}[\zeta] = \{a + b\zeta \mid a, b \in \mathbf{Z}\}.$$

Moreover, every element of $\mathbf{Z}[\zeta]$ can be written uniquely in the form $a + b\zeta$, since

$$\begin{aligned} a + b\zeta = a' + b'\zeta &\Longrightarrow a - a' = (b' - b)\zeta \\ &\Longrightarrow (b' - b)\zeta \in \mathbf{Z} \\ &\Longrightarrow b' - b = 0 \\ &\Longrightarrow a - a' = 0. \end{aligned}$$

Thus, we have proved

LEMMA 4: Every element of $\mathbf{Z}[\zeta]$ can be written uniquely in the form $a + b\zeta$, $a, b \in \mathbf{Z}$.

Proof of Theorem 3: If $a, b \in \mathbf{Q}$, define $\phi(a + b\zeta) = (a + b\zeta)(a + b\bar{\zeta}) = a^2 - ab + b^2$, where $\bar{\zeta}$ denotes the complex conjugate of ζ. Since $a + b\bar{\zeta} = \overline{a + b\zeta}$, we see that $\phi(a + b\zeta) = |a + b\zeta|^2 \geq 0$. Moreover, $\phi(a + b\zeta) = 0 \Longleftrightarrow a + b\zeta = 0$. Further, if $a, b \in \mathbf{Z}$, $\phi(a + b\zeta) = a^2 - ab + b^2 \in \mathbf{Z}$. We leave it as an exercise to verify that $\phi(\alpha\beta) = \phi(\alpha)\phi(\beta)$ for $\alpha, \beta \in \{a + b\zeta \mid a, b \in \mathbf{Q}\}$. Let us verify the division algorithm. This is where we use the special properties of ζ. Let $x = a + b\zeta$, $y = c + d\zeta \in \mathbf{Z}[\zeta]$, $y \neq 0$. Then, in $\mathbf{C}$, we have

$$\frac{x}{y} = \frac{a + b\zeta}{c + d\zeta} = \frac{(a + b\zeta)(c + d\bar{\zeta})}{|c + d\zeta|^2} = e + f\zeta,$$

where

$$e = \frac{ac - ad + bd}{c^2 - cd + d^2} \quad \text{and} \quad f = \frac{bc - ad}{c^2 - cd + d^2}$$

belong to $\mathbf{Q}$. There exist integers α and β such that

$$|e - \alpha| \leq \tfrac{1}{2}, \qquad |f - \beta| \leq \tfrac{1}{2}.$$

Set $q = \alpha + \beta\zeta$, $r = [(e - \alpha) + (f - \beta)\zeta]y$. Then $q \in \mathbf{Z}[\zeta]$ and

$$x = qy + r,$$

so that $r = x - qy \in \mathbf{Z}[\zeta]$. Finally,

$$\begin{aligned} \phi(r) &= \phi((e - \alpha) + (f - \beta)\zeta)\phi(y) \\ &= \{(e - \alpha)^2 - (e - \alpha)(f - \beta) + (f - \beta)^2\}\phi(y) \\ &\leq \{\tfrac{1}{4} + \tfrac{1}{4} + \tfrac{1}{4}\}\phi(y) \\ &< \phi(y). \end{aligned}$$

COROLLARY 5: If $\zeta = (-1 + \sqrt{3}i)/2$ is a primitive cube root of 1, then $\mathbf{Z}[\zeta]$ is a unique factorization domain.

5.4 *Exercises*

In this set of exercises we will explore some properties of the Gaussian integers $\mathbf{Z}[i] = \{a + bi \mid a, b \in \mathbf{Z}\}$.

1. If $\alpha = a + bi \in \mathbf{Z}[i]$, set $\phi(\alpha) = a^2 + b^2$.
   (a) Show that $\phi(\alpha\beta) = \phi(\alpha)\phi(\beta)$.
   (b) Show that $\phi(\alpha) = 0$ if and only if $\alpha = 0$.
2. Show that α is a unit in $\mathbf{Z}[i]$ if and only if $\phi(\alpha) = \pm 1$.
3. Show that the units of $\mathbf{Z}[i]$ are ± 1 and $\pm i$.
4. Show that $\mathbf{Z}[i]$ is a Euclidean domain. (Imitate the proof of Theorem 3.)
5. Show that if π is an irreducible element, then $\phi(\pi)$ is a power of a prime (of $\mathbf{Z}$).
6. Show that if π is an irreducible element, then π divides precisely one prime (of $\mathbf{Z}$).

*7. Let p be a prime.
   (a) If $p \equiv 3 \pmod 4$, then p is irreducible in $\mathbf{Z}[i]$.
   (b) If $p \equiv 1 \pmod 4$, then p is the product of two distinct irreducible elements in $\mathbf{Z}[i]$.
   (c) If $p = 2$, then $p = i\pi^2$ for π an irreducible element of $\mathbf{Z}[i]$.

5.5 *Proof of Fermat's Last Theorem for $n = 3$*

Let ζ be a primitive cube root of 1. In Section 5.4 we proved that $\mathbf{Z}[\zeta]$ is a UFD. Let us now make use of this fact in order to prove Fermat's last theorem for $n = 3$. Actually, we will prove somewhat more. We will prove

THEOREM 1: Let $x, y, z \in \mathbf{Z}[\zeta]$. If

$$x^3 + y^3 + z^3 = 0, \tag{1}$$

then one of x, y, z is 0.

In order to prove Theorem 1, it is necessary to develop further the arithmetic of $Z[\zeta]$. If $\alpha = a + b\zeta \in \mathbf{Z}[\zeta]$, let us define the norm of α, denoted $N(\alpha)$, by

$$\begin{aligned} N(\alpha) &= (a + b\zeta)(a + b\bar{\zeta}) \\ &= a^2 - ab + b^2. \end{aligned} \tag{2}$$

Then, $N(\alpha) = |a + b\zeta|^2 \geq 0$ and $N(\alpha) \in \mathbf{Z}$ by (2). Further, as we observed in the proof of Theorem 3 Section 5.4, we have

$$N(\alpha\beta) = N(\alpha)N(\beta), \tag{3}$$

$$N(\alpha) = 0 \text{ if and only if } \alpha = 0. \tag{4}$$

LEMMA 2: $N(\alpha) = 1$ if and only if α is a unit of $\mathbf{Z}[\zeta]$.

Proof: $\Rightarrow$ If $N(\alpha) = 1$, then $\alpha\bar{\alpha} = 1$, so that α is a unit. $\Leftarrow$ If α is a unit, then $\alpha\alpha^{-1} = 1$, so that $N(\alpha)N(\alpha^{-1}) = N(1) = 1$ [by (3)]. Therefore, $N(\alpha) = 1$, since $N(\alpha)$ is a nonnegative integer.

PROPOSITION 3: The units of $\mathbf{Z}[\zeta]$ are ± 1, $\pm\zeta$, and $\pm\zeta^2$.

Proof: By Lemma 2, $\alpha = a + b\zeta$ is a unit if and only if

$$a^2 - ab + b^2 = 1. \tag{5}$$

And it is easy to see that the only solutions of (5) in integers a and b are $(a, b) = (\pm 1, 0), (0, \pm 1). (\pm 1, \pm 1)$.

LEMMA 4: Let $\alpha \in \mathbf{Z}[\zeta]$. If $N(\alpha) = p$, where p is a prime, then α is irreducible in $\mathbf{Z}[\zeta]$.

Proof: Let $\alpha = \beta\gamma$, $\beta, \gamma \in \mathbf{Z}[\zeta]$. By (3), $N(\alpha) = N(\beta)N(\gamma)$, so that because p is prime, either $N(\beta) = 1$ or $N(\gamma) = 1$. Thus, by Lemma 2, either β or γ is a unit. Therefore, α is irreducible.

COROLLARY 5: Let $\lambda = 1 - \zeta$. Then λ is irreducible in $\mathbf{Z}[\zeta]$ and $\lambda \mid 3$.

Proof: $N(\lambda) = 1 \cdot 1 - 1(-1) + 1 \cdot 1 = 3$. Therefore, λ is irreducible by Lemma 4. Also, $(1 - \zeta)(2 + \zeta) = 3$, so that $\lambda \mid 3$.

We may introduce congruences in $\mathbf{Z}[\zeta]$ in much the same way that we introduced congruences in $\mathbf{Z}$. Let $\alpha, \beta, \eta \in \mathbf{Z}[\zeta]$. We say that α *is congruent to β modulo η*, denoted $\alpha \equiv \beta \pmod{\eta}$ if $\eta \mid \alpha - \beta$. The following results are proved in much the same way as the corresponding results for $\mathbf{Z}$: Let $\alpha \equiv \beta \pmod{\eta}$, $\gamma \equiv \delta \pmod{\eta}$. Then

$$\alpha + \gamma \equiv \beta + \delta \pmod{\eta}, \tag{6}$$

$$\alpha\gamma \equiv \beta\delta \pmod{\eta}. \tag{7}$$

Further, if $\alpha \equiv \beta \pmod{\eta}$, $\beta \equiv \gamma \pmod{\eta}$, then

$$\alpha \equiv \gamma \pmod{\eta}. \tag{8}$$

If $c \in \mathbf{Z}$, then $c \equiv 0, 1$ or $-1 \pmod 3$. Therefore, since $\lambda \mid 3$, $c \equiv 0, 1,$ or $-1 \pmod{\lambda}$. If $\alpha = a + b\zeta \in \mathbf{Z}[\zeta]$, then $\alpha = a + b + \lambda(-b)$, so that $\alpha \equiv a + b \pmod{\lambda}$. Therefore,

PROPOSITION 6: Let $\alpha \in \mathbf{Z}[\zeta]$. Then $\alpha \equiv 0, 1,$ or $-1 \pmod{\lambda}$.

If $\lambda \nmid \alpha$, then $\alpha \equiv \pm 1 \pmod{\lambda}$ by the preceding proposition. Therefore, if $\lambda \nmid \alpha$, we have $\alpha = \pm 1 + \kappa\lambda$ for some $\kappa \in \mathbf{Z}[\zeta]$, so that

$$\begin{aligned}\alpha^3 &= \pm 1 + 3\kappa\lambda \pm 3\kappa^2\lambda^2 + \kappa^3\lambda^3 \\ &= \pm 1 - \zeta^2\kappa\lambda^3 \mp \zeta^2\kappa^2\lambda^4 + \kappa^3\lambda^3 \qquad (\text{since } 3 = -\zeta^2\lambda^2) \\ &\equiv \pm 1 \pmod{\lambda^3}.\end{aligned} \tag{9}$$

Let us divide our proof of Theorem 1 into two cases:

Case 1: Equation (1) has no solution (x, y, z) such that $xyz \neq 0$ and $\lambda \nmid xyz$.

Case 2: Equation (1) has no solutions (x, y, z) such that $xyz \neq 0$ and $\lambda \mid xyz$.

Proof of Case 1: Assume that (x, y, z) is a solution of (1) such that $xyz \neq 0$ and $\lambda \nmid xyz$. Then none of x, y, z is divisible by λ. Therefore, by (9),

$$0 = x^3 + y^3 + z^3 \equiv \pm 1 \pm 1 \pm 1 \pmod{\lambda^3}.$$

If all the signs are the same, then $\pm 1 \pm 1 \pm 1 = \pm 3$ and $\pm 3 \equiv 0 \pmod{\lambda^3}$. But $\pm 3 = \mp\zeta^2\lambda^2$, so that ± 3 is not divisible by λ^3. On the other hand, if two of the signs are different, then these terms cancel and $\pm 1 \pm 1 \pm 1 = \pm 1$, and therefore $\pm 1 \equiv 0 \pmod{\lambda^3}$. But then $\lambda^3 \mid \pm 1$, which is impossible since λ is not a unit.

Proof in Case 2: Assume that (x, y, z) is a solution of (1) such that $xyz \neq 0$ and $\lambda \mid xyz$. Since λ is irreducible, λ divides one of x, y, z. If λ divides two of x, y, z, then it divides the third, since $x^3 + y^3 + z^3 = 0$. Without loss of generality, we may assume that x, y, and z have no common factors except units. For if η is a common factor, then $(x/\eta, y/\eta, z/\eta)$ is still a solution of (1). Henceforth, let us assume that x, y, and, z have no common factor. Then the reasoning above shows that λ divides precisely one of x, y, z. Without loss of generality, we may assume that $\lambda \mid z$. Then $\lambda \nmid x$, $\lambda \nmid y$, and $z = \lambda^s w$, for some $s \geq 1$, $w \in \mathbf{Z}[\zeta]$, $\lambda \nmid w$. Now since $\lambda \nmid x$, $\lambda \nmid y$, $\lambda \nmid w$, we have

$$x^3 + y^3 + \lambda^{3s}w^3 = 0, \qquad xyz \neq 0, \quad \lambda \nmid xyw. \tag{10}$$

At this point let us address a slightly more general problem. Let us prove

THEOREM 7: Let ϵ be a unit of $\mathbf{Z}[\zeta]$, $s \geq 1$. Then the equation

$$x^3 + y^3 + \epsilon\lambda^{3s}w^3 = 0 \tag{11}$$

has no solution with $x, y, w \in \mathbf{Z}[\zeta]$, $xyw \neq 0$, and $\lambda \nmid xyw$.

Proof: Let s denote the smallest positive integer for which there exist x, y, w, ϵ satisfying (11). Without loss of generality, assume that x, y, and w have no common factors except units. Since $s \geq 1$,

$$x^3 + y^3 \equiv x^3 + y^3 + \epsilon\lambda^{3s}w^3 \pmod{\lambda}$$
$$\equiv 0 \pmod{\lambda}.$$

Therefore, one of x, y is congruent to $+1 \pmod{\lambda}$ and the other is congruent to $-1 \pmod{\lambda}$. [Otherwise $\pm 2 \equiv 0 \pmod{\lambda}$, which is false.] Without loss of generality, assume that $x \equiv 1 \pmod{\lambda}$, $y \equiv -1 \pmod{\lambda}$. We assert that $s \geq 2$. Let $x = 1 + \lambda\alpha$, $y = -1 + \lambda\beta$, $\alpha, \beta \in \mathbf{Z}[\zeta]$. Then

$$\begin{aligned} x^3 + y^3 &= 3\lambda(\alpha + \beta) + 3\lambda^2(\alpha^2 - \beta^2) + \lambda^3(\alpha^3 + \beta^3) \\ &= -\zeta^2\lambda^3(\alpha + \beta) - \zeta^2\lambda^4(\alpha^2 - \beta^2) + \lambda^3(\alpha^3 + \beta^3) \qquad (\text{since } 3 = -\zeta^2\lambda^2) \\ &\equiv \lambda^3(-\zeta^2(\alpha + \beta) + (\alpha^3 + \beta^3)) \pmod{\lambda^4}. \end{aligned} \tag{12}$$

Moreover, since $\zeta \equiv 1 \pmod{\lambda}$,

$$-\zeta^2(\alpha + \beta) + (\alpha^3 + \beta^3) \equiv (\alpha^3 - \alpha) + (\beta^3 - \beta) \pmod{\lambda}. \tag{13}$$

And by (9), if $\lambda \nmid \gamma$, then

$$\gamma^3 \equiv \gamma \pmod{\lambda}. \tag{14}$$

But (14) also holds if $\lambda \mid \gamma$. Therefore, (13) and (14) imply that

$$-\zeta^2(\alpha + \beta) + (\alpha^3 + \beta^3) \equiv 0 + 0 \pmod{\lambda},$$

and thus we derive from (12) that $x^3 + y^3 \equiv 0 \pmod{\lambda^4}$. But since $x^3 + y^3 = \epsilon\lambda^{3s}(-w)^3$ and $\lambda \nmid w$, we see that $\lambda^4 \mid \lambda^{3s}$. Therefore, $3s \geq 4$ and $s \geq 2$.

Set

$$x' = \frac{x + \zeta y}{\lambda}, \qquad y' = \frac{\zeta(x + \zeta^2 y)}{\lambda}, \qquad z' = \frac{\zeta^2(x + y)}{\lambda}. \tag{15}$$

Since $x \equiv 1 \pmod{\lambda}$, $y \equiv -1 \pmod{\lambda}$, $\zeta \equiv 1 \pmod{\lambda}$, $\zeta^2 \equiv 1 \pmod{\lambda}$, we see that $x', y', z' \in \mathbf{Z}[\zeta]$. [For example, $x + \zeta y \equiv 1 + 1(-1) \pmod{\lambda}$ $\Rightarrow \lambda \mid x + \zeta y \Rightarrow x' \in \mathbf{Z}[\zeta]$.] Moreover, x', y', and z' are pairwise relatively prime. For example, if $\eta \mid x'$ and $\eta \mid y'$, then

$$\eta \left| \frac{x + \zeta y}{\lambda} - \frac{x + \zeta^2 y}{\lambda} \right. \Longrightarrow \eta \mid y,$$

$$\eta \left| \frac{\zeta(x + \zeta y)}{\lambda} - \frac{x + \zeta^2 y}{\lambda} \right. \Longrightarrow \eta \mid x.$$

But since $x^3 + y^3 = -\epsilon\lambda^{3s}w^3$, $\eta^3 \mid \lambda^{3s}w^3$, But $(\eta, \lambda) = 1$, $\lambda \nmid x$ and $\lambda \nmid y$. Therefore, $\eta^3 \mid z^3 \Rightarrow \eta \mid z$, using the fact that $\mathbf{Z}[\zeta]$ is a UFD. But then η is a common factor of x, y, and w, so that η is a unit and x' and y' are relatively prime. Similarly, $(x', z') = 1$. Note that

$$(x + \zeta y)(x + \zeta^2 y)(x + y) = \epsilon\lambda^{3s}(-w)^3;$$

this implies that

$$x'y'z' = \epsilon\lambda^{3(s-1)}(-w)^3. \tag{16}$$

Furthermore, since $1 + \zeta + \zeta^2 = 0$, we see that

$$x' + y' + z' = 0. \tag{17}$$

From (16), the fact that $\mathbf{Z}[\zeta]$ is a UFD, and the fact that x', y', z' are pairwise relatively prime, we see that there exist elements $a, b, c \in \mathbf{Z}[\zeta]$ and units $\epsilon_1, \epsilon_2, \epsilon_3$ such that

$$x' = \epsilon_1 a^3, \qquad y' = \epsilon_2 b^3, \qquad z' = \epsilon_3 c^3. \tag{18}$$

Since $s \geq 2$, and since x', y', and z' are pairwise relatively prime, (16) implies that λ divides precisely one of x', y', z'. Therefore, λ divides precisely one of a, b, c, say $\lambda \mid c$. Then (16) and (18) imply that

$$c = \lambda^{s-1} v, \qquad \lambda \nmid v. \tag{19}$$

Moreover, by (17), (18), and (19), we see that

$$a^3 + \epsilon_4 b^3 + \epsilon_5 \lambda^{3(s-1)} v^3 = 0, \tag{20}$$

where ϵ_4 and ϵ_5 are units of $\mathbf{Z}[\zeta]$. Since $s \geq 2$,

$$a^3 + \epsilon_4 b^3 \equiv 0 \pmod{\lambda^3}. \tag{21}$$

But by (9) and the fact that $\lambda \nmid a$, $\lambda \nmid b$, we see that

$$\pm 1 \pm \epsilon_4 \equiv 0 \pmod{\lambda^3}. \tag{22}$$

But $\epsilon_4 = \pm 1, \pm\zeta, \pm\zeta^2$. A simple verification shows that if (22) holds, then $\epsilon_4 = +1$ or -1. In the former case, (20) implies that

$$a^3 + b^3 + \epsilon_5 \lambda^{3(s-1)} v^3 = 0, \tag{23}$$

while in the latter case

$$a^3 + (-b)^3 + \epsilon_5 \zeta \lambda^{3(s-1)} v^3 = 0. \tag{24}$$

In either case, (23) and (24) contradict the original choice of s. Thus, a contradiction is established and Theorem 7 is proved.

5.6 Further Work on Fermat's Last Theorem

In Section 5.5 we showed how one could deduce the truth of Fermat's last theorem for $p = 3$ from the fact that $\mathbf{Z}[\zeta]$ is a UFD, where ζ is a primitive cube root of 1. Let us now cite some of the further work of Kummer and others, but without proofs. Using reasoning very similar to that used in Section 5.5, we could establish

THEOREM 1: Let ζ be a primitive pth root of 1. If $\mathbf{Z}[\zeta]$ is a UFD, then

$$x^p + y^p = z^p$$

has no solutions in integers x, y, z, $xyz \neq 0$.

It might be hoped that $\mathbf{Z}[\zeta]$ is a UFD for many values of p. Indeed, this is so for $p = 3, 5, 7, 11, 13, 17, 19$. However, it was recently shown that if $p > 19$, then $\mathbf{Z}[\zeta]$ is *never* a UFD. Thus, Kummer's original idea falls far short of proving the Fermat conjecture, even for "most" primes. However, Kummer made a much deeper analysis of the situation and got far better results.

If R is a ring, I an ideal of R, n a positive integer, let $I^n = I \cdot I \cdot \ldots \cdot I$ (n times), where the product on the right denotes the product of ideals.

Let p be an odd prime and let ζ be a primitive pth root of 1. Kummer made the following definition:

DEFINITION 2: The prime p is called a *regular prime* if, whenever A is an ideal of $\mathbf{Z}[\zeta]$ such that A^p is a principal ideal, we have that A is a principal ideal.

Then Kummer proved the following very deep result.

THEOREM 3: If p is a regular prime, then $x^p + y^p = z^p$ has no solutions in integers x, y, z, $xyz \neq 0$.

In order to appreciate this result, it is necessary to be able to determine whether a prime p is regular. An amazing criterion for regularity can be given in terms of Bernoulli numbers, and is due to Kummer. The Bernoulli numbers $B_m (m \geq 1)$ are defined inductively by the formulas

$$1 + \binom{2}{1} B_1 = 0$$

$$1 + \binom{3}{1} B_1 + \binom{3}{2} B_2 = 0$$

$$1 + \binom{4}{1} B_1 + \binom{4}{2} B_2 + \binom{4}{3} B_3 \pm 0$$

$$\vdots$$

$$1 + \binom{m}{1} B_1 + \binom{m}{2} B_2 + \cdots + \binom{m}{m-1} B_{m-1} = 0.$$

Thus, from the first formula, we can find B_1. Using the value of B_1, B_2 can be computed from the second formula, and so forth. Here $\binom{m}{n}$ denotes the binomial coefficient

$$\frac{m(m-1)\cdots(m-n+1)}{1 \cdot 2 \cdots n} \qquad (m \geq n).$$

Thus, we see that

$$\begin{array}{lll} B_1 = -\frac{1}{2} & B_5 = 0 & B_9 = 0 \\ B_2 = \frac{1}{6} & B_6 = \frac{1}{42} & B_{10} = \frac{5}{66}. \\ B_3 = 0 & B_7 = 0 & \\ B_4 = -\frac{1}{30} & B_8 = -\frac{1}{30} & \end{array}$$

Moreover, it can be shown that $B_{2m+1} = 0$ for $m \geq 1$.

The criterion for regularity given by Kummer is

THEOREM 4: Let $p \geq 3$ be a prime. Then p is regular if and only if the numerators of the Bernoulli numbers $B_2, B_4, \ldots, B_{p-3}$ are not divisible by p.

All primes ≤ 100, with the exception of 37, 59 and 67 are regular. It is known that infinitely many primes are not regular. Kummer asserted that infinitely many primes are regular. However, he gave no proof and his assertion is still unproved. Using Kummer's theory, it has been possible to establish Fermat's conjecture in all cases for $p \leq 4001$. If we ask for solutions x, y, z, of $x^p + y^p = z^p$ which satisfy $xyz \neq 0$ and $p \nmid xyz$, then it has been shown that there are none for $p \leq 253, 747, 889$. This was achieved through the work of many mathematicians, most notable among whom are D. Mirimanoff, H. S. Vandiver, and J. B. Rosser. They were able to show that if p is such that q^{p-1} is not congruent to 1 modulo p^2 for some prime q which is at most 43, then Fermat's equation has no solution (x, y, z) such that $xyz \neq 0$ and $p \nmid xyz$. This simple test, coupled with a computer, suffices to yield the result cited above.

5.7 *Unique Factorization in Polynomial Rings*

We have seen that if F is a field and X is an indeterminate over F, then $F[X]$ is a UFD. Thus, we can factorize polynomials in one variable with coefficients in a field into a product of irreducible polynomials in an essentially unique way. Can the same be said for polynomials in several variables? In other words, if $X_1, \ldots, X_n$ are indeterminates over F, is $F[X_1, \ldots, X_n]$ a UFD? The only way which we have at our disposal to prove that $F[X_1, \ldots, X_n]$ is a UFD is to verify that $F[X_1, \ldots, X_n]$ is a PID. However, we have seen that for $n = 2$, $F[X_1, \ldots, X_n]$ is not a PID. Therefore, the machinery developed earlier in the chapter is not sensitive enough to determine whether or not $F[X_1, \ldots, X_n]$ is a UFD. We will prove that it is, but we will require several new ideas, which are due to Gauss, who first exploited them in the early part of the nineteenth century. Our main result will be

THEOREM 1: Let R be a unique factorization domain and X an indeterminate over R. Then $R[X]$ is a unique factorization domain.

Before we begin proving Theorem 1, let us mention two easy consequences of it.

COROLLARY 2: Let R be a unique factorization domain and let $X_1, \ldots, X_n$ be indeterminates over R. Then $R[X_1, \ldots, X_n]$ is a unique factorization domain.

Proof: Induction on n. The result is true for $n = 1$ by Theorem 1. Assume the result for $n - 1$ ($n > 1$). Then $R[X_1, \ldots, X_{n-1}]$ is a UFD. Therefore, by Theorem 1, $R[X_1, \ldots, X_n] = R[X_1, \ldots, X_{n-1}][X_n]$ is a UFD.

COROLLARY 3: Let F be a field, $X_1, \ldots, X_n$ indeterminates over F. Then $F[X_1, \ldots, X_n]$ is a unique factorization domain.

Proof: F is a PID since its only ideals are (0) and $F = (1)$. Therefore, F is a UFD by Section 5.3. Therefore, $F[X_1, \ldots, X_n]$ is a UFD by Corollary 2.

To prove Theorem 1, it will be necessary to establish some preliminary machinery. Throughout the rest of this section, let R be a UFD. Let P be a set of irreducible elements of R such that (1) every irreducible element of R is an associate of some element of P and (2) no two elements of P are associates. Such a set P can be constructed as follows: Let P^* denote the set of all irreducible elements of R. Define an equivalence relation $\sim$ on P^* by setting $x \sim y$ if and only if x and y are associates. (Check that this defines an equivalence relation.) Then P can be gotten by choosing one element from each equivalence class of P^* with respect to $\sim$. Every element $x \in R^\times$ can be written in the form

$$x = \epsilon \prod_{\pi \in P} \pi^{a_\pi(x)}, \tag{1}$$

where $a_\pi(x)$ is a nonnegative integer, ϵ is a unit of R, $a_\pi(x) = 0$ for all but a finite number of π, and $\prod_{\pi \in P}$ denotes the product over all $\pi \in P$ for which $a_\pi(x) > 0$. The decomposition (1) follows from the fact that x can be written as a product of irreducible elements and every irreducible element of R is an associate of some $\pi \in P$. Since factorization in R is unique, the decomposition (1) is unique, up to rearrangements of the $\pi \in P$. (Here we use the fact that no two elements of P are associates.)

Let $a_1, \ldots, a_n \in R$. Then a *greatest common divisor* of $a_1, \ldots, a_n$ is an element of R such that

1. $c \mid a_1, \ldots, c \mid a_n$.
2. If $d \in R$ is such that $d \mid a_1, \ldots, d \mid a_n$, then $d \mid c$.

In case $n = 2$, this definition coincides with the definition of g.c.d. which we gave in Section 5.3. As in Section 5.3, we can prove that if d and d' are two g.c.d.s of $a_1, \ldots, a_n$, then d and d' are associates. In Section 5.3 we showed that if at least one of a and b is nonzero, then a and b have a g.c.d. in case R is a PID. We can generalize this statement

LEMMA 4: Let R be a UFD, $a_1, \ldots, a_n \in R$ not all 0. Then $a_1, \ldots, a_n$ have a g.c.d.

Proof: Let

$$a_i = \epsilon_i \prod_{\pi \in P} \pi^{a_\pi(a_i)}, \qquad \epsilon_i \text{ a unit of } R \quad (1 \leq i \leq n).$$

For each $\pi \in P$, let $a_\pi =$ the smallest of $a_\pi(a_1), \ldots, a_\pi(a_n)$. Then, since $a_\pi(a_i) = 0$ for all but a finite number of π, we see that $a_\pi = 0$ for all but a finite number of π, and thus we may set

$$c = \prod_{\pi \in P} \pi^{a_\pi}.$$

Then c is a g.c.d. of $a_1, \ldots, a_n$. We leave the details of this verification to the reader.

In Section 5.3. we showed that if R is a principal ideal domain, $a, b, \pi \in R$, π irreducible, then $\pi \mid ab$ implies that $\pi \mid a$ or $\pi \mid b$. Let us observe that this is true in any unique factorization domain.

PROPOSITION 5: Let R be a unique factorization domain, $a, b, \pi \in R$, π irreducible. If $\pi \mid ab$, then $\pi \mid a$ or $\pi \mid b$.

Proof: Since $\pi \mid ab$, there exists $k \in R$ such that $ab = k\pi$, Without loss of generality we may assume that $a \neq 0$, $b \neq 0$, $k \neq 0$. Since R is a UFD, we can write a and b as a product of irreducible elements. Since π appears in a decomposition of $k\pi$ into a product of irreducible elements, the uniqueness of factorization implies that π is an associate of either some irreducible element dividing a or some irreducible element dividing b. Thus, either $\pi \mid a$ or $\pi \mid b$.

If 1 is a g.c.d. of $a_1, \ldots, a_n \in R$, then we say that $a_1, \ldots, a_n$ are *relatively prime*. Moreover, if c is a g.c.d of $a_1, \ldots, a_n \in R$, then $a_i = k_i c$ $(1 \leq i \leq n)$ for some $k_i \in R$, and $k_1, \ldots, k_n$ are relatively prime. (Proof: Exercise.)

Let us now turn to the study of polynomials $f \in R[X]$. Let $f = a_0 + a_1 X + \cdots + a_n X^n$, $a_i \in R$, be a nonzero polynomial. If $a_0, \ldots, a_n$ are relatively prime, then we say that f is *primitive*. Let c be a g.c.d. of $a_0, \ldots, a_n$ and let $a_i = k_i c$ $(0 \leq i \leq n)$, where $k_i \in R$. Then $k_0, \ldots, k_n$ are relatively prime so that

$$f^* = k_0 + k_1 X + \ldots + k_n X^n$$

is primitive. Moreover,

$$f = cf^*.$$

Thus, we have proved

LEMMA 6: Let $f \in R[X]$ be nonzero and let c be a g.c.d. of the coefficients of f. Then f can be written in the form

$$f = cf^*,$$

where $f^* \in R[X]$ is primitive.

The element c of Lemma 6 is called the *content* of f. It is uniquely determined up to multiplication by units of R.

For example, consider the polynomial $4X^5 + 8X^3 + 4X^2 + 12 \in \mathbf{Z}[X]$. Then a g.c.d. of 4, 8, 4, and 12 is 4,

$$4X^5 + 8X^3 + 4X^2 + 12 = 4(X^5 + 2X^3 + X^2 + 3),$$

and $X^5 + 2X^3 + X^2 + 3$ is primitive.

The key to the proof of Theorem 1 is the following result, which is often referred to as *Gauss's lemma*.

THEOREM 7: Let $f, g \in R[X]$ be primitive. Then fg is primitive.

Proof: Let

$$f = a_0 + a_1X + \cdots + a_nX^n,$$
$$g = b_0 + b_1X + \cdots + b_mX^m,$$
$$fg = c_0 + c_1X + \cdots + c_{m+n}X^{m+n}.$$

Let π be an irreducible element of R. It suffices to prove that π does not divide all of $c_0, \ldots, c_{m+n}$. For then, if e is a g.c.d. of $c_0, \ldots, c_{m+n}$, then e is not divisible by any irreducible element of R and is thus a unit, so that $c_0, \ldots, c_{m+n}$ are relatively prime. Since f and g are primitive, π does not divide all of $a_0, \ldots, a_n$ and π does not divide all of $b_0, \ldots, b_m$. Let a_i and b_j be the first coefficients of f and g, respectively, which are not divisible by π. We assert that c_{i+j} is not divisible by π. Indeed, assume that $\pi \mid c_{i+j}$. Then

$$\pi \mid \underbrace{a_0b_{i+j} + \cdots + a_{i-1}b_{j+1}}_{\text{I}} + a_ib_j + \underbrace{a_{i+1}b_{j-1} + \cdots + a_{i+j}b_0}_{\text{II}}.$$

By the choice of a_i and b_j, we see that $a_0, \ldots, a_{i-1}, b_0, \ldots, b_{j-1}$ are all divisible by π, so that all terms of I and II are divisible by π. Thus,

$$\pi \mid a_ib_j,$$

which implies that $a_ib_j = \pi k$ for some $k \in R$. But π is irreducible in R, so that by the unique factorization in R and Proposition 5, π must divide either a_i or b_j, which is a contradiction to the choice of a_i and b_j. Thus, $\pi \nmid c_{i+j}$ as asserted.

Let F denote the quotient field of R. Every nonzero element x of F can be written in the form $x = a/b$, $a, b \in R^\times$. If λ is a g.c.d. of a and b, then $a^* = a/\lambda$, $b^* = b/\lambda$ belong to R and $x = a^*/b^*$, with a^* and b^* relatively prime. Thus, every nonzero $x \in F$ can be written in the form $x = a/b$, $a, b \in R$, a and b relatively prime. Therefore, if $f \in F[X]$ is nonzero, there exists $a^* \in F$ and $f^* \in R[X]$ such that $f = a^*f^*$. Moreover, it is clear that by choosing a^* appropriately, we may assume that f^* is primitive, by Lemma 6. Note that $F[X]$ is a UFD by Proposition 13 of Section 5.3.

COROLLARY 8: Let $f \in R[X]$ be primitive. Then f is irreducible in $R[X]$ if and only if f is irreducible in $F[X]$.

Proof: $\Rightarrow$ Assume that f is irreducible in $R[X]$ and that $f = gh, g, h \in F[X]$. By the above discussion, we can write $g = a^*g^*$, $h = b^*h^*$, where $g^*, h^* \in R[X]$, g^* and h^* are primitive, and $a^*, b^* \in F$. Then $f = (a^*b^*)(g^*h^*)$. By Theorem 7, g^*h^* is primitive. Therefore, since f is primitive, a^*b^* is a unit of R. But then, since $f = (a^*b^*)g^*h^*$ is irreducible, either g^* or h^* is a unit of $R[X]$. Thus, in particular either g^* or h^* is a constant polynomial, so that either g or h is a unit of $F[X]$. Thus, f is irreducible in $F[X]$.
$\Leftarrow$ Assume that f is irreducible in $F[X]$, and assume that $f = gh$, $g, h \in R[X]$. Then, by the irreducibility of f in $F[X]$, we see that one of g, h must be a unit of $F[X]$, that is, a constant polynomial. Say g is a constant polynomial. Then g is a unit of R, since $f = gh$ and f is primitive. Thus, f is irreducible in $R[X]$.

Proof of Theorem 1: Let $f \in R[X], f \neq$ a unit of $R[X]$. Let us first show that f can be written as a product of irreducible elements of $R[X]$. If f is a constant polynomial, then $f \in R$ and f can be written as a product of irreducible elements of R, since R is a UFD, and an irreducible element of R is an irreducible element of $R[X]$. (Why?) Thus, we may assume that $f \neq$ a constant polynomial. By Lemma 6, we can write

$$f = a^*f^*, \quad a^* \in R, \qquad f^* \in R[X], \quad f^* \text{ primitive.} \tag{2}$$

Since $F[X]$ is a UFD, we can write

$$f^* = f_1 \ldots f_n, \qquad f_i \in F[X], \quad f_i \text{ irreducible in } F[X]. \tag{3}$$

Now we can write $f_i = a_i f_i^*$, $f_i^* \in R[X]$, f_i^* primitive, $a_i \in F$. By Corollary 8, f_i^* is irreducible in $R[X]$. Moreover, by Theorem 7, $f_1^* \ldots f_n^*$ is primitive, so that by (3) and the fact that f^* is primitive, we see that

$$f^* = (a_1 \ldots a_n) f_1^* \ldots f_n^*$$

and $a_1 \ldots a_n$ is a unit ϵ of R. By (2),

$$f = (\epsilon a^*) f_1^* \ldots f_n^*.$$

Since R is a UFD, we can write $\epsilon a^* = \pi_1 \ldots \pi_m$, $\pi_i \in R$, π_i irreducible. But then, as observed above, π_i is irreducible in $R[X]$ and

$$f = \pi_1 \ldots \pi_m f_1^* \ldots f_n^*$$

is a decomposition of f into the product of irreducible elements of $R[X]$.

In order to complete the proof of Theorem 1, we must show that if

$$f = \pi_1 \ldots \pi_s = \lambda_1 \ldots \lambda_t$$

are two decompositions of f into a product of irreducible elements of $R[X]$, then $s = t$ and upon proper rearrangement of $\pi_1, \ldots, \pi_s$, we have π_i and λ_i are associates $(1 \leq i \leq s)$. This is proved using Proposition 5 plus the same reasoning as used in the proof of Theorem 12 of Section 5.3.

From the proof of Theorem 1, we glean the following useful fact:

COROLLARY 9: Let R be a unique factorization domain and let $f \in R[X]$ be a primitive polynomial. Further, let F denote the quotient field of R. Then the factorization of f into a product of irreducible factors in $F[X]$ can already be carried out in $R[X]$.

5.7 *Exercises*

1. Show that $\sim$ as defined in this section is an equivalence relation on P^*.
2. We showed (Exercise 9 of Section 2.4) that 3 is an irreducible element of $\mathbf{Z}[\sqrt{-5}]$. Show that $3 \mid (2+\sqrt{-5})(2-\sqrt{-5})$, but that $3 \nmid 2+\sqrt{-5}$, $3 \nmid 2-\sqrt{-5}$. Does this contradict Proposition 5?
3. Let $f \in \mathbf{Z}[X]$, $f \neq \mathbf{0}$. Define the *content* of f, denoted $C(f)$, to be the positive g.c.d. of all the coefficients of f. Show that $C(fg) = C(f)C(g)$.

5.8 *Factorization of Polynomials in* **Q** *[X]*

Since $\mathbf{Q}[X]$ is a UFD, every nonconstant polynomial $f \in \mathbf{Q}[X]$ can be factorized into a product of irreducible polynomials. This is guaranteed by our general theory. However, our general theory provides us with no way in which this factorization can be carried out in particular cases. No doubt, the reader has factored many polynomials in high school algebra. However, the methods used there are usually ad hoc, depending on a series of tricks. Moreover, these methods provide no real insight into the reducibility or irreducibility of a given polynomial.

In this section we will use the theory we have developed in order to provide a computational procedure for factorizing polynomials in $\mathbf{Q}[X]$. This beautiful development is due to the German mathematician Kronecker. Kronecker believed that mathematics should concern itself only with objects which can be computed or constructed in a finite number of steps. Thus, for Kronecker, a proof which asserts the existence of a mathematical object, without prescribing a recipe for constructing the object in a finite number of steps, is no proof at all! The developments of this section are very much in the spirit of Kronecker's philosophy.

First, let us make a few observations concerning polynomials. Let F be a field and let $\alpha_1, \ldots, \alpha_{n+1}, \beta_1, \ldots, \beta_{n+1}$ be elements of F with $\alpha_1, \ldots, \alpha_{n+1}$ distinct. The *Lagrange interpolation formula* (see Exercise 7 of Section 4.2) allows us to construct a polynomial $f \in F[X]$ of degree at most n such that

$$f(\alpha_i) = \beta_i \qquad (1 \leq i \leq n+1). \tag{1}$$

In fact, a polynomial f which works is given by

$$f = \sum_{i=1}^{n+1} \beta_i \frac{(X - \alpha_1)\cdots(X - \alpha_{i-1})(X - \alpha_{i+1})\cdots(X - \alpha_{n+1})}{(\alpha_i - \alpha_1)\cdots(\alpha_i - \alpha_{i-1})(\alpha_i - \alpha_{i+1})\cdots(\alpha_i - \alpha_{n+1})}. \tag{2}$$

LEMMA 1: Let $f \in F[X]$, $\alpha \in F$. If $f(\alpha) = 0$, then $X - \alpha \mid f$. Thus, if $\deg(f) = n$ and f has more than n zeros in F, then $f = \mathbf{0}$.

Proof: By the division algorithm in $F[X]$, there exist polynomials g and r such that

$$f = g(X - \alpha) + r, \qquad \deg(r) < \deg(X - \alpha) = 1. \tag{3}$$

Setting $X = \alpha$ in (3), we see that $r(\alpha) = 0$. But since $\deg(r) \leq 0$, r is a constant polynomial and $r = \mathbf{0}$. Thus, $f = g(X - \alpha)$ and $X - \alpha \mid f$. If $\deg(f) = n$ and f has zeros $\alpha_1, \ldots, \alpha_{n+1}$ in F, then $X - \alpha_i \mid f$ $(1 \leq i \leq n + 1)$, so that $(X - \alpha_1) \ldots (X - \alpha_{n+1}) \mid f$. Therefore, $f = \mathbf{0}$.

PROPOSITION 2: Let $\alpha_1, \ldots, \alpha_{n+1}, \beta_1, \ldots, \beta_{n+1} \in F$ with $\alpha_1, \ldots, \alpha_{n+1}$ distinct. Then there exists one and only one polynomial $f \in F[X]$ such that $\deg(f) \leq n$ and $f(\alpha_i) = \beta_i (1 \leq i \leq n + 1)$.

Proof: The existence of f is guaranteed by (2). Suppose that f and f^* satisfy the conditions of the proposition. Then $\deg(f - f^*) \leq n$ and $f - f^*$ has the $n + 1$ zeros $\alpha_1, \ldots, \alpha_{n+1}$. Therefore, by Lemma 1, $f - f^* = \mathbf{0}$.

Let us now describe Kronecker's factorization algorithm. Without loss of generality, let us restrict ourselves to consideration of a nonconstant primitive polynomial $f \in \mathbf{Z}[X]$ of degree n. Then the irreducible factors of f in $\mathbf{Q}[X]$ can be taken to be primitive polynomials in $\mathbf{Z}[X]$. If f is not irreducible in $\mathbf{Z}[X]$, then f has a factor of degree $\leq n/2$. Thus, in order to factor f, it suffices to determine whether f is irreducible and, if it is not, to find a factor of f of degree $\leq n/2$. For in the latter case, we can write

$$f = gh, \quad \deg(g) \leq \frac{n}{2}, \qquad g, h \in \mathbf{Z}[X],$$

and then we can repeat the procedure to determine whether g or h is irreducible. Eventually, we will get a factorization of f into primitive, irreducible polynomials in $\mathbf{Z}[X]$.

Suppose that $s =$ the largest positive integer $\leq n/2$, and suppose that g is a factor of f in $\mathbf{Z}[X]$, $\deg(g) \leq n/2$. Then for any integer α, $g(\alpha) \mid f(\alpha)$. In particular, if $f(\alpha) \neq 0$, then $g(\alpha)$ is a divisor of $f(\alpha)$ and there are only finitely many choices for $g(\alpha)$. Since f has at most n zeros in $\mathbf{Z}$, among the integers $1, 2, 3, \ldots, 2n$, we may select $s + 1$ of them, say $\alpha_0, \ldots, \alpha_s$, such that $f(\alpha_i) \neq 0$ $(0 \leq i \leq s)$. For each α_i, let us compute $S(\alpha_i) =$ the set of divisors of $f(\alpha_i)$. Then by our observations above, $g(\alpha_i) \in S(\alpha_i)$, $\deg(g) \leq s$. Moreover, if $\beta_0 \in$

$S(\alpha_0)$, $\beta_1 \in S(\alpha_1), \ldots, \beta_s \in S(\alpha_s)$, then there exists one and only one g such that $\deg(g) \leq s$ and $g(\alpha_i) = \beta_i$ $(0 \leq i \leq s)$, and this particular g can be calculated using (2). Thus, our plan of calculation is clear. For each $(s + 1)$-tuple $(\beta_0, \ldots, \beta_s)$, $\beta_i \in S(\alpha_i)$, we construct the unique g given by (2). Then using long divison, we determine whether g is a factor of f. If none of the factors thus constructed work, then f is irreducible.

This algorithm is rather clumsy to carry out manually, but it is rather easy to put on a computer. Nevertheless, let us carry out an easy example. Let $f = X^3 - 3X^2 + 4X + 1$. Then $s = 1$, and since $f(0) = 1 \neq 0, f(1) = 3 \neq 0$, we may set $\alpha_1 = 0$, $\alpha_2 = 1$. Then $S(\alpha_1) = \{1, -1\}$, $S(\alpha_2) = \{3, -3, 1, -1\}$. Note that if g corresponds to (β_1, β_2), then $-g$ corresponds to $(-\beta_1, -\beta_2)$, so that it is enough to check the cases

$$(\beta_1, \beta_2) = (1, 3), (1, -3), (1, 1), (1, -1).$$

In these cases, the value of g is, respectively,

$$2X + 1, \qquad 4X + 1, \qquad 1, \qquad -2X + 1.$$

However, none is a factor of f, so f is irreducible.

Kronecker's algorithm proceeds much more quickly if we can recognize irreducible polynomials easily. A simple test for irreducibility which is often applicable is the *Eisenstein irreducibility criterion.* (Eisenstein was a pupil of Gauss.)

THEOREM 3 (EISENSTEIN): Let $f = a_0 + a_1X + \cdots + a_nX^n \in \mathbf{Z}[X]$ be primitive and let p be a prime. Assume that $p \mid a_i$ $(0 \leq i \leq n - 1)$, $p \nmid a_n$, $p^2 \nmid a_0$. Then f is irreducible in $\mathbf{Z}[X]$.

Proof: Let us reason by contradiction. Assume that

$$f = (b_0 + \cdots + b_tX^t)(c_0 + \cdots + c_uX^u); \qquad b_i, c_i \in \mathbf{Z},$$

where $t \geq 1$, $u \geq 1$. Since $a_0 = b_0c_0$ and since $p^2 \nmid a_0$, we see that p cannot divide both of b_0 and c_0. On the other hand, since $p \mid a_0$, p divides one of b_0, c_0. Assume that $p \nmid b_0$ and $p \mid c_0$. Since f is primitive, not all c_i are divisible by p. Let c_v be the first c_i which is not divisible by p. Then,

$$a_v = b_vc_0 + b_{v-1}c_1 + \cdots + b_0c_v.$$

By assumption, $p \mid a_v$. Moreover, $p \mid c_0, p \mid c_1, \ldots, p \mid c_{v-1}$ by the choice of v. Therefore, $p \mid b_0c_v$, which contradicts the facts $p \nmid c_v$, $p \nmid b_0$.

EXAMPLE 1: $X^3 + 3X^2 + 9X + 6$ is irreducible in $\mathbf{Z}[X]$ (set $p = 3$). $f = X^4 + X^3 + X^2 + X + 1$ is irreducible in $\mathbf{Z}[X]$. Indeed, since

$$f(X + 1) = X^4 + 5X^3 + 10X^2 + 10X + 5,$$

we see that $f(X + 1)$ is irreducible in $\mathbf{Z}[X]$. Thus, f is irreducible in $\mathbf{Z}[X]$.

5.8 Exercises

1. Factor the following polynomials into irreducible factors in $\mathbf{Q}[X]$.
   (a) $X^3 + 2X + 3$.
   (b) $X^3 + 5X^2 + X + 2$.
   (c) $X^3 + 3X^2 + 9X + 3$.
   (d) $X^3 - 3X^2 - 3X + 1$.
   (e) $X^4 + X^3 + 2X^2 + 2X + 1$.
   (f) $X^5 + 5X^4 + 10X + 35$.

5.9 Prime Ideals and Maximal Ideals

Throughout this section let R be a commutative ring with identity.

DEFINITION 1: Let P be an ideal of R, $P \neq (1)$. We say that P is a *prime ideal* if, whenever $a, b \in R$ have the property that $ab \in P$, then either $a \in P$ or $b \in P$.

EXAMPLE 1: Let $R = \mathbf{Z}$ and let p be a prime. Then (p) is a prime ideal. For if $ab \in (p)$, then $p \mid ab$. Therefore, by Euclid's lemma, $p \mid a$ or $p \mid b$ and thus, either $a \in (p)$ or $b \in (p)$.

EXAMPLE 2: Example 1 suggests the following generalization. Let R be a UFD and let π be an irreducible element of R. Then (π) is a prime ideal. The proof is the same as Example 1, except Proposition 5 of Section 5.7 is used instead of Euclid's lemma.

EXAMPLE 3: Let $R = F[X, Y]$. We will prove below that (X, Y) is a prime ideal of R. (See Corollary 4).

EXAMPLE 4: Let $R = \mathbf{Z}$. Then (4) is *not* a prime ideal, since $2 \notin (4)$, but $2 \cdot 2 \in (4)$.

PROPOSITION 2: Let R be a principal ideal domain. Then the prime ideals of R are of the form (π), where π is an irreducible element of R or $\pi = 0$.

Proof: For let P be a prime ideal of R. Assume that $P \neq (0)$. Since P is a nontrivial ideal of R, $P = (a)$ for $a \in R^\times$, where a is a nonunit. Since R is a PID, R is a UFD, so that we may write $a = \pi_1 \ldots \pi_r$, where π_i is an irreducible element of R. Suppose that $r \geq 2$. Set $b = \pi_1$, $c = \pi_2 \ldots \pi_r$. Then $bc \in P$. Moreover, $b \notin P$ and $c \notin P$. (For example, if $b \in P$, then $b = da$ for some $d \in R$, so that $1 = dc$, and thus c is a unit, which is a contradiction.)

This contradicts the assumption that P is prime. Thus, $r = 1$ and $P = (\pi_1)$. On the other hand, if $P = (\pi)$, where π is irreducible, then P is prime by Example 2. The ideal (0) is prime since if $a \cdot b \in (0)$, then either $a = 0$ or $b = 0$ since R is an integral domain.

THEOREM 3: Let P be an ideal of R, a commutative ring with identity, $P \neq (1)$. Then P is a prime ideal if and only if R/P is an integral domain.

Proof: $\Rightarrow$ Assume that P is a prime ideal. It is clear that R/P is a commutative ring with identity. To show that R/P is an integral domain, we must show that if $a + P, b + P \in R/P$ are such that $(a + P)(b + P) = 0 + P$, then either $a + P = 0 + P$ or $b + P = 0 + P$. But $(a + P)(b + P) = ab + P$, so that if $(a + P)(b + P) = 0 + P$ then $ab \in P$. Therefore, since P is a prime ideal, either $a \in P$ or $b \in P$. Thus, either $a + P$ or $b + P$ equals $0 + P$.

$\Leftarrow$ Assume that R/P is an integral domain and that $a, b \in R$ are such that $ab \in P$. Then

$$\begin{aligned}(a + P)(b + P) &= ab + P \\ &= 0 + P.\end{aligned}$$

Therefore, since R/P is an integral domain, either $a + P$ or $b + P$ equals $0 + P$. Hence, either $a \in P$ or $b \in P$.

COROLLARY 4: Let F be a field, and let X and Y be indeterminates over F. Then (X, Y) is a prime ideal of $F[X, Y]$.

Proof: Consider the homomorphism

$$\psi: F[X, Y] \longrightarrow F$$

defined by mapping f into its value at (0, 0). It is easy to see that ψ is surjective and $\ker(\psi) = (X, Y)$. Therefore, $F[X, Y]/(X, Y) \approx$ F. But F is an integral domain, so that (X, Y) is a prime ideal by Theorem 3.

If R is a PID, we can rephrase the unique factorization in R in terms of prime ideals. This was the original motivation for introducing prime ideals into mathematics. We have

THEOREM 5: Let R be a principal ideal domain, A an ideal of R, $A \neq (0)$ or (1). Then A can be written as a product of prime ideals of R, and this expression is unique up to reordering of the prime ideals concerned.

Proof: Let $A = (a)$. Then, since $A \neq (0), (1)$, we see that $a \neq 0$ or a unit of R. Therefore, since R is a UFD, we may write

$$a = \pi_1 \ldots \pi_s,$$

$\pi_i (1 \leq i \leq s)$ irreducible. But then

$$A = (a) = (\pi_1) \dots (\pi_s)$$

is an expression of A as a product of prime ideals $(\pi_1), \dots, (\pi_s)$. Suppose that

$$\begin{aligned} A &= (\pi_1) \dots (\pi_s) \\ &= (\pi'_1) \dots (\pi'_t). \end{aligned} \tag{1}$$

[Every nonzero prime ideal of R is of the form (π) for π irreducible.] Then

$$\begin{aligned} (\pi_1 \dots \pi_s) &= (\pi'_1 \dots \pi'_t) \\ \Longrightarrow \pi_1 \dots \pi_s &= \epsilon \pi'_1 \dots \pi'_t, \end{aligned}$$

where ϵ is a unit of R. (See Exercise 1.) But since R is a UFD, $s = t$ and we may renumber $\pi_1, \dots, \pi_s$ so that π_i and π'_i are associates $(1 \leq i \leq s)$. Then $(\pi_i) = (\pi'_i)$ and the two decompositions of (1) are the same except for order of the prime ideals.

We have seen that there exist fairly simple integral domains (such as $\mathbf{Z}[\sqrt{-5}]$) for which unique factorization does not hold. However, it makes sense to inquire whether every nontrivial ideal of such a ring can be written uniquely as a product of prime ideals, where the uniqueness is taken to mean up to rearrangement. In essense, this is Kummer's idea. It is easy to see (Exercise 8) that if unique factorization among ideals holds and if the ring is a UFD, then the ring is a PID. Thus, in case all ideals are principal, unique factorization of ideals and unique factorization of elements are equivalent. However, if there exist nonprincipal ideals, then it is possible for unique factorization of ideals to hold even though unique factorization of elements does not. An integral domain in which every nontrivial ideal can be expressed uniquely as a product of prime ideals is called a *Dedekind domain*. The rings $\mathbf{Z}[\sqrt{-5}]$ and $\mathbf{Z}[\zeta]$ are examples of Dedekind domains. We cannot pursue the theory of Dedekind domains any further in this introductory book. Their theory properly belongs to the subject of *algebraic number theory*. The interested reader will find the beautiful book *The Theory of Algebraic Numbers* by H. Pollard (Wiley, New York, 1956) an accessible introduction, at the same level as this book.

Having introduced the prime ideals of a ring, let us now introduce another important class of ideals—the maximal ideals. An ideal A of R is said to be *proper* if $A \neq R$.

DEFINITION 6: An ideal M of R is said to be a *maximal ideal* if (1) M is a proper ideal and (2) if A is a proper ideal containing M, then $M = A$.

We can reformulate the definition of a maximal ideal in a more convenient way:

LEMMA 7: Let M be a proper ideal of R. Then if M is a maximal ideal for all $x \in R - M$, we have $M + (x) = R$.

Proof: $\Rightarrow$ Let $x \in R - M$. Then $A = M + (x)$ is an ideal containing M. If A is proper, then $M = A$, since M is maximal. But this is impossible, since $x \in A - M$. Therefore, $M + (x) = R$.
$\Leftarrow$ Let A be a proper ideal containing M, and let $x \in A - M$. Then $M + (x) = R$. Thus, $A = R$, which is a contradiction to the assumption that A is proper. But this implies that $A - M = \varnothing$ and $M \supseteq A$. Therefore, since $M \subseteq A$, we have $M = A$ and M is maximal.

EXAMPLE 5: Let R be a PID, π an irreducible element of R. Then (π) is a maximal ideal. For if $x \in R - (\pi)$, then x is not divisible by π. Thus, since π is irreducible, x and π are relatively prime. By Exercise 2, there exist $a, b \in R$ such that $a\pi + bx = 1$. Thus, $1 \in (\pi) + (x)$, and

$$(\pi) + (x) = (1) = R,$$

so that (π) is a maximal ideal. In particular, if p is a prime, then (p) is a maximal ideal of $\mathbf{Z}$, and if $f \in F[X]$ is irreducible, then (f) is a maximal ideal of $F[X]$.

THEOREM 8: M is a maximal ideal of R if and only if R/M is a field.

Proof: $\Rightarrow$ Assume that M is maximal. Then M is proper, so that R/M is not the trivial ring. Let $x + M \in R/M$, $x + M \neq 0 + M$. We must show that $x + M$ is a unit in R/M. But $x \notin M$, so that by Lemma 7, $M + (x) = R$. In particular, $1 \in M + (x)$, so that there exists $a \in R$ such that $1 - ax \in M$. But then $(a + M)(x + M) = 1 + M$ and $x + M$ is a unit in R/M.
$\Leftarrow$ Assume that R/M is a field. Then R/M is not the trivial ring, so that $R/M \neq \{M\}$ and M is proper. Let $x \in R - M$. Then $x + M \neq 0 + M$, so that there exists $a + M \in R/M$ such that $(a + M)(x + M) = 1 + M$. But then $1 - ax \in M$ and $1 \in (x) + M$, so that $M + (x) = R$. Thus, M is maximal by Lemma 7.

COROLLARY 9: If M is a maximal ideal, then M is a prime ideal.

Proof: If M is maximal, then R/M is a field by Theorem 8. But a field is an integral domain, so that M is prime by Theorem 3.

We have seen that every maximal ideal is a prime. However, the converse is not true. For example, let $R = F[X, Y]$. Then (X) is a prime ideal since X is irreducible in R. However, (X) is not a maximal ideal since $(X) \subseteq (X, Y) \subseteq R$ and (X, Y) is a proper ideal such that $(X, Y) \neq (X)$.

A maximal ideal cannot be properly contained in any nontrivial ideal.

Thus, it is natural to ask if every ideal which is not the entire ring is contained in some maximal ideal. This is a highly nontrivial question. Consider the following example. Let $C[0, 1]$ denote the set of all continuous real-valued functions on the closed interval $[0, 1]$, and let sum and product of functions be defined in the usual way, so that $C[0, 1]$ becomes a commutative ring with identity. For each integer j $(j = 1, 2, 3, \ldots)$ let $I_j = X^{1/2^j} C[0, 1]$. Then I_j is a principal ideal and

$$I_1 \subsetneq I_2 \subsetneq I_3 \subsetneq \ldots.$$

Therefore, it is clear that none of the I_j is a maximal ideal. Is there a maximal ideal containing I_j? It is not obvious, but in fact there is. For let I denote the set of all functions in $C[0, 1]$ which are zero at $x = 0$. Then it is easy to see that I is an ideal of $C[0, 1]$. Moreover, I is a maximal ideal, being the kernel of the surjective homomorphism

$$C[0, 1] \longrightarrow \mathbf{R},$$
$$f \longrightarrow f(0).$$

Thus, since $\mathbf{R}$ is a field, I is a maximal ideal by Theorem 8. And it is clear that I contains all the ideals I_j.

THEOREM 10: Let I be a proper ideal of the commutative ring with identity R. Then I is contained in some maximal ideal of R.

The proof of Theorem 10 will not be given, since it rests on a very subtle set-theoretic result known as Zorn's lemma, which, in turn, is a consequence of the *axiom of choice*, a set-theory axiom which is very convenient to assume and which actually has entered into a number of our arguments in this book, albeit in a very subtle way. Since a discussion of these delicate matters of set theory would take us rather far afield, it seems best for us to gloss over them in this first course.

5.9 Exercises

1. Let R be an integral domain, $a, b \in R$. Suppose that $(a) = (b)$. Show that $a = \epsilon b$ for ϵ a unit of R.
2. Let R be a PID and let $a, b \in R$ be nonzero. Show that if a and b are relatively prime, then there exist $x, y \in R$ such that $ax + by = 1$.

*3. Let F be a field and let $X_1, \ldots, X_n$ be indeterminates over F. Further let $a_1, \ldots, a_n \in F$.
   (a) Show that
$$(X_1 - a_1, \ldots, X_k - a_k) \qquad (k \leq n)$$
   is a prime ideal of $F[X_1, \ldots, X_n]$.

(b) Show that

$$(X_1 - a_1, \ldots, X_n - a_n)$$

is a maximal ideal of $F[X_1, \ldots, X_n]$.

4. For each of the following ideals of **Z**, determine whether the ideal is prime:
   (a) (2). (b) (3, 6).
   (c) (32, 6). (d) (9).

5. For each of the following ideals of $F[X, Y]$, determine whether the ideal is prime or maximal (or both).
   (a) $(X - 3)$. (b) $(X - 2, X - 1)$.
   (c) $(X - 4)$. (d) $(X - 3, Y - 2)$.

6. Let P be a prime ideal of R and let A and B be ideals of R such that $P \supseteq AB$. Show that $P \supseteq A$ or $P \supseteq B$.

7. Let F be a field, X an indeterminate over F, $f \in F[X]$, $f \neq 0$. Then $F[X]/(f)$ is an integral domain if and only if f is irreducible in $F[X]$.

8. Let R be a ring in which every nonzero proper ideal can be uniquely written as a product of prime ideals, where uniqueness is up to reordering. Show that R is a UFD if and only if R is a PID.

9. Let $C[0, 1]$ denote the ring of all continuous functions (real-valued) on the closed interval [0, 1].
   (a) Let $a \in [0, 1]$, and let $F_a = \{f \in C[0, 1] \mid f(a) = 0\}$. Prove that F_a is a maximal ideal of $C[0, 1]$.
   (b) Show that every maximal ideal of $C[0, 1]$ is of the form F_a for some a.
   (c) Show that $C[0, 1]$ has no irreducible elements.
   (d) What are the units of $C[0, 1]$?

10. Let R be a ring, X an indeterminate over R, and let $R[[X]]$ denote the set of all formal sums $a_0 + a_1X + \ldots$ $(a_i \in R)$. Thus, $R[[X]]$ consists of "infinite polynomials" with coefficients in R. Define addition and multiplication in $R[[X]]$ just as for polynomials. Prove that with respect to these operations, $R[[X]]$ becomes a ring, called the *ring of formal power series with coefficients in R*.

11. Let the notations be the same as in Exercise 10, and let F be a field, $f = a_0 + a_1X + \cdots \in F[[X]]$.
    (a) Show that f is a unit of $F[[X]]$ if and only if $a_0 \neq 0$.
    (b) Show that every ideal of $F[[X]]$ is of the form $X^nF[[X]]$ for some $n \geq 0$.

6 VECTOR SPACES

In the preceding chapters we have studied various algebraic structures, including groups, rings, and fields. We will now introduce a new algebraic structure, which will play an important role in our development of the theory of equations later. This algebraic structure—the vector space—is motivated by the consideration of vectors in physics. However, its importance is felt in many branches of mathematics and in many disciplines to which mathematics is applied.

6.1 The Concept of a Vector Space

In physics one often meets physical quantities which have both a magnitude and a direction. For example, a force is described by its magnitude and by the direction in which it is applied. Such quantities are called *vectors* by the physicist. For simplicity, let us confine our discussion to vectors in two dimensions. A convenient representation of a vector quantity can be obtained by drawing an arrow from the origin of a rectangular coordinate system in the direction of the vector, where the length of the arrow equals the magnitude of the vector (see Figure 6.1). It is but a short abstract jump to replace the arrow by the ordered pair (a, b) of real numbers, giving the coordinates of the endpoint of the arrow. Certainly, all the information conveyed by the arrow is provided by the ordered pair (a, b). Thus, for the mathematician, the vector of the physicist (at least in two dimensions) is an ordered pair of real numbers.

Suppose that forces F_1 and F_2 are applied to a body. Experiment has

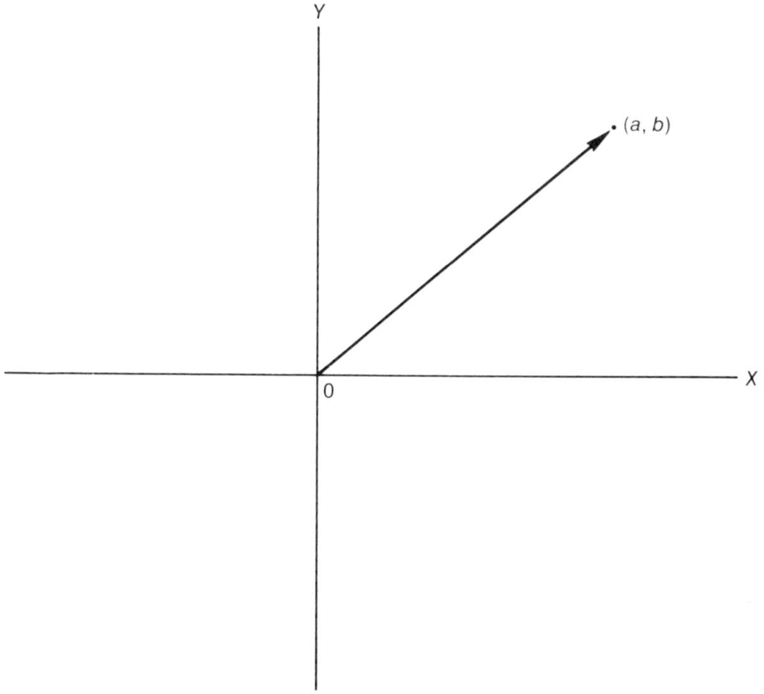

FIGURE 6-1: A Vector in the Plane.

shown the physicist that the resulting force, which is called the *sum* of F_1 and F_2, is the same as the force described by constructing a parallelogram with F_1 and F_2 as sides and taking the diagonal of this parallelogram (see Figure 6.2). Thus, if F_1 and F_2 correspond to the ordered pairs (a, b), (c, d), respectively, then a casual reference to the figure should convince the reader that $F_1 + F_2$ corresponds to $(a + c, b + d)$.

The mathematician interprets addition of vectors as follows: Let V_2 denote the set of all ordered pairs (a, b) $(a,b \in \mathbf{R})$. Define an operation $+$, called *addition*, on V_2 by setting

$$(a, b) + (c, d) = (a + c, b + d).$$

It is trivial to verify that with respect to this operation, V_2 is an abelian group. The identity element is $(0, 0)$, whereas the inverse of (a, b) is $(-a, -b)$.

In addition to the operation $+$ among vectors, there is another natural operation present. If F is a vector and α is a real number, then it is possible to "multiply" α and F as follows: aF is the vector whose magnitude is $|\alpha|\cdot|F|$, where $|F|$ denotes the magnitude of F. The direction of αF is the same as the direction of F if $\alpha \geq 0$, and is the opposite direction to F if $\alpha < 0$ (see Figure 6.3.) Thus, if F corresponds to (a, b), then αF corresponds to $(\alpha a, \alpha b)$. To

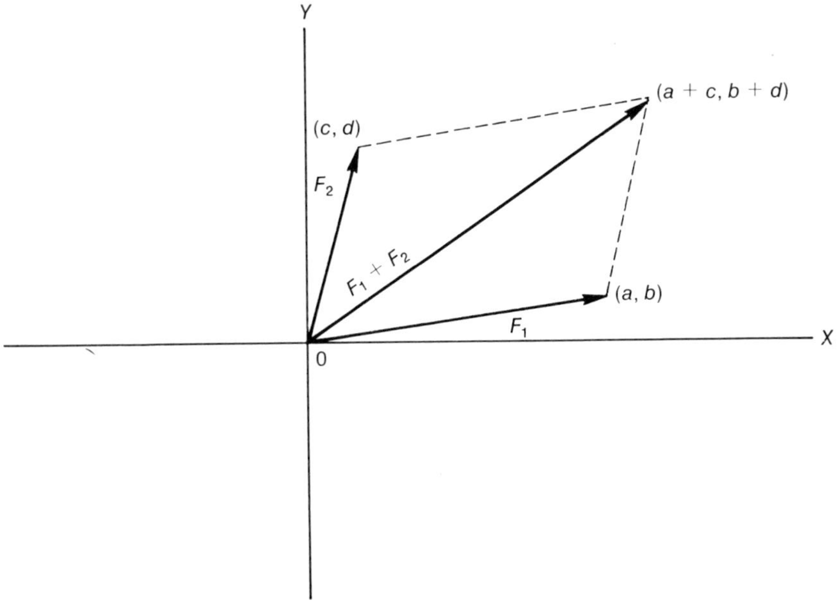

FIGURE 6-2: Addition of Vectors.

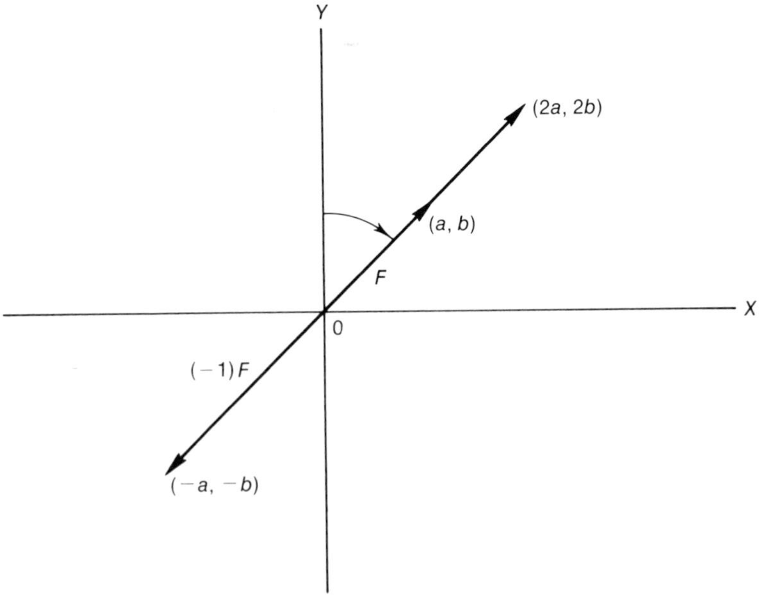

FIGURE 6-3: Scalar Multiplication.

distinguish the real numbers from the vectors, the real numbers are called *scalars*. Thus, we have defined a multiplication which allows taking the product of a scalar and a vector. This multiplication is called *scalar multiplication* and has the following properties: If $\alpha, \beta \in \mathbf{R}$, $v, w \in V_2$, then

$$(\alpha + \beta)\cdot v = \alpha\cdot v + \beta\cdot v, \tag{1}$$

$$\alpha\cdot(v + w) = \alpha\cdot v + \alpha\cdot w, \tag{2}$$

$$(\alpha\beta)\cdot v = \alpha\cdot(\beta v), \tag{3}$$

$$1\cdot v = v. \tag{4}$$

For example, let us prove (1): Let $v = (a, b)$. Then

$$\begin{aligned}(\alpha + \beta)\cdot v &= (\alpha + \beta)\cdot(a, b)\\ &= ((\alpha + \beta)a, (\alpha + \beta)b)\\ &= (\alpha a + \beta a, \alpha b + \beta b)\\ &= (\alpha a, \alpha b) + (\beta a, \beta b)\\ &= \alpha\cdot(a, b) + \beta\cdot(a, b)\\ &= \alpha\cdot v + \beta\cdot v.\end{aligned}$$

We leave the proofs of (2)–(4) as exercises.

We have just created a new algebraic system. The vectors in two dimensions form an abelian group with respect to addition, but they also admit multiplication by a set of elements known as scalars, such that the properties (1)–(4) are satisfied. Such an algebraic system is called a *vector space*. Let us now formally define this concept.

DEFINITION 1: Let F be a field. A *vector space over* F is a nonempty set V plus two functions $+: V \times V \to V$, $\cdot: F \times V \to V$, called *vector addition* and *scalar multiplication*, respectively, such that the following properties are satisfied:

V_1. With respect to vector addition, V is an abelian group.
V_2. For all $\alpha \in F$, $\mathbf{v}, \mathbf{w} \in V$, we have $\alpha\cdot(\mathbf{v} + \mathbf{w}) = \alpha\cdot\mathbf{v} + \alpha\cdot\mathbf{w}$.
V_3. For all $\alpha, \beta \in F$, $\mathbf{v} \in V$, we have $(\alpha + \beta)\cdot\mathbf{v} = \alpha\cdot\mathbf{v} + \beta\cdot\mathbf{v}$.
V_4. For all $\alpha, \beta \in F$, $\mathbf{v} \in V$, we have $\alpha\cdot(\beta\cdot\mathbf{v}) = (\alpha\cdot\beta)\cdot\mathbf{v}$.
V_5. For all $\mathbf{v} \in V$, we have $1\cdot\mathbf{v} = \mathbf{v}$.

The elements of V are called *vectors* and the elements of F are called *scalars*.

EXAMPLE 1: $F = \mathbf{R}$, $V = V_2$. Then V_2 is a vector space over $\mathbf{R}$.

EXAMPLE 2: Let F be any field. $V = \{(a_1, \ldots, a_n) \mid a_i \in F, 1 \leq i \leq n\}$. Define vector addition in V by

$$(a_1, \ldots, a_n) + (b_1, \ldots, b_n) = (a_1 + b_1, \ldots, a_n + b_n).$$

Define scalar multiplication by

$$\alpha \cdot (a_1, \ldots, a_n) = (\alpha a_1, \ldots, \alpha a_n) \qquad (\alpha \in F).$$

Then with respect to these operations, V becomes a vector space over F, denoted F^n.

EXAMPLE 3: Let F be any field, $V = F[X]$, where X is an indeterminate over F. Define vector addition to be the usual addition of polynomials and define scalar multiplication as follows: Let $\alpha \in F, f = a_0 + a_1X + \cdots + a_nX^n \in F[X]$. Set

$$\alpha \cdot f = \alpha a_0 + (\alpha a_1)X + \cdots + (\alpha a_n)X^n.$$

Then with respect to these operations, $F[X]$ becomes a vector space over F.

EXAMPLE 4: Let F be any field, $V = \{f \in F[X] \mid \deg(f) \leq n\}$, and let vector addition and scalar multiplication be defined as in Example 3. Then V is a vector space over F.

EXAMPLE 5: Let $\mathcal{C}[0, 1]$ denote the set of all real-valued functions which are continuous on the closed interval $[0, 1]$. Let $F = \mathbf{R}$, $V = \mathcal{C}[0, 1]$. Define vector addition by

$$(f + g)(x) = f(x) + g(x) \qquad (x \in [0, 1]),$$

and define scalar multiplication by

$$(\alpha f)(x) = \alpha f(x) \qquad (x \in [0, 1]).$$

Then $\mathcal{C}[0, 1]$ is a vector space over $\mathbf{R}$.

EXAMPLE 6: Let E be a field, F a subfield of E. Let $V = E$ and define vector addition to be the addition in E. Let scalar multiplication be the multiplication in E. Then E is a vector space over F.

Thus, we see from the above examples that a great variety of mathematical objects qualify as vector spaces. The notion of a vector space is one of the fundamental unifying notions in modern mathematics, both pure and applied, and finds applications in fields as diverse as psychology and quantum mechanics.

Some Remarks: 1. If V is a vector space with vector addition $+$, then V is an abelian group with respect to $+$. In order to minimize confusion, we will always use additive notation with respect to this group, The identity element will be denoted $\mathbf{0}$, whereas the inverse of $\mathbf{v}$ will be denoted $-\mathbf{v}$. Note that $\mathbf{0}$ is not the same as $0 \in F$!

2. Vectors will always be denoted by lowercase Roman boldface letters, for example, $\mathbf{v}, \mathbf{w}, \mathbf{x}, \mathbf{y}$. Scalars will be denoted by lowercase Greek letters, for example, $\alpha, \beta, \gamma, \delta$.

3. We will usually omit the $\cdot$ signifying scalar multiplication and will write $\alpha\mathbf{v}$ instead of $\alpha\cdot\mathbf{v}$.

4. The reason for the axiom $1\cdot\mathbf{v} = \mathbf{v}$ ($\mathbf{v} \in V$) in our definition of a vector space is to eliminate degenerate examples for which $\alpha\mathbf{v} = \mathbf{0}$ for all $\mathbf{v} \in V$, which satisfy the remaining axioms.

Three elementary, but useful, facts concerning vector spaces are

$$0\mathbf{v} = \mathbf{0} \qquad (v \in V), \tag{5}$$

$$(-1)\mathbf{v} = -\mathbf{v} \qquad (v \in V), \tag{6}$$

$$\alpha(-\mathbf{v}) = (-\alpha)\mathbf{v} = -(a\mathbf{v}) \qquad (\alpha \in F, v \in V). \tag{7}$$

We leave these as exercises for the reader.

6.1 *Exercises*

1. Determine the following vectors in $\mathbf{R}^3$:
   (a) $5\cdot(0, 2, 7) + 3\cdot(0, 1, -1)$
   (b) $(-6)(-\mathbf{v}) - 2\mathbf{w}$, where $\mathbf{v} = (-1, 1, 1)$, $\mathbf{w} = (1, 0, 0)$
   (c) $a_1(1, 0, 0) + a_2(0, 1, 0) + a_3(0, 0, 1)$ $(a_i \in \mathbf{R})$.
2. Let F be a field and let $\mathbf{e}_i = (0, \ldots, 0, 1, 0, \ldots, 0)$, where the 1 is in the ith component. Let $\mathbf{v} \in F^n$.
   (a) Show that $\mathbf{v}$ can be written in the form
   $$\mathbf{v} = a_1\mathbf{e}_1 + a_2\mathbf{e}_2 + \cdots + a_n\mathbf{e}_n.$$
   (b) Show that $a_1, \ldots, a_n$ in part (a) are uniquely determined by $\mathbf{v}$.
3. Verify that Examples 1–6 satisfy the axioms for a vector space.
4. Prove properties (5)–(7).
5. Let S be a set and let V denote the set of all functions $f: S \longrightarrow \mathbf{R}$. Define addition in V by setting $(f + g)(s) = f(s) + g(s)$ $(f,g \in V, s \in S)$. If $f \in V$, and $\alpha \in R$, define the product αf by $(\alpha f)(s) = \alpha f(s)$. Show that with respect to these operations, V forms a vector space over $\mathbf{R}$.
6. Let V denote the set of all real numbers of the form $a + b\sqrt{2}$, $a,b \in \mathbf{Q}$. Show that with respect to the usual operations of addition and multiplication of real numbers, V forms a vector space over $\mathbf{Q}$.

6.2 *Subspaces and Quotient Spaces*

Let V be a vector space over the field F. Let us pursue the analogy with the theories of groups and rings and define subspaces and quotient spaces of V.

DEFINITION 1: A *subspace* of V is a subset of V which is a vector space over F with respect to the operations of vector addition and scalar multiplication of V.

It is clear that W is a subspace of V if and only if W is a subgroup of the additive group of V and W is closed with respect to multiplication by scalars. Thus, by Proposition 2 of Section 3.3 we have

PROPOSITION 2: Let $W \subseteq V$. Then W is a subspace of $V \leftrightarrow$ *For all* $\mathbf{w}, \mathbf{w}' \in W$, $\alpha \in F$, *we have* $\mathbf{w} - \mathbf{w}' \in W$, $\alpha\mathbf{w} \in W$.

EXAMPLE 1: Let $V = F^2$, $W = \{(a, 0) \mid a \in F\}$. Then

$$(a, 0) - (a', 0) = (a - a', 0) \in W,$$
$$\alpha(a, 0) = (\alpha a, 0) \in W,$$

so that by Proposition 2, W is a subspace of V.

EXAMPLE 2: Let $V = \mathcal{C}[0, 1]$, $W = \{f \in V \mid f(0) = 0\}$. Then W is a subspace of V.

EXAMPLE 3: Let $V = F^3$, $W = \{(a, b, 3a) \mid a, b \in F\}$. Then W is a subspace of V.

Let S be an arbitrary subset of V, and let $[S]$ denote the set of all sums of the form

$$\alpha_1\mathbf{s}_1 + \cdots + \alpha_m\mathbf{s}_m, \tag{1}$$

where $\alpha_1, \ldots, \alpha_m \in F$ and $\mathbf{s}_1, \ldots, \mathbf{s}_m \in S$. From Proposition 2, it is immediate that $[S]$ is a subspace of V. Moreover, $[S]$ contains S. If W is a subspace of V which contains S, then W contains all sums of the form (1) and thus $W \supseteq [S]$. Therefore, $[S]$ is the smallest subspace of V which contains S. We call $[S]$ the subspace *generated* by S. If W is a subspace of V and S is a subset of V such that $W = [S]$, then we say that S is a set of generators for W. For example, in Example 1, $W = [(1, 0)]$; in example 3, $W = [S]$, where $S = \{(1, 0, 3), (0, 1, 0)\}$. Note that $F^n = [S]$, where $S = \{\mathbf{e}_1, \ldots, \mathbf{e}_n\}$ and

$$\begin{aligned}
\mathbf{e}_1 &= (1, 0, 0, \ldots, 0)\\
\mathbf{e}_2 &= (0, 1, 0, \ldots, 0)\\
&\vdots\\
\mathbf{e}_n &= (0, 0, 0, \ldots, 1).
\end{aligned}$$

Indeed, if $\mathbf{v} = (a_1, \ldots, a_n)$, then

$$\mathbf{v} = a_1\mathbf{e}_1 + a_2\mathbf{e}_2 + \cdots + a_n\mathbf{e}_n,$$

so that $F^n \subseteq [S]$. And it is clear that $[S] \subseteq F^n$. Thus, $F^n = [S]$.

Let V be a vector space over F and let W be a subspace of V. Then W is a subgroup of the additive group of V. Since the additive group of V is abelian, W is a normal subgroup of V and thus the quotient group V/W is defined. The quotient group V/W consists of cosets of the form $\mathbf{v} + W$ ($\mathbf{v} \in V$) and the group operation is defined by

$$(\mathbf{v} + W) + (\mathbf{v}' + W) = (\mathbf{v} + \mathbf{v}') + W.$$

If $\alpha \in F$, let us define

$$\alpha(\mathbf{v} + W) = \alpha\mathbf{v} + W. \tag{2}$$

Let us first show that the above definition is consistent. Indeed, if $\mathbf{v} + W = \mathbf{v}' + W$, then $\mathbf{v} - \mathbf{v}' \in W$. But since W is a subspace, $\alpha(\mathbf{v} - \mathbf{v}') \in W$, and thus, since $\alpha(\mathbf{v} - \mathbf{v}') = \alpha\mathbf{v} + \alpha(-\mathbf{v}') = \alpha\mathbf{v} - \alpha\mathbf{v}'$ [by Equation (7), Section 6.1], we have that

$$\begin{aligned} \alpha\mathbf{v} - \alpha\mathbf{v}' \in W &\Longrightarrow \alpha\mathbf{v} + W = \alpha\mathbf{v}' + W, \\ &\Longrightarrow \alpha(\mathbf{v} + W) = \alpha(\mathbf{v}' + W). \end{aligned}$$

Thus, the operation (2) is consistent. It is now elementary to check that with respect to the group operation in V/W and the scalar multiplication defined by (2), V/W is a vector space over F. The vector space V/W is called the *quotient space of V modulo W*.

EXAMPLE 4: Let $V = F^2$, $W = \{(a, 0) \mid a \in F\}$. Then, since

$$(a, b) = (a, 0) + (0, b),$$

we see that every element of V/W can be written in the form

$$(0, b) + W.$$

Moreover, no two of these elements are equal, since

$$\begin{aligned} (0, b) + W = (0, b') + W &\Longrightarrow (0, b - b') \in W \\ &\Longrightarrow (0, b - b') = (a, 0) \qquad \text{for some } a \in F \\ &\Longrightarrow a = b - b' = 0 \\ &\Longrightarrow b = b'. \end{aligned}$$

Thus, $V/W = \{(0, b) + W \mid b \in F\}$.

EXAMPLE 5: Let $V = F^3$, $W = \{(a, b, 0) \mid a, b \in F\}$. Then

$$V/W = \{(0, 0, c) + W \mid c \in F\}$$

6.2 Exercises

1. Let $a_1, \ldots, a_n \in F$, where F is a field. Show that
$$W = \{\mathbf{v} = (x_1, \ldots, x_n) \in F^n \mid a_1x_1 + \cdots + a_nx_n = 0\}$$
is a subspace of F^n.

2. Let F be a field and let $F[X]$ be the ring of polynomials over F in the indeterminate X. Regard $F[X]$ as a vector space over F. Describe $[S]$, where S is the following:
   (a) $S = \{0\}$.
   (b) $S = \{1, X\}$.
   (c) $S = \{1, X, \ldots, X^n\}$.
   (d) $S = \{f(X)\}$, where $f(X) \in F[X]$ is a fixed polynomial.
   (e) $S = \varnothing$.
3. Let F be a field and let $\mathbf{e}_1, \ldots, \mathbf{e}_n$ be as given in the text. Show that no proper subset S of $\{\mathbf{e}_1, \ldots, \mathbf{e}_n\}$ generates F^n.
4. Let $m \leq n$, and let $\{\mathbf{e}_1, \ldots, \mathbf{e}_n\}$ be as given in the text. Describe $F^n/[S]$, where $S = \{\mathbf{e}_1, \ldots, \mathbf{e}_m\}$.
5. Let $V = F^n$ $(n \geq 2)$ and let $W = [(a, b, a, \ldots, a)]$, where $a, b \in F$ are given. Describe V/W.
6. Give an example of a vector space V and a subgroup W of the additive group of V such that W is not a subspace of V.

6.3 Basis and Dimension

In the preceding section, we observed that $S = \{\mathbf{e}_1, \mathbf{e}_2, \ldots, \mathbf{e}_n\}$, where

$$\begin{aligned} \mathbf{e}_1 &= (1, 0, 0, \ldots, 0) \\ \mathbf{e}_2 &= (0, 1, 0, \ldots, 0) \\ &\vdots \\ \mathbf{e}_n &= (0, 0, 0, \ldots, 1) \end{aligned}$$

is a set of generators of the vector space F^n. Thus, every element $\mathbf{v}$ of F^n can be written in the form

$$\mathbf{v} = \sum_{i=1}^{n} \alpha_i \mathbf{e}_i, \qquad \alpha_i \in F. \tag{1}$$

Actually, the set S of generators has a very special property: The elements α_i in (1) are uniquely determined. Indeed,

$$\sum_{i=1}^{n} \alpha_i \mathbf{e}_i = (\alpha_1, \alpha_2, \ldots, \alpha_n),$$

so that if (1) holds, and if $\mathbf{v} = (v_1, \ldots, v_n)$, then $\alpha_i = v_i$ $(1 \leq i \leq n)$.

DEFINITION 1: Let V be a vector space over F. *A basis of* V is a subset $\{\mathbf{e}_i\}$ of V (finite or infinite) with the property that every element $\mathbf{v}$ of V can be uniquely written in the form

$$\mathbf{v} = \sum_i \alpha_i \mathbf{e}_i.$$

(Here, we always assume that if $\{\mathbf{e}_i\}$ is infinite, then $\alpha_i = 0$ for all but a finite number of i, so that the sum is a finite sum.)

EXAMPLE 1: $\{\mathbf{e}_1, \ldots, \mathbf{e}_n\}$ is a basis of F^n.

EXAMPLE 2: $\{\mathbf{e}_1 - \mathbf{e}_2, \mathbf{e}_1 + \mathbf{e}_2\}$ is a basis of F^2 since if $\mathbf{v} = (v_1, v_2)$, then

$$\mathbf{v} = \frac{v_1 - v_2}{2}(\mathbf{e}_1 - \mathbf{e}_2) + \frac{v_1 + v_2}{2}(\mathbf{e}_1 + \mathbf{e}_2),$$

and if $\mathbf{v} = \alpha(\mathbf{e}_1 - \mathbf{e}_2) + \beta(\mathbf{e}_1 + \mathbf{e}_2)$, then $\mathbf{v} = (\alpha + \beta, \beta - \alpha)$, so that $\alpha = (v_1 - v_2)/2$, $\beta = (v_1 + v_2)/2$. Thus, every element of F^2 can be represented in one and only one way in the form $\alpha(\mathbf{e}_1 - \mathbf{e}_2) + \beta(\mathbf{e}_1 + \mathbf{e}_2)$, so that $\{\mathbf{e}_1 - \mathbf{e}_2, \mathbf{e}_1 + \mathbf{e}_2\}$ is a basis of F^2.

The main goal of this section is to show that any two bases of a vector space V have the same number of elements, in the following sense: If one basis of V has a finite number of elements, then every basis has a finite number of elements, and every basis has the same number. If one basis V is infinite, then every basis is infinite. Assuming this result for the moment, let us make the following definition.

DEFINITION 2: Let V be a vector space. If V has a finite set of generators, then we say that V is *finite-dimensional* and we define its *dimension*, denoted $\dim_F V$, to be the minimum possible number of elements in such set. If V does not have a finite set of generators, then we say that V is *infinite-dimensional*, and we define $\dim_F V$ to be ∞.

For example, let $V = \{\mathbf{0}\}$. Since $S = \varnothing$ is a set of generators for V, and since S is clearly as small as possible a set of generators, we see that $\dim_F V = 0$.

Assume now that $V \neq \{\mathbf{0}\}$ and that V is finite-dimensional. Let $S = \{\mathbf{v}_1, \ldots, \mathbf{v}_n\}$ be a set of generators having as few elements as possible. (Since $V \neq \{\mathbf{0}\}$, S is nonempty.) Then, by definition of $\dim_F V$, we have $\dim_F V = n$. We will now show that $\{\mathbf{v}_1, \ldots, \mathbf{v}_n\}$ *is a basis of* V. Indeed, since S generates V, every element $\mathbf{v}$ of V can be written in the form

$$\mathbf{v} = \alpha_1\mathbf{v}_1 + \cdots + \alpha_n\mathbf{v}_n \qquad (\alpha_i \in F).$$

We assert that this representation is unique. For if $\mathbf{v} = \beta_1\mathbf{v}_1 + \cdots + \beta_n\mathbf{v}_n$ with $\alpha_1 \neq \beta_1$, say, then $\alpha_1 - \beta_1 \neq 0$ and

$$\mathbf{v}_1 = \gamma_2\mathbf{v}_2 + \cdots + \gamma_n\mathbf{v}_n,$$

where $\gamma_i = (\alpha_1 - \beta_1)^{-1}(\beta_i - \alpha_i)$ $(2 \leq i \leq n)$. But if $\mathbf{w} \in V$, we may write

$$\begin{aligned}\mathbf{w} &= \delta_1\mathbf{v}_1 + \cdots + \delta_n\mathbf{v}_n, \qquad \delta_i \in F\\ &= (\delta_2 + \delta_1\gamma_2)\mathbf{v}_2 + \cdots + (\delta_n + \delta_1\gamma_n)\mathbf{v}_n,\end{aligned}$$

so that $\mathbf{w}$ can be written as a linear combination of $\mathbf{v}_2, \ldots, \mathbf{v}_n$. Then, since $\mathbf{w} \in V$ is arbitrary, $\{\mathbf{v}_2, \ldots, \mathbf{v}_n\}$ is a set of generators for V, which contradicts the hypothesis that S is as small as possible. Thus, the representation is unique, and S is a basis for V. We have shown that the dimension of V can be computed as the number of elements in a set of generators that is as small as possible and such a set of generators is a basis for V. We will show next, that any two bases of V have the same number of elements. Thus, we will have proved that $\dim_F V =$ *the number of elements in a basis for* V. By the way, we should remark that it is a consequence of our above reasoning that every *nonzero finite—dimensional vector space has a basis.* (Just take as basis any set of generators that is as small as possible.) For example, $\dim_F F^n = n$, since $\{\mathbf{e}_1, \ldots, \mathbf{e}_n\}$ is a basis for F^n.

In order to establish the fact that any two bases have the same number of elements, we must first explore some preliminary ideas.

DEFINITION 3: Let $S \subseteq V$ be a nonempty subset of V. Then S is said to be *linearly dependent* if there exist distinct elements $\mathbf{s}_1, \ldots, \mathbf{s}_n$ belonging to S and $\alpha_1, \ldots, \alpha_n$ belonging to F, not all 0, such that

$$\alpha_1\mathbf{s}_1 + \alpha_2\mathbf{s}_2 + \cdots + \alpha_n\mathbf{s}_n = \mathbf{0}.$$

If S is not linearly dependent, then we say that S is *linearly independent.*

EXAMPLE 3: Let $V = \mathbf{R}^2$, $S = \{\mathbf{e}_1, 2\mathbf{e}_1\}$. Then S is linearly dependent, since

$$(-2)\cdot\mathbf{e}_1 + 1\cdot(2\mathbf{e}_1) = \mathbf{0}.$$

EXAMPLE 4: Let V be arbitrary and let $S = \{\mathbf{0}\}$. Then S is linearly dependent, since $1\cdot\mathbf{0} = \mathbf{0}$.

EXAMPLE 5: Let V be arbitrary and let S be a basis of V. Then S is linearly independent, since if

$$\alpha_1\mathbf{s}_1 + \alpha_2\mathbf{s}_2 + \cdots + \alpha_n\mathbf{s}_n = \mathbf{0}, \qquad \alpha_i \in F, \qquad \mathbf{s}_i \in S,$$

then, since

$$\mathbf{0} = 0\cdot\mathbf{s}_1 + 0\cdot\mathbf{s}_2 + \cdots + 0\cdot\mathbf{s}_n,$$

the fact that $\mathbf{0}$ is uniquely representable in terms of the basis implies that $\alpha_1 = \alpha_2 = \cdots = \alpha_n = 0$.

PROPOSITION 4: Let V be a vector space and let $S \subseteq V$ be a nonempty subset. Then S is a basis for V if and only if S is a set of generators for V and S is linearly independent.

Proof: $\Longrightarrow$ If S is a basis, then S is clearly a set of generators for V. Moreover, S is linearly independent by Example 5.

$\Leftarrow$ Let $\mathbf{v} \in V$. Since S is a set of generators, there exist $\alpha_s(s \in S)$ belonging to F such that

$$\mathbf{v} = \sum_{s \in S} \alpha_s \mathbf{s}, \tag{2}$$

where all but a finite number of the α_s are 0. Let us show that the representation (2) is unique. Indeed, if

$$\mathbf{v} = \sum_{s \in S} \beta_s \mathbf{s}, \qquad \beta_s \in F,$$

then

$$\mathbf{0} = \mathbf{v} - \mathbf{v} = \sum_{s \in S} (\alpha_s - \beta_s)\mathbf{s}.$$

But since S is linearly independent, we see that $\alpha_s - \beta_s = 0$ for all s. Thus, $\alpha_s = \beta_s$ for all $s \in S$ and the representation (2) is unique. Thus, S is a basis of V.

THEOREM 5: Let V be a vector space. Then any two bases of V contain the same number of elements.

Proof: Let us consider the special case where V has a finite basis $\{\mathbf{e}_1, \ldots, \mathbf{e}_n\}$. We will prove that if $\{\mathbf{f}_i\}_{i \in I}$ is any other basis of V, then it is finite and contains n elements. Consideration of the special case suffices, for then if V has one infinite basis, then the special case implies that all bases must be infinite. Since $\{\mathbf{f}_i\}_{i \in I}$ is linearly independent, all the $\mathbf{f}_i$ are nonzero. Let $\mathbf{f}_1$ be one of the $\mathbf{f}_i$. Since $\{\mathbf{e}_1, \ldots, \mathbf{e}_n\}$ is a basis of V, there exist $\alpha_1, \ldots, \alpha_n \in F$ such that

$$\mathbf{f}_1 = \alpha_1\mathbf{e}_1 + \cdots + \alpha_n\mathbf{e}_n.$$

Since $\mathbf{f}_1 \neq \mathbf{0}$, not all α_j are zero. Let us renumber the $\mathbf{e}_j$ so that $\alpha_1 \neq 0$. Then, since F is a field,

$$\mathbf{e}_1 = \alpha_1^{-1}\mathbf{f}_1 - \alpha_1^{-1}\alpha_2\mathbf{e}_2 - \cdots - \alpha_1^{-1}\alpha_n\mathbf{e}_n. \tag{3}$$

Let us show that $\{\mathbf{f}_1, \mathbf{e}_2, \ldots, \mathbf{e}_n\}$ is a basis of V. Let $\mathbf{v} \in V$. Then there exist $\beta_1, \ldots, \beta_n \in F$ such that

$$\mathbf{v} = \beta_1\mathbf{e}_1 + \cdots + \beta_n\mathbf{e}_n.$$

Therefore, by (3),

$$\mathbf{v} = \beta_1'\mathbf{f}_1 + \beta_2'\mathbf{e}_2 + \cdots + \beta_n'\mathbf{e}_n.$$

Thus, $\{\mathbf{f}_1, \mathbf{e}_2, \ldots, \mathbf{e}_n\}$ is a set of generators for V. If $\{\mathbf{f}_1, \mathbf{e}_2, \ldots, \mathbf{e}_n\}$ is linearly dependent, then there exist $\gamma_1, \ldots, \gamma_n \in F$ such that

$$\gamma_1\mathbf{f}_1 + \gamma_2\mathbf{e}_2 + \cdots + \gamma_n\mathbf{e}_n = \mathbf{0}, \tag{4}$$

where $\gamma_1, \ldots, \gamma_n$ are not all 0. If $\gamma_1 = 0$, then one of $\gamma_2, \ldots, \gamma_n$ is nonzero and $\gamma_2\mathbf{e}_2 + \cdots + \gamma_n\mathbf{e}_n = \mathbf{0}$, which contradicts the fact that $\mathbf{e}_2, \ldots, \mathbf{e}_n$ are linearly independent. Therefore, $\gamma_1 \neq 0$. Without loss of generality, we may multiply (4) by γ_1^{-1} and assume that $\gamma_1 = 1$. Then

$$\alpha_1\mathbf{e}_1 + (\alpha_2 + \gamma_2)\mathbf{e}_2 + \cdots + (\alpha_n + \gamma_n)\mathbf{e}_n = \mathbf{0}, \qquad \alpha_1 \neq 0,$$

which contradicts the fact that $\{\mathbf{e}_1, \ldots, \mathbf{e}_n\}$ are linearly independent. The contradiction proves that $\{\mathbf{f}_1, \mathbf{e}_2, \ldots, \mathbf{e}_n\}$ is linearly independent, so that by Proposition 4, $\{\mathbf{f}_1, \mathbf{e}_2, \ldots, \mathbf{e}_n\}$ is a basis of V. If $n = 1$, then $\{\mathbf{f}_1\}$ is a basis of V. Therefore, $\{\mathbf{f}_i\}_{i \in I}$ consists of a single element and the theorem is correct. Thus, assume that $n > 1$. If $\{\mathbf{f}_i\}_{i \in I}$ consists only of the single element $\mathbf{f}_1$, then $\{\mathbf{f}_1\}$ is a basis of V, which contradicts the fact that $\{\mathbf{f}_1, \mathbf{e}_2, \ldots, \mathbf{e}_n\}$ is a basis of V and $n > 1$. Thus, $\{\mathbf{f}_i\}_{i \in I}$ consists of at least two elements. Let $\mathbf{f}_2 \in \{\mathbf{f}_i\}_{i \in I}$, $\mathbf{f}_2 \neq \mathbf{f}_1$. Since $\{\mathbf{f}_1, \mathbf{e}_2, \ldots, \mathbf{e}_n\}$ is a basis of V, there exist $\delta_1, \ldots, \delta_n \in F$ such that

$$\mathbf{f}_2 = \delta_1\mathbf{f}_1 + \delta_2\mathbf{e}_2 + \cdots + \delta_n\mathbf{e}_n,$$

$\delta_1, \ldots, \delta_n$ not all 0. If $\delta_2 = \cdots = \delta_n = 0$, then $\mathbf{f}_2 - \delta_1\mathbf{f}_1 = 0$, which contradicts the fact that $\mathbf{f}_1$ and $\mathbf{f}_2$ are linearly independent. Thus, one of $\delta_2, \ldots, \delta_n$ is nonzero. Without loss of generality, assume that $\delta_2 \neq 0$. Repeating the above argument, we see that $\{\mathbf{f}_1, \mathbf{f}_2, \mathbf{e}_3, \ldots, \mathbf{e}_n\}$ is a basis of V. Continuing in this way, we find $\mathbf{f}_1, \ldots, \mathbf{f}_n$ in $\{\mathbf{f}_i\}_{i \in I}$ such that $\{\mathbf{f}_1, \ldots, \mathbf{f}_n\}$ is a basis of V. But since $\{\mathbf{f}_i\}_{i \in I}$ is also a basis for V, we see that $\{\mathbf{f}_i\}_{i \in I}$ consists of precisely n elements. For if $\mathbf{f}_{n+1} \in \{\mathbf{f}_i\}_{i \in I}$ is distinct from $\mathbf{f}_1, \ldots, \mathbf{f}_n$, then $\mathbf{f}_{n+1} = \eta_1\mathbf{f}_1 + \eta_2\mathbf{f}_2 + \cdots + \eta_n\mathbf{f}_n$ $(\eta_i \in F)$ since $\{\mathbf{f}_1, \ldots, \mathbf{f}_n\}$ is a basis of V. But this contradicts the fact that $\{\mathbf{f}_1, \ldots, \mathbf{f}_{n+1}\}$ are linearly independent. This completes the proof of the theorem.

Remarks: 1. It is possible to strengthen Theorem 5 in case V has infinite dimension. Namely, it is possible to show that not only are any two bases of V infinite, but, given any two bases of V, there exists a bijection of one onto the other.

2. It is not clear from what we have said that any vector space has a basis. In order to prove this assertion, it is necessary to appeal to Zorn's lemma. If V is finite-dimensional, then the existence of a basis of V was proved. In most of what follows, we will be concerned only with finite-dimensional vector spaces, so we will not give a proof of the existence of a basis in the most general case.

6.3 *Exercises*

1. Let F be a field.
   (a) Find a basis of $F[X]$ over F.
   (b) What is $\dim_F F[X]$?
2. Let $\mathbf{C}$ (respectively, $\mathbf{R}$) denote the field of complex (respectively, real) numbers.
   (a) Show that $\mathbf{C}$ is a vector space over $\mathbf{R}$ if vector addition is taken to be addition in $\mathbf{C}$ and scalar multiplication is taken to be multiplication in $\mathbf{C}$.
   (b) Show that $\{1, i\}$ is a basis of $\mathbf{C}$ over $\mathbf{R}$.

3. Let $F_n[X]$ denote the F-vector space of all polynomials of degree $\leq n$.
   (a) Find two different bases for $F_n[X]$.
   (b) Determine $\dim_F F_n[X]$.
4. Let E and F be fields $E \supseteq F$. Define vector addition in E to be the field addition in E and define scalar multiplication $a \cdot \alpha$ ($a \in F$, $\alpha \in E$) to be the field product of a and α.
   (a) Show that with respect to these operations, E is a vector space over F.
   **(b) Show that $\dim_{\mathbf{Q}}(\mathbf{R})$ is infinite.

6.4 *Linear Transformations*

In this section we will again pursue the analogy between vector spaces, on the one hand, and groups and rings, on the other, by defining an analogue of the homomorphisms of group theory and ring theory. In the case of vector spaces, the homomorphisms are called *linear transformations* (or *linear operators*).

DEFINITION 1: Let V and W be vector spaces over the same field F. A *linear transformation* from V to W is a function $T: V \longrightarrow W$ such that for all $\mathbf{v},\mathbf{v}' \in V$ and all $\alpha \in F$, we have

$$T(\mathbf{v} + \mathbf{v}') = T(\mathbf{v}) + T(\mathbf{v}'), \tag{1}$$

$$T(\alpha\mathbf{v}) = \alpha T(\mathbf{v}). \tag{2}$$

Note that a linear transformation $T: V \longrightarrow W$ is a group homomorphism from the additive group of V to the additive group of W. Thus, in particular, if $\mathbf{v},\mathbf{v}' \in V$, then

$$T(\mathbf{v} - \mathbf{v}') = T(\mathbf{v}) - T(\mathbf{v}'). \tag{3}$$

Let $\mathbf{v}_1, \ldots, \mathbf{v}_n \in V, \alpha_1, \ldots, \alpha_n \in F$. By repeated application of (1) and (2), we see that if $T: V \longrightarrow W$ is a linear transformation, then

$$T(\alpha_1\mathbf{v}_1 + \cdots + \alpha_n\mathbf{v}_n) = \alpha_1 T(\mathbf{v}_1) + \cdots + \alpha_n T(\mathbf{v}_n). \tag{4}$$

EXAMPLE 1: Let $\alpha_1, \ldots, \alpha_n \in F$. Then the mapping $T: F^n \longrightarrow F^1$ defined by $T((a_1, \ldots, a_n)) = \alpha_1 a_1 + \cdots + \alpha_n a_n$ is a linear transformation.

EXAMPLE 2: Let V and W be finite-dimensional vector spaces over the field F and let $\{\mathbf{e}_1, \ldots, \mathbf{e}_n\}, \{\mathbf{f}_1, \ldots, \mathbf{f}_m\}$ be bases for V and W, respectively. Let $T: V \longrightarrow W$ be a linear transformation. We may describe the effect of T as follows: Since $T(\mathbf{e}_i) \in W$, we have

$$T(\mathbf{e}_i) = a_{1i}\mathbf{f}_1 + \cdots + a_{mi}\mathbf{f}_m \qquad (1 \leq i \leq n),$$

where $a_{ij} \in F$. The set $\{a_{ij}\}$, which consists of mn elements of F, completely determines T. For if $\mathbf{v} \in V$, then

$$\mathbf{v} = \alpha_1\mathbf{e}_1 + \cdots + \alpha_n\mathbf{e}_n, \qquad \alpha_i \in F.$$

Therefore, by (4),

$$\begin{aligned} T(\mathbf{v}) &= \alpha_1 T(\mathbf{e}_1) + \cdots + \alpha_n T(\mathbf{e}_n) \\ &= \alpha_1(a_{11}\mathbf{f}_1 + \cdots + a_{m1}\mathbf{f}_m) + \cdots + \alpha_n(a_{1n}\mathbf{f}_1 + \cdots + a_{mn}\mathbf{f}_m) \qquad (5) \\ &= (a_{11}\alpha_1 + \cdots + a_{1n}\alpha_n)\mathbf{f}_1 + \cdots + (a_{m1}\alpha_1 + \cdots + a_{mn}\alpha_n)\mathbf{f}_n. \end{aligned}$$

Thus, the image of any vector $\mathbf{v}$ with respect to T is completely determined by the set $\{a_{ij}\}$. Conversely, given any set $\{a_{ij}\}$ of mn elements of F, formula (5) defines a linear transformation of V into W.

EXAMPLE 3: Let $V = W = F[X]$, considered as a vector space over F. If

$$f = a_0 + a_1 X + \cdots + a_n X^n \in F[X],$$

let us define the *formal derivative* of f, denoted Df, by

$$Df = 1 \cdot a_1 + \cdots + n \cdot a_n X^{n-1}.$$

Then $D(f + g) = Df + Dg$ and $D(\alpha f) = \alpha Df$ for all $f, g \in F[X], \alpha \in F$. Thus, $D: F[X] \to F[X]$ is a linear transformation.

EXAMPLE 4: Let $\mathcal{C}[0, 1]$ = the vector space over $\mathbf{R}$ of all continuous functions $f: [0, 1] \to \mathbf{R}$. If $f \in \mathcal{C}[0, 1]$, set

$$T(f) = \int_0^1 f(x)\,dx.$$

Then, by the properties of the integral proved in calculus, we have

$$\begin{aligned} T(f + g) &= \int_0^1 (f(x) + g(x))\,dx \\ &= \int_0^1 f(x)\,dx + \int_0^1 g(x)\,dx \\ &= T(f) + T(g), \\ T(\alpha f) &= \int_0^1 \alpha f(x)\,dx \\ &= \alpha \int_0^1 f(x)\,dx \\ &= \alpha T(f) \end{aligned}$$

for all $f, g \in \mathcal{C}[0, 1], \alpha \in \mathbf{R}$. Thus, $T: \mathcal{C}[0, 1] \to \mathbf{R}^1$ is a linear transformation.

As in group theory, and ring theory let us define the *kernel* of the linear transformation $T: V \to W$, denoted $\ker(T)$, by

$$\ker(T) = \{\mathbf{v} \in V \mid T(\mathbf{v}) = \mathbf{0}\}. \qquad (6)$$

PROPOSITION 2: Let $T: V \to W$ be a linear transformation.

(1) $T(V)$ is a subspace of W.

(2) $\ker(T)$ is a subspace of V.

Proof: (1) Let $\mathbf{w},\mathbf{w}' \in T(V)$, $\alpha \in F$. Then there exist $\mathbf{v},\mathbf{v}' \in V$ such that $\mathbf{w} = T(\mathbf{v})$, $\mathbf{w}' = T(\mathbf{v}')$. Then, by (3),

$$\mathbf{w} - \mathbf{w}' = T(\mathbf{v}) - T(\mathbf{v}') = T(\mathbf{v} - \mathbf{v}') \in T(V),$$

since $\mathbf{v} - \mathbf{v}' \in V$. Similarly, by (2),

$$\alpha\mathbf{w} = \alpha T(\mathbf{v}) = T(\alpha\mathbf{v}) \in T(V),$$

since $\alpha\mathbf{v} \in V$. Thus, by Proposition 6.2.2, $T(V)$ is a subspace of W. (2) Let $\mathbf{v},\mathbf{v}' \in \ker(T)$, $\alpha \in F$. Then by (3) and the fact that $\mathbf{v},\mathbf{v}' \in \ker(T)$,

$$\begin{aligned} T(\mathbf{v} - \mathbf{v}') &= T(\mathbf{v}) - T(\mathbf{v}') \\ &= 0 - 0 \\ &= 0 \\ &\Longrightarrow \mathbf{v} - \mathbf{v}' \in \ker(T). \end{aligned}$$

And, by (2),

$$\begin{aligned} T(\alpha\mathbf{v}) &= \alpha T(\mathbf{v}) = \alpha\mathbf{0} = \mathbf{0} \\ &\Longrightarrow \alpha\mathbf{v} \in \ker(T). \end{aligned}$$

Thus, $\ker(T)$ is a subspace of V.

Let $T: V \longrightarrow W$ be a linear transformation. Since T is a homomorphism of additive groups, we have, by Corollary 5 of Section 3.6, the following result.

PROPOSITION 3: T is an isomorphism $\Longleftrightarrow \ker(T) = \{\mathbf{0}\}$.

In Chapter 3 we remarked that one of the fundamental goals of the theory of finite groups is to make a list of all nonisomorphic finite groups, this goal being beyond the present state of knowledge. In the case of vector spaces, however, the situation is far different. The following theorem gives a complete classification of all finite-dimensional vector spaces over a field F.

THEOREM 4: Let V be a finite-dimensional vector space over a field F and let $n = \dim_F V$. Then $V \approx F^n$.

Proof: Let $\{\mathbf{e}_1, \ldots, \mathbf{e}_n\}$ be a basis of V. Then every vector $\mathbf{v} \in V$ can be expressed uniquely in the form

$$\mathbf{v} = \alpha_1\mathbf{e}_1 + \cdots + \alpha_n\mathbf{e}_n.$$

Define the function $T: V \longrightarrow F^n$ by

$$T(\alpha_1\mathbf{e}_1 + \cdots + \alpha_n\mathbf{e}_n) = (\alpha_1, \ldots, \alpha_n).$$

It is easy to check that T is a linear transformation and T is clearly surjective.

If $\mathbf{v} \in \ker(T)$, then

$$\begin{aligned} T(\mathbf{v}) = \mathbf{0} &\Longrightarrow (\alpha_1, \ldots, \alpha_n) = (0, \ldots, 0) \\ &\Longrightarrow \alpha_1 = \alpha_2 = \cdots = \alpha_n = 0 \\ &\Longrightarrow \mathbf{v} = \mathbf{0}. \end{aligned}$$

Thus, by Proposition 3, T is an isomorphism. Therefore, $V \approx F^n$.

THEOREM 5: Let V and W be vector spaces over F, and let $T: V \longrightarrow W$ be a linear transformation.

(1) $\dim_F T(V) \leq \dim_F(V)$

(2) If T is an isomorphism, $\dim_F(T(V)) = \dim_F(V)$.

Proof: (1) Without loss of generality, let us assume that V is finite-dimensional, and let $\{\mathbf{e}_1, \ldots, \mathbf{e}_n\}$ be a basis of V. Then

$$V = \{\alpha_1\mathbf{e}_1 + \cdots + \alpha_n\mathbf{e}_n \mid \alpha_i \in F\},$$

so that

$$T(V) = \{\alpha_1 T(\mathbf{e}_1) + \cdots + \alpha_n T(\mathbf{e}_n) \mid \alpha_i \in F\},$$

and thus $\{T(\mathbf{e}_1), \ldots, T(\mathbf{e}_n)\}$ is a set of generators for $T(V)$. In particular, $\dim_F(T(V)) \leq n$.

(2) Let the notation be as in the proof of (1). It suffices to show that $\{T(\mathbf{e}_1), \ldots, T(\mathbf{e}_n)\}$ is a basis for $T(V)$. Since we have already shown that this set generates $T(V)$, it suffices to show that $\{T(\mathbf{e}_1), \ldots, T(\mathbf{e}_n)\}$ is linearly independent. If $\alpha_1, \ldots, \alpha_n \in F$ are such that

$$\mathbf{0} = \alpha_1 T(\mathbf{e}_1) + \cdots + \alpha_n T(\mathbf{e}_n),$$

then, by (4), we have

$$\begin{aligned} &\mathbf{0} = T(\alpha_1\mathbf{e}_1 + \cdots + \alpha_n\mathbf{e}_n) \\ &\Longrightarrow \mathbf{0} = \alpha_1\mathbf{e}_1 + \cdots + \alpha_n\mathbf{e}_n && \text{(since } T \text{ is an isomorphism)} \\ &\Longrightarrow \alpha_1 = \cdots = \alpha_n = 0 && \text{(since } \{\mathbf{e}_1, \ldots, \mathbf{e}_n\} \text{ is linearly independent)} \\ &\Longrightarrow \{T(\mathbf{e}_1), \ldots, T(\mathbf{e}_n)\} && \text{is linearly independent.} \end{aligned}$$

Theorem 5 has an interesting application to the theory of simultaneous linear equations. Let us consider the following system of equations:

$$\begin{aligned} a_{11}x_1 + a_{12}x_2 + \cdots + a_{1m}x_m &= b_1 \\ a_{21}x_1 + a_{22}x_2 + \cdots + a_{2m}x_m &= b_2 \\ &\;\;\vdots \\ a_{n1}x_1 + a_{n2}x_2 + \cdots + a_{nm}x_m &= b_n, \end{aligned} \tag{7}$$

where a_{ij}, b_i all belong to a given field F. A *solution* of the system (7) is an

m-tuple $(x_1, \ldots, x_m) \in F^m$ for which Equations (7) hold. A given system may or may not have a solution. The reader will find in the exercises examples of the various phenomena which can occur. In what follows we will study *homogeneous systems*—that is, systems of the form (7), in which $b_1 = b_2 = \cdots = b_n = 0$. Such a system always has at least one solution, the *zero solution* $(x_1, \ldots, x_m) = (0, \ldots, 0)$. The following result guarantees the existence of nonzero solutions—that is, solutions for which at least one x_i is nonzero.

THEOREM 6: Let

$$\begin{array}{c} a_{11}x_1 + a_{12}x_2 + \cdots + a_{1m}x_m = 0 \\ \vdots \qquad\qquad\qquad \vdots \\ a_{n1}x_1 + a_{n2}x_2 + \cdots + a_{nm}x_m = 0 \end{array} \tag{8}$$

be a homogeneous system with coefficients a_{ij} belonging to the field F. If $n < m$, then the system always has a nonzero solution.

Proof: Consider the linear transformation $T: F^m \longrightarrow F^n$ defined by $T((x_1, \ldots, x_m)) = (a_{11}x_1 + \cdots + a_{1m}x_m, \ldots, a_{n1}x_1 + \cdots + a_{nm}x_m) \in F^n$. Note that $(x_1, \ldots, x_m)$ is a solution of the system (8) $\longleftrightarrow$

$$T((x_1, \ldots, x_m)) = (0, \ldots, 0)$$
$$\longleftrightarrow (x_1, \ldots, x_m) \in \ker(T).$$

Assume that (8) has only the zero solution. Then $\ker(T) = \{\mathbf{0}\}$. Therefore, T is an isomorphism. By Theorem 5,

$$\dim_F(T(F^m)) = \dim_F(F^m) = m. \tag{9}$$

However, since $T(F^m) \subseteq F^n$, we see that $\dim_F(T(F^m)) \leq n$. Therefore, by (9), $m \leq n$, which contradicts the assumption that $n < m$. Thus, (8) has nonzero solutions.

6.4 Exercises

1. Verify that the mappings of Examples 1–4 define linear transformations.
2. Determine the kernel of each of the linear transformations of Examples 1–4.
3. Let F be a field and let $T: F^2 \longrightarrow F^2$ be the linear transformation defined by
$$T((a_1, a_2)) = (x_{11}a_1 + x_{12}a_2, x_{21}a_1 + x_{22}a_2) \qquad (x_{ij} \in F).$$
Show that T is an isomorphism if and only if $x_{11}x_{22} - x_{12}x_{21} \neq 0$.

*4. Let V and W be vector spaces over F, and let $T: V \longrightarrow W$ be a linear transformation. Show that
$$\dim_F V = \dim_F(\ker(T)) + \dim_F(T(V)).$$

5. Let V and W be vector spaces over F. Let $\mathbf{e}_1, \ldots, \mathbf{e}_n$ be a basis of V, and let $\mathbf{f}_1, \ldots, \mathbf{f}_n$ be arbitrary elements of W. Show that there exists one and only one linear transformation of V into W such that $\mathbf{e}_i$ is mapped into $\mathbf{f}_i$ $(1 \leq i \leq n)$.

6. Show that every linear transformation of F^n into F^m is of the form

$$T((x_1, \ldots, x_n)) = (a_{11}x_1 + \cdots + a_{1n}x_n, \ldots, a_{m1}x_1 + \cdots + a_{mn}x_n),$$

where $a_{ij} \in F$.

7. Let F be a field, and suppose that the system of linear equations

$$\begin{aligned} a_{11}x_1 + \cdots + a_{1n}x_n &= b_1 \\ a_{21}x_1 + \cdots + a_{2n}x_n &= b_2 \\ &\vdots \\ a_{m1}x_1 + \cdots + a_{mn}x_n &= b_m \end{aligned}$$

$(a_{ij}, b_k \in F)$ is consistent—that is, there exists a solution $(x_1, \ldots, x_n)$ in F^n. Show that any solution of the system is of the form $(x_1 + x_1', \ldots, x_n + x_n')$, where $(x_1', \ldots, x_n')$ is a solution of the corresponding homogeneous system—that is, the system with all b_i replaced by 0.

8. Formulate and prove an analogue of the first isomorphism theorems of group theory and ring theory in the setting of vector spaces.

7 THE THEORY OF FIELDS

7.1 Introduction

In this chapter let us begin to study the theory of fields, with an eye toward the Galois theory of equations, which will be taken up in Chapter 9. As a consequence of the theory which we will develop in this chapter, we will be able to attack the classical geometrical problems concerning constructions with compass and straightedge. In particular, we will show that it is impossible to trisect the general angle or to duplicate the cube, using only compass and straightedge.

If E is a field, then a *subfield of* E is a subset of E which is also a field with respect to the operations of E. For example, $\mathbf{Q}$ is a subfield of $\mathbf{R}$. If F is a subfield of E, then we say that E is an *extension field of* F (or simply E is an *extension of* F). Thus, for example, $\mathbf{R}$ is an extension of $\mathbf{Q}$. Our point of view in this chapter will be roughly as follows: For a given field F, we will study the properties of extensions of F. Just why we want to proceed in this way will be discussed below. But let us first consider some nontrivial examples of extensions.

Consider the set of all real numbers of the form

$$\frac{a_0 + a_1\sqrt{2} + a_2\sqrt{2}^2 + \cdots + a_r\sqrt{2}^r}{b_0 + b_1\sqrt{2} + b_2\sqrt{2}^2 + \cdots + b_s\sqrt{2}^s}, \qquad a_i, b_j \in \mathbf{Q}, \qquad (1)$$

where the denominator is nonzero. It is reasonably apparent that the set of all such quotients is a field containing $\mathbf{Q}$ and $\sqrt{2}$. Let us denote this field by $\mathbf{Q}(\sqrt{2})$. If F is a subfield of $\mathbf{R}$ which contains $\mathbf{Q}$ and $\sqrt{2}$, then F must contain every real number of the form (1), so that $F \supseteq \mathbf{Q}(\sqrt{2})$. Therefore,

$\mathbf{Q}(\sqrt{2})$ is the smallest subfield of $\mathbf{R}$ containing $\mathbf{Q}$ and $\sqrt{2}$. In particular, $\mathbf{Q}(\sqrt{2})$ is an extension of $\mathbf{Q}$. Let us examine this extension more closely. Since $\sqrt{2}^2 = 2$, we have $\sqrt{2}^3 = 2\sqrt{2}, \sqrt{2}^4 = 2^2, \ldots$. Therefore every quotient of the form (1) can be written in the form

$$\frac{\alpha + \beta\sqrt{2}}{\gamma + \delta\sqrt{2}}, \qquad \alpha, \beta, \gamma, \delta \in \mathbf{Q}. \tag{2}$$

But since $(\gamma + \delta\sqrt{2})^{-1} = (\gamma - \delta\sqrt{2})/(\gamma^2 - 2\delta^2)$, we see that every element of $\mathbf{Q}(\sqrt{2})$ is of the form

$$a + b\sqrt{2}, \qquad a, b \in \mathbf{Q}. \tag{3}$$

Conversely, every element of the form (3) belongs to $\mathbf{Q}(\sqrt{2})$. Therefore,

$$\mathbf{Q}(\sqrt{2}) = \{a + b\sqrt{2} \mid a, b \in \mathbf{Q}\}. \tag{4}$$

Let us generalize the above example into a general procedure for constructing extensions. Let F be a field, E an extension of F, S a set of elements of E. Further, let $\mathfrak{C}$ denote the collection of all subfields of E which contain F and S. Then

$$\bigcap_{G \in \mathfrak{C}} G$$

is a subfield of E containing F and S, and is the smallest such subfield. Let us denote this subfield by $F(S)$. Then $F(S)$ is called the field obtained by *adjoining the elements of S to F*. In the above example, $F = \mathbf{Q}$, $E = \mathbf{R}$, $S = \{\sqrt{2}\}$. In case $S = \{a_1, \ldots, a_n\}$, we will write $F(a_1, \ldots, a_n)$ instead of $F(\{a_1, \ldots, a_n\})$. A field of the form $F(S)$ is an extension of F. Moreover, if E is an extension of F, then $E = F(S)$ for $S = E$. Therefore, every extension of F can be obtained by adjoining a set of elements to F. And thus, in order to study the properties of extensions of F, we must examine this process of adjunction more closely.

PROPOSITION 1: Let F be a field, E an extension of F, S a subset of E. If $S = S_1 \cup S_2$, then

$$F(S) = F(S_1)(S_2).$$

That is, $F(S)$ can be obtained by adjoining the elements of S_2 to $F(S_1)$.

Proof: $F(S_1)(S_2)$ is a subfield of E containing $F(S_1)$ and S_2, so that $F(S_1)(S_2)$ contains F, S_1, and S_2. In particular, $F(S_1)(S_2)$ is a subfield of E containing F and $S_1 \cup S_2 = S$. Therefore, $F(S_1)(S_2) \supseteq F(S)$. Since $S \supseteq S_1$, $S \supseteq S_2$, we see that $F(S) \supseteq F(S_1)$, $F(S) \supseteq S_2$. Therefore, $F(S) \supseteq F(S_1)(S_2)$. Thus, we have $F(S) = F(S_1)(S_2)$.

If $S = \{a_1, \ldots, a_n\}$, then Proposition 1 implies that $F(a_1, \ldots, a_n) = F(a_1, \ldots, a_{n-1})(a_n)$. Therefore, we can reduce the process of adjoining n elements to F to n successive adjunctions of a single element: First we adjoin a_1 to F; then we adjoin a_2 to $F(a_1)$; then we adjoin a_3 to $F(a_1, a_2)$;

and so on. An extension of F of the form $F(a)$ is called a *simple extension of F*. We have just shown that we can adjoin a set of n elements to F by forming n consecutive simple extensions. Thus, the simple extensions of a field F should be studied more closely. For now, let us be content to observe that $F(a)$ consists of all quotients of the form

$$\frac{a_0 + a_1 a + \cdots + a_r a^r}{b_0 + b_1 a + \cdots + b_s a^s}, \qquad a_i, b_j \in F, \tag{5}$$

where the denominator is nonzero. The proof is identical to the argument we used for $\mathbf{Q}(\sqrt{2})$ above. We will examine the simple extensions more closely in Section 7.2.

We still must supply an answer to the question: Why should we study extensions of a field? Let us answer this question by constructing a very important class of extensions of $\mathbf{Q}$. If $f \in \mathbf{Q}[X]$ is a nonconstant, monic† polynomial of degree n, then we will show in Chapter 10 that f has zeros $\alpha_1, \ldots, \alpha_n$ in $\mathbf{C}$ such that

$$f = (X - \alpha_1)(X - \alpha_2) \cdots (X - \alpha_n).$$

Let us form the extension $\mathbf{Q}(\alpha_1, \alpha_2, \ldots, \alpha_n)$ of $\mathbf{Q}$. One of our major tasks is to get information about $\alpha_1, \ldots, \alpha_n$. We will accomplish this by studying the algebraic properties of the extension $\mathbf{Q}(\alpha_1, \alpha_2, \ldots, \alpha_n)$.

In studying field extensions, it will be necessary to study isomorphisms of one field to another. Therefore, let us clarify the notion of isomorphism we have in mind. If E_1 and E_2 are fields, then an *isomorphism* $f: E_1 \longrightarrow E_2$ is a ring isomorphism of E_1 into E_2. Thus, an isomorphism $f: E_1 \longrightarrow E_2$ is injective and preserves sums, differences, and products. But then it is easy to see that if $a \in E_1^\times$, we have $f(a^{-1}) = f(a)^{-1}$.

If E_1 and E_2 are extensions of the same field F, then an *F-isomorphism of E_1 into E_2* is an isomorphism $f: E_1 \longrightarrow E_2$ such that $f(x) = x$ for all $x \in F$. We say that E_1 and E_2 are *F-isomorphic* if there exists a surjective F-isomorphism $f: E_1 \longrightarrow E_2$. In this case, we write $E_1 \approx_F E_2$.

PROPOSITION 2: Let E_1 and E_2 be fields, $f: E_1 \longrightarrow E_2$ a ring homomorphism. Then either $f(x) = 0$ for all $x \in E_1$ or f is an isomorphism.

Proof: Since E_1 is a field, the only ideals of E_1 are $\{0\}$ and E_1 (Example 3 of Section 5.2). But $\ker(f)$ is an ideal of E_1. If $\ker(f) = E_1$, then $f(x) = 0$ for all $x \in E_1$. If $\ker(f) = \{0\}$, then f is an isomorphism.

7.1 Exercises

1. Show that every element of $\mathbf{Q}(\sqrt{2})$ can be written uniquely in the form $a + b\sqrt{2}$, $a, b \in \mathbf{Q}$.

† That is, the coefficient of the highest power of X appearing in f is 1.

2. Describe the elements of $\mathbf{Q}(\sqrt{5})$ and $\mathbf{Q}(\sqrt{7})$.
3. Show that every element of $\mathbf{Q}(\sqrt[3]{5})$ is of the form $a + b\sqrt[3]{5} + c\sqrt[3]{5^2}$, $a, b, c \in \mathbf{Q}$.
4. Show that every element of $\mathbf{Q}(\sqrt{2}, \sqrt{3})$ is of the form $a + b\sqrt{2} + c\sqrt{3} + d\sqrt{6}$, $a, b, c, d \in \mathbf{Q}$.
5. Let X be an indeterminate over F, and let $F(X)$ be the quotient field of $F[X]$.
   (a) Show that $F(X)$ is an extension field of F gotten by adjoining X to F.
   (b) Show that $F(X^n) \neq F(X)$ for $n > 1$. [Here $F(X^n)$ means the extension gotten by adjoining X^n to F.]
   (c) Show that $F(X^2, X^3) = F(X)$.
6. Let E be an extension of F. Show that E is a vector space over F, where the product $\alpha \cdot \mathbf{v}$ of a scalar $\alpha \in F$ and a vector $\mathbf{v} \in E$ is just the product of α and $\mathbf{v}$ as elements of E.
7. Show that $\{1, \sqrt{2}\}$ is a basis of $\mathbf{Q}(\sqrt{2})$ over $\mathbf{Q}$.
8. Suppose that E is a simple extension of F. Is E necessarily finite-dimensional over F? (*Hint*: Look at Exercise 5.)
9. Let $f\colon \mathbf{Q}(\sqrt{2}) \longrightarrow \mathbf{Q}(\sqrt{2})$ be defined by $f(a + b\sqrt{2}) = a - b\sqrt{2}$. Show that f is a $\mathbf{Q}$-isomorphism.
10. Let E and E' be extensions of $\mathbf{Q}$ and let $f\colon E \longrightarrow E'$ be an isomorphism. Show that f is a $\mathbf{Q}$-isomorphism.
11. Let $f\colon \mathbf{C} \longrightarrow \mathbf{C}$ be defined by $f(x) = \bar{x}$. Show that F is an $\mathbf{R}$ isomorphism.
12. Let E be an extension of F and let $f\colon E \longrightarrow E$ be an F-isomorphism. Then show that $\{x \in E \mid f(x) = x\}$ is an extension field of F.
13. Show that $\mathbf{Q}(\sqrt{2})$ is not $\mathbf{Q}$-isomorphic to $\mathbf{Q}(\sqrt{3})$. [*Hint*: Suppose that $f\colon \mathbf{Q}(\sqrt{2}) \longrightarrow \mathbf{Q}(\sqrt{3})$ is a surjective $\mathbf{Q}$-isomorphism. Then what is $f(\sqrt{2})$?]

7.2 Algebraic and Transcendental Elements

Throughout this section let F be a field and E an extension field of F. Let us begin by classifying the elements of E into two broad categories with respect to F. Let $\alpha \in E$. We say that α is *algebraic over* F if there exists a nonzero polynomial $f \in F[X]$ such that $f(\alpha) = 0$. If α is not algebraic over F, then we say that α is *transcendental over* F. In order to get some feel for this classification, let us study some examples.

EXAMPLE 1: $\alpha \in F \Rightarrow \alpha$ is algebraic over F. For α is a zero of $X - \alpha \in F[X]$.

EXAMPLE 2: Let $F = \mathbf{Q}$, $E = \mathbf{C}$. Then $\alpha_1 = \sqrt{2}$, $\alpha_2 = \sqrt[3]{7}$, $\alpha_3 = \sqrt{2} + \sqrt[3]{7}$ are all algebraic over $\mathbf{Q}$. For α_1 is a zero of $X^2 - 2$; α_2 is a zero of $X^3 - 7$; α_3 is a zero of $X^6 - 6X^4 - 14X^3 + 12X^2 - 84X + 41$. (Check it!)

EXAMPLE 3: Let $F = \mathbf{Q}, E = \mathbf{C}$. Then $\pi = 3.1415926 \cdots$, $e = 2.718281 \cdots$ are transcendental over $\mathbf{Q}$. We cannot come near to proving these two assertions. Both are celebrated results of nineteenth-century mathematics. It was long suspected that neither π nor e can satisfy a polynomial equation with rational coefficients. However, the transcendentality of π was not proved until 1882, when it was settled by Lindemann. The transcendentality of e is somewhat easier and was first proved by Hermite in 1873.

EXAMPLE 4: The theory of transcendental numbers is a vast realm of mathematics in which research is still going on in many parts of the mathematical community. It is an extremely difficult branch of mathematics, which relies for many of its proofs on analysis, especially on functions of a complex variable. Some apparently very simple questions on transcendence cannot be answered at the present time. For example, it is known that e^π is transcendental over $\mathbf{Q}$, but it is not known whether π^e is transcendental. Although we cannot go into an extended discussion of the theory of transcendental numbers, let us cite one of the crowning achievements of the theory, the Gelfond–Schneider theorem:

THEOREM 1: Let $\alpha \neq 0, 1$ be algebraic over $\mathbf{Q}$. Let β be algebraic over $\mathbf{Q}$, $\beta \notin \mathbf{Q}$. Then α^β is transcendental over $\mathbf{Q}$.

Thus, for example, $2^{\sqrt{2}}$ is transcendental over $\mathbf{Q}$. The statement of this theorem was first guessed at by Leonhard Euler in the eighteenth century and was posed by Hilbert as one of his famous 23 problems at the International Congress of Mathematicians in Paris in 1900. A proof of the result was discovered almost simultaneously in 1934 by Gelfond and Schneider, working independently.

We have not given any examples of transcendental elements which we could actually show to be transcendental. There is one obvious example, however:

EXAMPLE 5: Let F be a field, X an indeterminate over F. Let $F(X)$ be the quotient field of $F[X]$. Then $F(X)$ is an extension field of F, so we may set $E = F(X)$. Then $X \in E$ is transcendental over F. For if X satisfies a polynomial equation with coefficients in F, then

$$c_nX^n + c_{n-1}X^{n-1} + \cdots + c_0 = 0, \qquad c_i \in F$$

$$\Longrightarrow c_0 = c_1 = \cdots = c_n = 0 \quad \text{since } X \text{ is an indeterminate over } F.$$

Thus, X cannot be the zero of any nonzero polynomial with coefficients in F.

In the remainder of this book we will have little to say concerning transcendental elements. This is for at least two reasons. First, the theory of transcendental elements is far too difficult for a first course. Second, we are striving toward a theory of polynomial equations. And the zeros of a polynomial are algebraic over any field containing the coefficients of the polynomial.

Let $\alpha \in E$ be algebraic over F. Our first task is to single out a special polynomial having α as a zero. Let $f \in F[X]$ have α as a zero. Since $F[X]$ is a UFD, we may write $f = p_1, \ldots, p_r$, where $p_i \in F[X]$ is irreducible. Then

$$0 = f(\alpha) = p_1(\alpha) \ldots p_r(\alpha).$$

But since E is an integral domain and $p_i(\alpha) \in E$ $(1 \leq i \leq r)$, we see that $p_i(\alpha) = 0$ for some i. Thus, α is a zero of some irreducible polynomial $p \in F[X]$. Without loss of generality, we may normalize p to be monic. We assert that p is the unique monic, irreducible polynomial in $F[X]$ having α as a zero. Indeed, if q were another such polynomial, then p and q are relatively prime, so that there exist polynomials $a, b \in F[X]$ such that $ap + bq = 1$. But then

$$0 = a(\alpha)p(\alpha) + b(\alpha)q(\alpha) = 1,$$

which is a contradiction. Then polynomial p is called the *irreducible polynomial of α over F* and is denoted $\text{Irr}_F(\alpha, X)$. Moreover, the manner in which we constructed p implies

LEMMA 2: Let $\alpha \in E$ be algebraic over F, and let $f \in F[X]$ be such that $f(\alpha) = 0$. Then f is divisible by $\text{Irr}_F(\alpha, X)$.

EXAMPLE 6: $X^2 - 2$ is monic and irreducible in $\mathbf{Q}[X]$. Therefore, $\text{Irr}_{\mathbf{Q}}(\sqrt{2}, X) = X^2 - 2$.

EXAMPLE 7: Let $d \in \mathbf{Z}$ not be a perfect square. Then $X^2 - d$ is monic and irreducible in $\mathbf{Q}[X]$, so that $\text{Irr}_{\mathbf{Q}}(\sqrt{d}, X) = X^2 - d$.

EXAMPLE 8: Let $\alpha \in F$. Then $\text{Irr}_F(\alpha, X) = X - \alpha$.

As the reader may have already guessed, it is often relatively easy to exhibit a polynomial $f \in F[X]$ for which $f(\alpha) = 0$. The difficult part of constructing $\text{Irr}_F(\alpha, X)$ is to make sure that f is irreducible. There is a reasonable way of going about this, at least in the case $F = \mathbf{Q}$. If we can exhibit some $f \in F[X]$ such that $f(\alpha) = 0$, then we can try to factor f and determine an irreducible factor of f which has α as a zero. The irreducibility tests developed in Chapter 5 are useful in proving that a given factor of f is irreducible. Particular attention should be called to the Eisenstein irreducibility criterion, which should prove helpful in solving several of the exercises for this section.

If E is an extension of the field F, then E may be regarded as a vector space over F. The product $\alpha \cdot \mathbf{v}$ of a scalar $\alpha \in F$ and a vector $\mathbf{v} \in E$ is just the product of α and $\mathbf{v}$, considered as elements of E. The dimension of E over F is called the *degree of E over F*, denoted $\deg(E/F)$.

For example, we showed in Section 7.1 that

$$\mathbf{Q}(\sqrt{2}) = \{a + b\sqrt{2} \mid a, b \in \mathbf{Q}\}.$$

Therefore, $\{1, \sqrt{2}\}$ generates $\mathbf{Q}(\sqrt{2})$ over $\mathbf{Q}$. Moreover, $\{1, \sqrt{2}\}$ is linearly independent over $\mathbf{Q}$, since if $a + b\sqrt{2} = 0$ with a, b not both 0, we see that $b \neq 0$, so that $\sqrt{2} = -a/b \in \mathbf{Q}$, which contradicts Theorem 1 of Section 4.6. Thus, we see that $\{1, \sqrt{2}\}$ is a basis of $\mathbf{Q}(\sqrt{2})$ over $\mathbf{Q}$ and that $\deg(\mathbf{Q}\sqrt{2})/\mathbf{Q}) = 2$. Note that $\operatorname{Irr}_{\mathbf{Q}}(\sqrt{2}, X) = X^2 - 2$ and that $\deg(\operatorname{Irr}_{\mathbf{Q}}(\sqrt{2}, X)) = \deg(\mathbf{Q}\sqrt{2})/\mathbf{Q})$. This phenomenon is not accidental, as the following result shows.

THEOREM 3: Let α be algebraic over F, $n = \deg(\operatorname{Irr}_F(\alpha, X))$. Then

(1) $\deg(F(\alpha)/F) = n$.
(2) $\{1, \alpha, \alpha^2, \ldots, \alpha^{n-1}\}$ is a basis for $F(\alpha)$ over F.

Proof: Since $\{1, \alpha, \alpha^2, \ldots, \alpha^{n-1}\}$ contains n elements, it is clear that part (2) implies part (1). To prove part (2), we must show that every $\beta \in F(\alpha)$ can be written uniquely in the form

$$\beta = a_0 \cdot 1 + a_1 \cdot \alpha + \cdots + a_{n-1}\alpha^{n-1} \qquad (a_i \in F). \tag{1}$$

The uniqueness is easy. For if $\beta = a'_0 \cdot 1 + a'_1\alpha + \cdots + a'_{n-1}\alpha^{n-1}$, then

$$(a_0 - a'_0) + (a_1 - a'_1)\alpha + \cdots + (a_{n-1} - a'_{n-1})\alpha^{n-1} = 0.$$

But then the polynomial $f = (a_0 - a'_0) + (a_1 - a'_1)X + \cdots + (a_{n-1} - a'_{n-1})X^{n-1}$ has α as a zero. Therefore, by Lemma 2, f is divisible by $\operatorname{Irr}_F(\alpha, X)$, where the latter has degree n. Thus, we are forced to conclude that $f = \mathbf{0}$, which implies that $a_0 = a'_0, a_1 = a'_1, \ldots, a_{n-1} = a'_{n-1}$, which gives us the uniqueness of the representation (1). Let us now show that every $\beta \in F(\alpha)$ can be written in the form (1). Let $F[\alpha]$ denote the smallest subring of $F(\alpha)$ containing F and α. Then $F[\alpha]$ consists of all sums of the form

$$a_0 + a_1\alpha + \cdots + a_m\alpha^m \qquad (a_i \in F). \tag{2}$$

We assert that every sum of the form (2) can be rewritten in the form

$$b_0 + b_1\alpha + \cdots + b_{n-1}\alpha^{n-1} \qquad (b_i \in F). \tag{3}$$

It clearly suffices to show that α^r $(r \geq n)$ can be so written. Let us proceed by induction on r. Let

$$\operatorname{Irr}_F(\alpha, X) = X^n + c_{n-1}X^{n-1} + \cdots + c_0.$$

Then $\alpha^n = -c_{n-1}\alpha^{n-1} - c_{n-2}\alpha^{n-2} - \cdots - c_0$, so α^n can be written in the form (3). Thus, the assertion is true for $r = n$. Assume that $r > n$ and that α^{r-1} can be written in the form (3). Then

$$\begin{aligned}\alpha^r = \alpha \cdot \alpha^{r-1} &= \alpha \cdot (b_0 + b_1\alpha + \cdots + b_{n-1}\alpha^{n-1}) \\ &= b_0\alpha + b_1\alpha^2 + \cdots + b_{n-1}\alpha^n \\ &= b'_0 + b'_1\alpha + \cdots + b'_{n-1}\alpha^{n-1},\end{aligned}$$

since α^n can be written in the form (3). Thus, the induction is complete. We have shown that

$$F[\alpha] = \{a_0 + a_1\alpha + \cdots + a_{n-1}\alpha^{n-1} \mid a_i \in F\}. \tag{4}$$

It is clear that $F[\alpha]$ contains both F and α and $F[\alpha] \subseteq F(\alpha)$. Therefore, if we prove that $F[\alpha]$ is a field, then we can conclude that $F[\alpha] \supseteq F(\alpha)$ and therefore $F[\alpha] = F(\alpha)$. Therefore, by (4), we conclude that every $\beta \in F(\alpha)$ can be written in the form (2). Thus, we must show that $F[\alpha]$ is a field. Let

$$\psi_\alpha : F[X] \longrightarrow F[\alpha]$$

be the homomorphism "evaluation at α." Then it is easy to see that ψ_α is surjective. Moreover, $\ker(\psi_\alpha) = g \cdot F[X]$, where $g = \mathrm{Irr}_F(\alpha, X)$. Therefore,

$$F[X]/g \cdot F[X] \approx F[\alpha].$$

However, since g is irreducible, $g \cdot F[X]$ is a maximal ideal (Example 5 of Section 5.9) and thus $F[X]/g \cdot F[X]$ is a field by Theorem 8 of Section 5.9. Thus, $F[\alpha]$ is a field, since the image of a field is under an isomorphism again a field.

In Theorem 3, we described the simple extension $F(\alpha)$ in case α is algebraic over F. Let us now carry out the corresponding task in case α is transcendental over F. The clue is provided by Example 5.

PROPOSITION 4: Let $F(\alpha)$ be a simple extension of F and let α be transcendental over F. Then the mapping

$$\psi : F(X) \longrightarrow F(\alpha),$$

$$\frac{p(X)}{q(X)} \longrightarrow \frac{p(\alpha)}{q(\alpha)} \qquad (p(X), q(X) \in F[X], q(X) \neq \mathbf{0})$$

is a surjective F-isomorphism. In particular, $F(X)$ and $F(\alpha)$ are F-isomorphic.

Proof: First note that $q(\alpha) \neq 0$ since α is transcendental over F and $q(X) \neq \mathbf{0}$. Thus, $p(\alpha)/q(\alpha)$ makes sense. Second, let us show that ψ is well defined: If $p_1(X)/q_1(X) = p_2(X)/q_2(X)$, then

$$p_1(X)q_2(X) = q_1(X)p_2(X),$$

$$\Longrightarrow p_1(\alpha)q_2(\alpha) = q_1(\alpha)p_2(\alpha),$$

$$\Longrightarrow \frac{p_1(\alpha)}{q_1(\alpha)} = \frac{p_2(\alpha)}{q_2(\alpha)}.$$

Thus, ψ is well defined. It is easy to check that ψ is a ring homomorphism. Therefore, by the last Proposition 2 Section 7.1, ψ is an isomorphism since ψ is not identically zero. By (5) of Section 7.1, ψ is surjective.

We showed in Theorem 3 that if α is algebraic over F, then $\deg(F(\alpha)/F) = \deg(\mathrm{Irr}_F(\alpha, X))$. In particular, $\deg(F(\alpha)/F)$ is finite. The situation is quite different if α is transcendental over F, however. Indeed, the set

$$\{1, \alpha, \alpha^2, \dots\} = S$$

is then an infinite, linearly independent subset of $F(\alpha)$. For if

$$\sum_{i \geq 0} a_i \alpha^i = 0, \quad a_i \in F, \qquad a_i \text{ not all } 0,$$

is a linear relation among the elements of S, then α is the zero of some nonzero polynomial with coefficients in F, a contradiction to the assumption that α is transcendental over F. Thus, if α is transcendental over F, $\deg(F(\alpha)/F)$ is infinite. If E is an extension field of F, then we say that E is a *finite extension of F* (or *that E/F is finite*) if $\deg(E/F)$ is finite. Then we have proved the following result:

THEOREM 5: $F(\alpha)/F$ is finite if and only if α is algebraic over F.

7.2 Exercises

1. Show that the following complex numbers are algebraic over $\mathbf{Q}$:
   (a) $\sqrt{5}$. (b) $\sqrt{d}$, $d \in \mathbf{Z}$. (c) $\sqrt{11} + 1$.
   (d) $\sqrt[3]{2} + 3\sqrt{2}$. (e) $i + 1$. (f) $\dfrac{-1 + \sqrt{-3}}{2}$.
2. Let α be algebraic over F and let $f = \mathrm{Irr}_F(\alpha, X)$. If $a \in F$, show that $\alpha + a$ is algebraic over F and that $\mathrm{Irr}_F(\alpha + a, X) = f(X - a)$.
3. Let $d \in \mathbf{Z}$.
   (a) Show that $X^2 - d$ is irreducible in $\mathbf{Q}[X]$ if $d \neq x^2$, $x \in \mathbf{Z}$ and is reducible otherwise.
   (b) Show that $\deg(\mathbf{Q}(\sqrt{d})/\mathbf{Q}) = 1$ if $d = x^2$ $(x \in \mathbf{Z})$, $= 2$ if $d \neq x^2$ $(x \in \mathbf{Z})$.
   (c) Find a basis for $\mathbf{Q}(\sqrt{d})$ over $\mathbf{Q}$.
4. Show that $\sqrt{2}^{\sqrt{3}}$ is transcendental over $\mathbf{Q}$.
5. Is $\sqrt{2}^{\sqrt{3}} + 1$ transcendental over $\mathbf{Q}$? (*Hint*: Use Exercise 2.)
*6. Is $\sqrt{2}^{\sqrt{2}} + \sqrt{2}$ transcendental over $\mathbf{Q}$?
7. Let p be an odd prime and let ζ be a primitive pth root of 1.
   (a) Show that ζ is algebraic over $\mathbf{Q}$.
   (b) Show that $\mathrm{Irr}_{\mathbf{Q}}(\zeta, X) = X^{p-1} + X^{p-2} + \cdots + X + 1$. [*Hint*: $\zeta^p = 1$. To show that $f = X^{p-1} + X^{p-2} + \cdots + X + 1$ is irreducible over $\mathbf{Q}$, show that $f(X + 1)$ is irreducible over $\mathbf{Q}$ by the Eisenstein irreducibility criterion.]
8. Let α be a complex zero of the polynomial $X^4 - 5X + 10$.
   (a) Describe $\mathbf{Q}(\alpha)$.
   (b) Compute $\deg(\mathbf{Q}(\alpha)/\mathbf{Q})$.

7.3 Algebraic Extensions

Let F be a field and let E be an extension of F. We say that E is an *algebraic extension of F* if every $\alpha \in E$ is algebraic over F. One of the main results of this section asserts that if α is algebraic over F, then $F(\alpha)$ is an algebraic extension of F. [The reader should not be duped into thinking this is a trivial result! For example, let $F = \mathbf{Q}$, $E = F(\sqrt[3]{2})$. By Theorem 3 of Section 7.2, every $\alpha \in E$ is of the form $\alpha = a_1 + a_2\sqrt[3]{2} + a_3\sqrt[3]{2}^2$. The reader should try to show that all such α are algebraic over $\mathbf{Q}$.] Let us begin with a modest result.

PROPOSITION 1: Let E/F be finite. Then E is algebraic over F.

Proof: Let $\alpha \in E$, $n = \deg(E/F)$. The $n + 1$ elements

$$1, \alpha, \ldots, \alpha^n$$

of E must be linearly dependent over F, since $\dim_F(E) = n$. Therefore, there exist $c_i \in F$ $(0 \leq i \leq n)$, c_i not all 0, such that

$$c_n\alpha^n + c_{n-1}\alpha^{n-1} + \cdots + c_0 = 0.$$

Therefore, α is algebraic over F.

The next two results are elementary, but of critical importance.

PROPOSITION 2: $\deg(E/F) = 1$ if and only if $E = F$.

Proof: $\Leftarrow$ Obvious.

$\Rightarrow$. Suppose that $\deg(E/F) = 1$. Then we can choose a basis of E over F consisting of one element α, and every $\beta \in E$ is of the form $a\alpha$ for some $a \in F$. In particular, $1 = a_0\alpha$ for some $a_0 \in F$, so that $\alpha = a_0^{-1} \in F$. Therefore, $a\alpha \in F \Rightarrow E \subseteq F$. Thus, since $F \subseteq E$, we have $E = F$.

THEOREM 3: Let $E \subseteq F \subseteq G$ be three fields. Assume that $\deg(F/E)$ and $\deg(G/F)$ are finite. Then $\deg(G/E)$ is finite and

$$\deg(G/E) = \deg(G/F)\cdot\deg(F/E).$$

Further, if $\{\alpha_1, \ldots, \alpha_n\}$ is a basis of G over F and $\{\beta_1, \ldots, \beta_m\}$ is a basis of F over E, then $\{\alpha_i\beta_j\}$ $(1 \leq i \leq n, 1 \leq j \leq m)$ is a basis of G over E.

Proof: Let $x \in G$. Then there exist elements $a_i \in F$ $(1 \leq i \leq n)$ such that

$$x = \sum_{i=1}^{n} a_i\alpha_i. \tag{1}$$

However, since $\{\beta_1, \ldots, \beta_m\}$ is a basis of F over E, for each i $(1 \leq i \leq n)$, there exist elements $b_{ij} \in E$ $(1 \leq j \leq m)$ such that

$$a_i = \sum_{j=1}^{m} b_{ij}\beta_j \qquad (1 \leq i \leq m). \tag{2}$$

Combining (1) and (2) we see that

$$x = \sum_{i=1}^{n} \sum_{j=1}^{m} b_{ij}\beta_j\alpha_i.$$

Therefore, every element $x \in G$ is expressible as a linear combination of the elements $\alpha_i\beta_j$. Moreover, this expression is unique, since

$$\sum_{i=1}^{n} \sum_{j=1}^{m} b_{ij}\beta_j\alpha_i = \sum_{i=1}^{n} \sum_{j=1}^{m} b'_{ij}\beta_j\alpha_i,$$

$$\Longrightarrow \sum_{i=1}^{n} \sum_{j=1}^{m} c_{ij}\alpha_i\beta_j = 0 \qquad \text{where } c_{ij} = b_{ij} - b'_{ij}.$$

But

$$\sum_{i=1}^{n} \sum_{j=1}^{m} c_{ij}\alpha_i\beta_j = 0 \Longrightarrow \sum_{i=1}^{n}\Big(\sum_{j=1}^{m} c_{ij}\beta_j\Big)\alpha_i = 0$$

$$\Longrightarrow \sum_{j=1}^{m} c_{ij}\beta_j = 0 \qquad (1 \leq i \leq n)$$

since the α_i are linearly independent over F. But then

$$c_{ij} = 0 \qquad (1 \leq i \leq n, 1 \leq j \leq m)$$

since the β_j are linearly independent over E. Therefore,

$$b_{ij} = b'_{ij}$$

for all i, j.

In what follows, let the elements $\alpha_1, \alpha_2, \ldots, \alpha_n, \beta$ be drawn from some extension E of F.

COROLLARY 4: If α_1 and α_2 are algebraic over F, then $F(\alpha_1, \alpha_2)$ is of finite degree over F and is therefore algebraic over F.

Proof: By Theorem 5 of Section 7.2, $F(\alpha_1)$ is finite over F. Moreover, α_2 algebraic over F implies that α_2 is algebraic over $F(\alpha_1)$. [For if α_2 is the zero of a polynomial $f \in F[X]$, then it certainly is the zero of some polynomial in $F(\alpha_1)[X]$, namely f.] Therefore, by Proposition 1 of Section 7.1, $F(\alpha_1)(\alpha_2) = F(\alpha_1, \alpha_2)$ is finite over $F(\alpha_1)$. Thus, by Theorem 3, $F(\alpha_1, \alpha_2)$ is finite over F, so that by Proposition 1, $F(\alpha_1, \alpha_2)$ is algebraic over F.

COROLLARY 5: If $\alpha_1, \ldots, \alpha_n$ are algebraic over F, then $F(\alpha_1, \ldots, \alpha_n)$ is of finite degree over F and is therefore algebraic over F.

Proof: Apply Corollary 4 and induction on n.

COROLLARY 6: Let α and β be algebraic over F. Then $\alpha \pm \beta, \alpha \cdot \beta, \alpha/\beta$ $(\beta \neq 0)$ are algebraic over F.

Proof: $F(\alpha, \beta)$ is algebraic over F, and $\alpha \pm \beta, \alpha \cdot \beta, \alpha/\beta$ $(\beta \neq 0)$ belong to $F(\alpha, \beta)$.

7.3 Exercises

1. (a) Show that $X^2 - 3$ is irreducible in $\mathbf{Q}(\sqrt{2})[X]$.
   (b) Compute the degree of $\mathbf{Q}(\sqrt{2}, \sqrt{3})$ over $\mathbf{Q}$.
   (c) Exhibit a basis of $\mathbf{Q}(\sqrt{2}, \sqrt{3})$ over $\mathbf{Q}$.
2. Let $n_1, n_2, \ldots, n_t$ be distinct integers.
   (a) Show that $\deg(\mathbf{Q}\sqrt{n_1}, \sqrt{n_2}, \ldots, \sqrt{n_t})/\mathbf{Q}) \leq 2^t$.
   (b) Can strict inequality hold in (a)?
3. Let ζ denote the primitive eighth root of 1 given by $\zeta = \cos(2\pi/8) + i \sin(2\pi/8)$.
   (a) Show that $(\zeta + \zeta^{-1})^2 = 2$.
   (b) Show that $\mathbf{Q}(\sqrt{2}) \subseteq \mathbf{Q}(\zeta)$.
   (c) Compute $\deg(\mathbf{Q}(\zeta)/\mathbf{Q}(\sqrt{2}))$.
4. Let $\bar{\mathbf{Q}} = \{x \in \mathbf{C} \mid x \text{ is algebraic over } \mathbf{Q}\}$.
   (a) Show that $\bar{\mathbf{Q}}$ is a field containing $\mathbf{Q}$. ($\bar{\mathbf{Q}}$ is called the field of algebraic numbers.)
   *(b) Show that if $x \in \mathbf{C}$ and x is algebraic over $\bar{\mathbf{Q}}$, then $x \in \bar{\mathbf{Q}}$.
   *(c) Show that $\deg(\bar{\mathbf{Q}}/\mathbf{Q})$ is infinite. [*Hint*: Let p_n denote the nth prime $(n = 1, 2, 3, \ldots)$, and let $K_n = \mathbf{Q}(\sqrt{p_1}, \ldots, \sqrt{p_n})$. Show that $K_n \subseteq \bar{\mathbf{Q}}$ and that $\deg(K_n/\mathbf{Q}) = 2^n$. Therefore, conclude that $\deg(\bar{\mathbf{Q}}/\mathbf{Q}) \geq 2^n$ for every $n \geq 1$.]
   (d) Is an algebraic extension always finite? (We proved that the converse is true.)
5. Let $K_{p^n} = \mathbf{Q}(\zeta_{p^n})$, where p is a prime, $n \geq 0$, and ζ_{p^n} is a primitive p^nth root of 1.
   (a) Show that $K_p \subseteq K_{p^2} \subseteq K_{p^3} \subseteq \ldots$.
   (b) Show that
   $$K = \bigcup_{n=1}^{\infty} K_{p^n}$$
   is a field.
   (c) If $n \geq 1$, show that $\text{Irr}_{K_{p^n}}(\zeta_{p^{n+1}}, X) = X^p - \zeta_{p^n}$.
   (d) Show that $\deg(K_{p^{n+1}}/K_{p^n}) = p \quad (n \geq 1)$.
6. Let E be an algebraic extension of F, and let A be a subring of E such that $E \supseteq A \supseteq F$. Show that A is a field.
7. Does the conclusion of Exercise 6 hold if we do not assume that E is algebraic over F?
8. Let F be a field, $E = F(\alpha, \beta)$ an algebraic extension of F. Let $f = \text{Irr}_F(\alpha, X)$, $g = \text{Irr}_F(\beta, X)$, and assume that $\deg(f)$ and $\deg(g)$ are relatively prime to one another.
   (a) Show that g is irreducible in $F(\alpha)[X]$.
   (b) Show that $\deg(E/F) = m \cdot n$, where $m = \deg(f)$, $n = \deg(g)$.

7.4 The Complex Numbers Revisited

Let us reexamine our construction of **C** from a somewhat different viewpoint. If $x \in \mathbf{R}$, then $x^2 \geq 0$, and therefore $x^2 + 1 \neq 0$. Thus, the polynomial $X^2 + 1$ has no zeros in **R**. In particular, $X^2 + 1$ is irreducible in $\mathbf{R}[X]$. In **C**, $X^2 + 1$ has two zeros—$+i$ and $-i$, and in $\mathbf{C}[X]$,

$$X^2 + 1 = (X - i)(X + i).$$

Since every complex number is of the form $a + bi$, it is clear that $\mathbf{C} = \mathbf{R}(i)$. In other words, **C** is obtained from **R** by adjoining a zero of a polynomial of $\mathbf{R}[X]$ which is irreducible. Let us examine this statement a little more closely and hopefully achieve a better understanding of what it means to adjoin a root of a polynomial to a field.

Suppose we were faced with the following question: *Given* **R** *and the irreducible polynomial* $X^2 + 1 \in \mathbf{R}[X]$, *construct an extension field of* **R** *which contains a zero of* $X^2 + 1$. Of course, we solved this problem by our construction of **C** in Chapter 4. However, that construction was rather ad hoc. Let us try to get at it using a more general recipe, which was already introduced in Theorem 3 of Section 7.2.

Since $X^2 + 1$ is irreducible in $\mathbf{R}[X]$, $I = (X^2 + 1)\mathbf{R}[X]$ is a maximal ideal of $\mathbf{R}[X]$. Therefore, by Theorem 8 of Section 5.9,

$$\mathbf{R}[X]/(X^2 + 1)\mathbf{R}[X]$$

is a field. Also,

$$\mathbf{R}_0 = \{a + I \mid a \in \mathbf{R}\}$$

is a subfield of $\mathbf{R}[X]/(X^2 + 1)\mathbf{R}[X]$ which is isomorphic to **R** under the isomorphism $a \longrightarrow a + I$ $(a \in \mathbf{R})$. Thus, by identifying **R** and $\mathbf{R}_0$, we may view $\mathbf{R}[X]/(X^2 + 1)\mathbf{R}[X]$ as an extension field of **R**. Consider the element $X + I$ of $\mathbf{R}[X]/(X^2 + 1)\mathbf{R}[X]$ and note that

$$\begin{aligned}(X + I)^2 &= X^2 + I \\ &= -1 + (X^2 + 1) + I \\ &= -1 + I.\end{aligned}$$

Therefore, $X + I$ is a zero of the polynomial $Y^2 + 1$. In other words, we have just constructed an "abstract" field extension of **R** which contains a zero of $Y^2 + 1$. How does this extension compare with **C**? By the Euclidean algorithm, every polynomial $f(X) \in \mathbf{R}[X]$ can be uniquely written in the form

$$f(X) = a + bX + (X^2 + 1)q(X), \qquad q(X) \in \mathbf{R}[X].$$

Therefore, every coset in $\mathbf{R}[X]/(X^2 + 1)\mathbf{R}[X]$ can be uniquely written in the form

$$a + bX + I.$$

Moreover,

$$[a + bX + I] + [c + dX + I] = (a + c) + (b + d)X + I$$

$$\begin{aligned}[a + bX + I]\cdot[c + dX + I] &= ac + (bc + ad)X + bdX^2 + I \\ &= ac + (bc + ad)X + bd(X^2 + 1) - bd + I \\ &= (ac - bd) + (bc + ad)X + I.\end{aligned}$$

Therefore, the mapping

$$\mathbf{C} \longrightarrow \mathbf{R}[X]/(X^2 + 1)\mathbf{R}[X],$$

$$(a, b) \longrightarrow a + bX + I$$

is a surjective isomorphism! Thus, we have realized the complex numbers as $\mathbf{R}[X]/(X^2 + 1)\mathbf{R}[X]$. In the various constructions of $\mathbf{C}$ which we have given, the last one, in some sense, is the most natural, as we shall see when we discuss splitting fields in the next section.

7.4 Exercises

1. Using the reasoning of this section, establish the following $\mathbf{Q}$-isomorphisms:

$$\mathbf{Q}(\sqrt{3}) \approx \mathbf{Q}[X]/(X^2 - 3)\mathbf{Q}[X]$$
$$\mathbf{Q}(i) \approx \mathbf{Q}[X]/(X^2 + 1)\mathbf{Q}[X]$$
$$\mathbf{Q}(\sqrt[3]{2}) \approx \mathbf{Q}[X]/(X^3 - 2)\mathbf{Q}[X].$$

2. Show that $\mathbf{Q}(\sqrt[3]{2}) \neq \mathbf{Q}(\sqrt[3]{2}\,\omega)$, $\omega =$ a primitive cube root of 1, but nevertheless, $\mathbf{Q}(\sqrt[3]{2})$ is $\mathbf{Q}$-isomorphic to $\mathbf{Q}(\sqrt[3]{2}\,\omega)$.
3. Let $f \in \mathbf{R}[X]$. Show that if $\alpha \in \mathbf{C}$ is a zero of f, then $\bar{\alpha}$ is a zero of f.

7.5 Splitting Fields

In Section 7.4 we constructed the complex numbers $\mathbf{C}$ by adjoining to $\mathbf{R}$ a zero of the polynomial $X^2 + 1$ in $\mathbf{R}[X]$. Let us now take up the general problem of adjoining to F a zero of some polynomial $f \in F[X]$, where F is an arbitrary field. Throughout this section, let

$$f = X^n + a_{n-1}X^{n-1} + \cdots + a_0, \qquad a_i \in F.$$

Let us first obtain some information about what zeros f can have in an extension E of F. (The case $E = F$ is not excluded.) Our first result is a generalization of a well-known result from high school algebra.

PROPOSITION 1: Let $\alpha \in E$. Then α is a zero of f if and only if $X - \alpha$ divides f in $E[X]$; that is, $f = (X - \alpha)g$ for some $g \in E[X]$.

Proof: By the division algorithm in $E[X]$, we may write

$$f = (X - \alpha)g + r, \qquad g, r \in E[X],$$

where $r = \mathbf{0}$ or $0 \leq \deg(r) < 1$. In either case, r is a constant polynomial, say $r = \beta \in E$, so that $f = (X - \alpha)g + \beta$. Evaluating f at α, we see that

$$f(\alpha) = \beta,$$

and therefore $f = (X - \alpha)g + f(\alpha)$. Thus, $f(\alpha) = 0$ if and only if $X - \alpha$ divides f in $E[X]$.

COROLLARY 2: f has at most n zeros in E.

Proof: Let $\alpha_1, \ldots, \alpha_k$ be distinct zeros of f in E. Then, in $E[X]$, f is divisible by $(X - \alpha_1)(X - \alpha_2) \cdots (X - \alpha_k)$. Therefore, $n \geq k$.

COROLLARY 3: Suppose that $n > 1$ and that f is irreducible in $E[X]$. Then f has no zeros in E.

Proof: For if f has a zero α in E, then f is divisible by $X - \alpha$ in $E[X]$, so that f is reducible in $E[X]$ (since $n > 1$).

Let us now prove the main result of this section.

THEOREM 4: There exists an extension E of F which contains a zero of f.

Proof: If f_1 is an irreducible factor of f in $F[X]$, then every zero of f_1 is also a zero of f. Therefore, without loss of generality, let us assume that f is irreducible in $F[X]$. Then $I = f \cdot F[X]$ is a maximal ideal of $F[X]$, so that by Theorem 8 of Section 5.9, $F[X]/I$ is a field. Since f is irreducible, f is not a constant. Therefore, the mapping

$$F \longrightarrow F[X]/I,$$
$$a \longrightarrow a + I$$

is an isomorphism. (We leave the verification as an exercise.) Thus, let us identify F with the subring $\{a + I \mid a \in F\}$ of $F[X]/I$, so that we can view $F[X]/I$ as an extension of F. Let $E = F[X]/I$ and set $\alpha = X + I \in E$. Then

$$\begin{aligned} f(\alpha) &= f(X) + I \\ &= 0 + I \qquad [\text{since } f(X) = f \in I], \end{aligned}$$

so that α is a zero of f in E.

As an easy consequence of Theorem 4 we deduce

COROLLARY 5: There exists an extension E of F such that, in $E[X]$, $f(X)$ can be written in the form

$$f(X) = \prod_{i=1}^{n} (X - \alpha_i), \qquad \alpha_i \in E.$$

Proof: By Theorem 4, there exists an extension E_1 of F in which $f(X)$ has a zero α_1. By Proposition 1, $f(X)$ is divisible by $X - \alpha_1$ in $E_1[X]$. Therefore, there exists a monic polynomial $g_1(X) \in E_1[X]$ such that $f(X) = (X - \alpha_1)g_1(X)$. Applying the same reasoning to $g_1(X)$ as we just applied to $f(X)$, we see that there exists an extension E_2 of E_1, $\alpha_2 \in E_2$, and $g_2(X)$ such that $g_1(X) = (X - \alpha_2)g_2(X)$. Therefore, $f(X) = (X - \alpha_1)(X - \alpha_2)g_2(X)$. Proceeding in this way, we can construct an extension E_n of F such that

$$f(X) = g_n(X)\prod_{i=1}^{n}(X - \alpha_i), \qquad g_n(X) \in E_n[X].$$

But since $n = \deg f(X)$, $\deg g_n(X) = 0$, so that $g_n(X) \in E_n$. However, because f is monic, $g_n(X) = 1$. Therefore, we may set $E = E_n$.

Let E denote an extension of F such that

$$f(X) = \prod_{i=1}^{n}(X - \alpha_i), \qquad \alpha_i \in E.$$

The subfield $F(\alpha_1, \ldots, \alpha_n)$ of E gotten by adjoining all zeros of $f(X)$ to F is called a *splitting field* of f. It is clear that $F(\alpha_1, \ldots, \alpha_n)$ is the smallest subfield of E in which $f(X)$ splits into a product of linear factors, whence the term "splitting field."

A simple argument utilizing Corollary 5 enables us to prove

COROLLARY 6: Let $f_1(X), f_2(X), \ldots, f_r(X)$ be monic polynomials belonging to $F[X]$ such that $\deg f_i(X) \geq 1$ $(1 \leq i \leq r)$. Then there exists an extension E of F such that, in $E[X]$, each polynomial $f_i(X)$ splits into a product of linear factors.

Proof: Exercise.

7.5 Exercises

1. Find the splitting fields over $\mathbf{Q}$ of the following polynomials:
   (a) $X^2 - 2$.
   (b) $X^4 + 1$.
   (c) $X^2 - 1$.
   (d) $X^4 - 16$.
   (e) $X^5 - 7$.
   (f) $X^2 + X + 1$.
   (g) $X^3 + 3X^2 + 3X - 4$.
   (h) $X^6 + X^3 + 1$.
2. Let p be a prime. Determine the splitting field of $1 + X + \cdots + X^{p-1}$ over $\mathbf{Q}$. [*Hint*: $X^p - 1 = (X - 1)(X^{p-1} + \cdots + 1)$.]
3. Let F be the splitting field over $\mathbf{Q}$ of $f \in \mathbf{Q}[X]$. Assume that $\deg(F/\mathbf{Q})$ is a prime and let α be any zero of f. Show that $F = \mathbf{Q}(\alpha)$, provided $\alpha \notin \mathbf{Q}$.
4. Let $f = X^2 + aX + b \in \mathbf{C}[X]$.
   (a) Show that f factors into linear factors in $\mathbf{C}$.
   (b) Determine the zeros of f in $\mathbf{C}$.
   (c) Let $F = \mathbf{Q}(a, b)$. Show that the splitting field of f over F is $F(\sqrt{\Delta})$, where $\Delta = a^2 - 4b$ and $\sqrt{\Delta}$ denotes one of the square roots of Δ in $\mathbf{C}$.

5. Let $f = X^3 + aX^2 + bX + c \in \mathbf{Q}[X]$. Show that the splitting field of f over $\mathbf{Q}$ is of degree 1, 2, 3, or 6 over $\mathbf{Q}$.
6. Let $f = X^3 + aX^2 + bX + c \in \mathbf{Q}[X]$ be irreducible.
   (a) Show that f has either three real zeros or a real zero and a pair of complex-conjugate zeros. [*Hint*: Apply Exercise 3 of Section 7.4.]
   (b) Let $E \subseteq \mathbf{C}$ be a splitting field of f over $\mathbf{Q}$. Show that $\deg(E/\mathbf{Q}) = 3$ or 6.
7. (a) Show that $X^2 + X + 1$ is irreducible in $\mathbf{Z}_2[X]$.
   (b) Prove that there exists a field with four elements.
   (c) Write down tables showing the laws of addition and multiplication for the field of (b).
8. Prove Corollary 6.

7.6 Some Examples

Let us now give some examples of splitting fields. Throughout the section let F be a field, $f \in F[X]$, E_f = a splitting field of f over F.

EXAMPLE 1: $F = \mathbf{Q}, f = X^2 + aX + b \ (a, b \in \mathbf{Q})$.

The zeros of f in $\mathbf{C}$ are given by the quadratic formula

$$\frac{-a + \sqrt{a^2 - 4b}}{2}, \qquad \frac{-a - \sqrt{a^2 - 4b}}{2},$$

where $\sqrt{a^2 - 4b}$ denotes one of the square roots of $a^2 - 4b$ in $\mathbf{C}$. Therefore, since $a \in \mathbf{Q}$, we see that

$$E_f = \mathbf{Q}(\sqrt{a^2 - 4b}). \tag{1}$$

Let us analyze the situation further: We consider two cases.

CASE 1: f reducible in $\mathbf{Q}[X]$.

In this case, the zeros of f are contained in $\mathbf{Q}$ and $E_f = \mathbf{Q}$ and $\deg(E_f/\mathbf{Q}) = 1$.

CASE 2: f irreducible in $\mathbf{Q}[X]$.

Note that by (1), E_f can be obtained from $\mathbf{Q}$ by adjoining either one of the zeros of f. Therefore, by Theorem 3 of Section 7.2,

$$\deg(E_f/\mathbf{Q}) = \deg(f) = 2.$$

EXAMPLE 2: $F = \mathbf{Q}, f = X^n - 1$.

The zeros of f in $\mathbf{C}$ are just the nth roots of 1. Let ζ be a primitive nth root of 1. Then the zeros of f are

$$1, \zeta, \zeta^2, \ldots, \zeta^{n-1}.$$

Since all these roots are contained in $\mathbf{Q}(\zeta)$, we see that

$$E_f = \mathbf{Q}(\zeta).$$

The field E_f is called the nth *cyclotomic field.* The term "cyclotomic" means "circle dividing," and the reason for the terminology is that the field E_f was first considered in connection with the problem of constructing a regular polygon of n sides using only a ruler and compass. We will come to this application in the next section.

Let us determine the degree of $\mathbf{Q}(\zeta)$ over $\mathbf{Q}$. The best way to do this is to determine $\operatorname{Irr}_{\mathbf{Q}}(\zeta, X)$. For by Theorem 3 of Section 7.2,

$$\deg(\mathbf{Q}(\zeta)/\mathbf{Q}) = \deg(\operatorname{Irr}_{\mathbf{Q}}(\zeta, X)). \tag{2}$$

Set

$$\Phi_n(X) = \prod_{\substack{i=0 \\ (i,n)=1}}^{n-1} (X - \zeta^i). \tag{3}$$

Then $\Phi_n(X)$ is called the nth *cyclotomic polynomial.* It first appears that $\Phi_n(X)$ has complex coefficients. But actually the coefficients of $\Phi_n(X)$ are rational and, in fact, are even integers! Note the zeros of $\Phi_n(X)$ are just the $\phi(n)$ distinct primitive nth roots of 1. Moreover, if η is an nth root of 1, let d be the order of η as an element of the group X_n of nth roots of 1. Then d is the smallest positive integer such that $\eta^d = 1$ and η is a primitive dth root of 1. Moreover, since the order of X_n is n and the order of an element divides the order of the group, we see that $d \mid n$. Thus, every nth root of 1 is a primitive dth root of unity for some uniquely determined d which divides n. Therefore,

$$\begin{aligned} X^n - 1 &= \prod_{\eta \in X_n} (X - \eta) \\ &= \prod_{\substack{d \mid n \\ d \geq 1}} \Big[\prod_{\substack{\eta \text{ is a primitive} \\ d\text{th root of } 1}} (X - \eta) \Big] \\ &= \prod_{d \mid n} \Phi_d(X). \end{aligned} \tag{4}$$

Let us use formula (4) to compute $\Phi_n(X)$ for the first few n. For $n = 1$, (4) reads

$$\Phi_1(X) = X - 1. \tag{5}$$

For $n = 2$, formula (4) reads

$$\Phi_1(X) \cdot \Phi_2(X) = X^2 - 1.$$

Therefore, by (5),

$$\Phi_2(X) = X + 1. \tag{6}$$

For $n = 3$, (4) reads

$$X^3 - 1 = \Phi_1(X) \cdot \Phi_3(X),$$

and thus by (5),

$$\Phi_3(X) = \frac{X^3 - 1}{X - 1} = X^2 + X + 1. \tag{7}$$

For $n = 4$, (4) reads

$$X^4 - 1 = \Phi_1(X) \cdot \Phi_2(X) \cdot \Phi_4(X),$$

and thus by (5) and (6),

$$\Phi_4(X) = \frac{X^4 - 1}{(X - 1)(X + 1)} = X^2 + 1. \tag{8}$$

From these few computations, it becomes clear that if we have computed $\Phi_d(X)$ for all $d < n$, then we may use (4) to compute $\Phi_n(X)$ from

$$\Phi_n(X) = (X^n - 1)/ \prod_{\substack{d \mid n \\ 1 \leq d < n}} \Phi_d(X). \tag{9}$$

We observe that $\Phi_1(X)$ has rational coefficients. Moreover, if $\Phi_d(X)$ is assumed to have rational coefficients for all $d < n$, then (9) implies that $\Phi_n(X)$ has rational coefficients, since $\Phi_n(X)$ is computed from $\Phi_d(X)$ $(d < n)$ using only rational operations. Thus, by induction,

$$\Phi_n(X) \in \mathbf{Q}[X] \qquad (n \geq 1). \tag{10}$$

A more careful analysis of the algorithm for computing Φ_n shows that $\Phi_n \in \mathbf{Z}[X]$. (See Exercise 5.) It is clear that $\Phi_n(X)$ is a monic polynomial having ζ as a zero. Thus, if we knew that $\Phi_n(X)$ were irreducible in $\mathbf{Q}[X]$, we could deduce that

$$\begin{aligned} \Phi_n(X) &= \operatorname{Irr}_{\mathbf{Q}}(\zeta, X), \\ \Longrightarrow \deg(\mathbf{Q}(\zeta)/\mathbf{Q}) &= \deg(\operatorname{Irr}_{\mathbf{Q}}(\zeta, X)) \\ &= \deg(\Phi_n(X)) \\ &= \phi(n), \end{aligned}$$

where $\phi(n)$ denotes Euler's ϕ-function. The proof of the irreducibility of $\Phi_n(X)$ is a rather delicate affair and we reserve its proof to Appendix A. But for the moment, let us record

THEOREM 1: If ζ is a primitive nth root of 1, then $\deg(\mathbf{Q}(\zeta)/\mathbf{Q}) = \phi(n)$.

EXAMPLE 3: $F = \mathbf{Q}, f = X^3 - 2$.

Let $\sqrt[3]{2}$ denote the real cube root of 2. Then the zeros of $X^3 - 2$ in $\mathbf{C}$ are all the cube roots of 2, given by

$$\sqrt[3]{2}, \quad \sqrt[3]{2}\,\omega, \quad \sqrt[3]{2}\,\omega^2,$$

where $\omega = (-1 + \sqrt{3}\,i)/2$ is a primitive cube root of 1. Moreover,

$$\begin{aligned} E_f &= \mathbf{Q}(\sqrt[3]{2}, \quad \sqrt[3]{2}\,\omega, \quad \sqrt[3]{2}\,\omega^2) \\ &= \mathbf{Q}(\sqrt[3]{2}, \omega). \end{aligned} \tag{11}$$

Since $X^3 - 2$ is irreducible in $\mathbf{Q}[X]$, by Eisenstein's criterion,

$$\deg(\mathbf{Q}(\sqrt[3]{2})/\mathbf{Q}) = 3. \tag{12}$$

Moreover, ω is a zero of $X^2 + X + 1 \in \mathbf{Q}(\sqrt[3]{2})[X]$, $\mathrm{Irr}_{\mathbf{Q}(\sqrt[3]{2})}(\omega, X) \mid X^2 + X + 1$, and therefore

$$\deg(\mathbf{Q}(\sqrt[3]{2}, \omega)/\mathbf{Q}(\sqrt[3]{2})) \leq 2.$$

Thus, by Theorem 3 of Section 7.3,

$$\begin{aligned} \deg(\mathbf{Q}(\sqrt[3]{2}, \omega)/\mathbf{Q}) &= \deg(\mathbf{Q}(\sqrt[3]{2}, \omega)/\mathbf{Q}(\sqrt[3]{2}))\cdot\deg(\mathbf{Q}(\sqrt[3]{2})/\mathbf{Q}) \\ &\leq 2\cdot 3 = 6 \end{aligned} \tag{13}$$

Note, however, that $X^2 + X + 1$ is irreducible over $\mathbf{Q}$ by Example 1. Therefore,

$$\deg(\mathbf{Q}(\omega)/\mathbf{Q}) = 2. \tag{14}$$

But by Theorem 3 of Section 7.3 and (14),

$$\begin{aligned} &\deg(\mathbf{Q}(\omega)/\mathbf{Q}) \mid \deg(\mathbf{Q}(\sqrt[3]{2}, \omega)/\mathbf{Q}) \\ &\Longrightarrow 2 \mid \deg(\mathbf{Q}(\sqrt[3]{2}, \omega)/\mathbf{Q}). \end{aligned} \tag{15}$$

Similarly,

$$\begin{aligned} &\deg(\mathbf{Q}(\sqrt[3]{2})/\mathbf{Q}) \mid \deg(\mathbf{Q}(\sqrt[3]{2}, \omega)/\mathbf{Q}) \\ &\Longrightarrow 3 \mid \deg(\mathbf{Q}(\sqrt[3]{2}, \omega)/\mathbf{Q}) \end{aligned} \tag{16}$$

by (15). By (15) and (16), we see that

$$\begin{aligned} &6 \mid \deg(\mathbf{Q}(\sqrt[3]{2}, \omega)/\mathbf{Q}) \\ &\Longrightarrow \deg(\mathbf{Q}(\sqrt[3]{2}, \omega)/\mathbf{Q}) \geq 6. \end{aligned} \tag{17}$$

Finally, by (13) and (17),

$$\deg(\mathbf{Q}(\sqrt[3]{2}, \omega)/\mathbf{Q}) = 6. \tag{18}$$

EXAMPLE 4: Let $F = \mathbf{Q}, f = X^n - a$ ($n \geq 1, a \in \mathbf{Q}$).

Let $a^{1/n}$ be any nth root of a in $\mathbf{C}$ and let ζ be a primitive nth root of 1. The zeros of f are just the nth roots of a and therefore are

$$a^{1/n}, a^{1/n}\zeta, a^{1/n}\zeta^2, \ldots, a^{1/n}\zeta^{n-1}.$$

Thus,

$$\begin{aligned} E_f &= \mathbf{Q}(a^{1/n}, a^{1/n}\zeta, \ldots, a^{1/n}\zeta^{n-1}) \\ &= \mathbf{Q}(a^{1/n}, \zeta). \end{aligned}$$

Calculating the degree of E_f over $\mathbf{Q}$ is a rather intricate problem, and we will not consider its complete solution. However, let us say what we can without undue expenditure of effort. Note that

$$E_f \supseteq \mathbf{Q}(\zeta) \supseteq \mathbf{Q}$$

and

$$\begin{aligned} \deg(E_f/\mathbf{Q}) &= \deg(E_f/\mathbf{Q}(\zeta))\cdot\deg(\mathbf{Q}(\zeta)/\mathbf{Q}) \\ &= \phi(n)\deg(E_f/\mathbf{Q}(\zeta)) \end{aligned} \tag{19}$$

by Theorem 1. Therefore, in order to compute $\deg(E_f/\mathbf{Q})$, we must compute $\deg(E_f/\mathbf{Q}(\zeta))$. And, in turn, to compute $\deg(E_f/\mathbf{Q}(\zeta))$, we must compute

$$\deg(\mathrm{Irr}_{\mathbf{Q}(\zeta)}(a^{1/n}, X)).$$

Since $X^n - a$ is polynomial in $\mathbf{Q}(\zeta)[X]$ having $a^{1/n}$ as a zero, we see that

$$\mathrm{Irr}_{\mathbf{Q}(\zeta)}(a^{1/n}, X) \mid X^n - a,$$

and therefore

$$\begin{aligned} \deg(E_f/\mathbf{Q}(\zeta)) &= \deg(\mathrm{Irr}_{\mathbf{Q}(\zeta)}(a^{1/n}, X)) \\ &\leq \deg(X^n - a) \\ &= n. \end{aligned}$$

Thus, finally, by (19), we see that

$$\deg(E_f/\mathbf{Q}) \leq n\phi(n). \tag{20}$$

In special cases, we can compute $\deg(E_f/\mathbf{Q})$ exactly. For example,

THEOREM 2: Let p be a prime and suppose that $f = X^p - a$ is irreducible in $\mathbf{Q}[X]$. Then

$$\deg(E_f/\mathbf{Q}) = p(p - 1).$$

Proof: Since p is prime, we have

$$\phi(p) = p - 1, \tag{21}$$

$$(p - 1, p) = 1. \tag{22}$$

By equations (19) and (21), $p - 1 \mid \deg(E_f/\mathbf{Q})$. Since $X^p - a$ is irreducible in $\mathbf{Q}[X]$,

$$\deg(\mathbf{Q}(a^{1/p})/\mathbf{Q}) = p.$$

Therefore, since

$$\begin{aligned} \deg(E_f/\mathbf{Q}) &= \deg(E_f/\mathbf{Q}(a^{1/p}))\cdot\deg(\mathbf{Q}(a^{1/p})/\mathbf{Q}) \\ &= p\cdot\deg(E_f/\mathbf{Q}(a^{1/p})), \end{aligned}$$

we see that

$$p \mid \deg(E_f/\mathbf{Q}).$$

Thus, by (22), we have $p(p - 1) \mid \deg(E_f/\mathbf{Q})$. However, by (20), we see that $\deg(E_f/\mathbf{Q}) \leq p(p - 1)$, and hence the proof of the theorem is complete.

EXAMPLE 5: Let F be a subfield of $\mathbf{C}$ which contains the nth roots of 1 for some $n \geq 1$, and let $f = X^n - a$ $(a \in F)$. If $a^{1/n}$ is one nth root of a in $\mathbf{C}$, then the zeros of f are

$$a^{1/n}, a^{1/n}\zeta, \ldots, a^{1/n}\zeta^{n-1},$$

where ζ is a primitive nth root of 1. Therefore, since $\zeta \in F$,

$$\begin{aligned} E_f &= F(a^{1/n}, a^{1/n}\zeta, \ldots, a^{1/n}\zeta^{n-1}) \\ &= F(a^{1/n}). \end{aligned}$$

The extension $F(a^{1/n})/F$ is a typical example of what is called a *Kummer extension.*

7.6 Exercises

1. Compute $\Phi_n(X)$ for $n \leq 10$.
2. Let p be a prime, $n \geq 1$. Show that

$$\Phi_{p^n}(X) = \frac{X^{p^n} - 1}{X^{p^{n-1}} - 1} = 1 + X^{p^{n-1}} + X^{2p^{n-1}} + \cdots + X^{(p-1)p^{n-1}}.$$

3. Show that $\Phi_{p^n}(X)$ is irreducible, where p is a prime. [*Hint*: Show that $\Phi^{p^n}(X + 1)$ is irreducible by using the binomial theorem and the Eisenstein irreducibility criterion.]
4. Let $n \geq 1$ satisfy $(n, \phi(n)) = 1$ and let $x^n - a (a \in \mathbf{Q})$ be irreducible in $\mathbf{Q}[X]$. Show that the splitting field of $X^n - a$ over $\mathbf{Q}$ has degree $n\phi(n)$. [*Hint:* Apply Exercise 8 of Section 7.3.]

7.7 Constructions with Straightedge and Compass

We may not appear to have accomplished anything of significance yet concerning the theory of fields, but we have enough machinery to apply to several significant problems. In this section we will consider the problem of constructing geometric figures using only a compass and a straightedge. (For our purposes, a straightedge is a device which can be used only to draw a line between two points in the plane. It has no measurement capabilities.) Such constructions were first considered by the Greek geometers and are probably familiar to the student from his high school course in geometry. The Greeks posed three very interesting questions concerning constructibility which they were unable to solve:

PROBLEM 1 (TRISECTING AN ANGLE): Given an arbitrary angle θ, construct the angle $\theta/3$, using only straightedge and compass.

PROBLEM 2 (DUPLICATING A CUBE): Given an arbitrary cube C of volume V, construct a cube having volume $2V$, using only straightedge and compass.

PROBLEM 3 (SQUARING A CIRCLE): Given an arbitrary circle C of area A, construct a square having area A, using only straightedge and compass.

Our plan in this section is to first discuss the general problem of constructibility. From this discussion will emerge a powerful theory which will provide immediate solutions to Problems 1 and 2 and will point the way to a solution of Problem 3. It will turn out that all three problems have no solution. We will then turn our attention to the construction of regular polygons.

Let us first lay down the ground rules for our investigation. Let us restrict ourselves to plane constructions. Further, we will identify the Euclidean plane with the complex numbers by identifying the point with coordinates (a, b) with the complex number $a + bi$. Finally, let us assume that in addition to a compass and straightedge, we are given the line segment connecting (0, 0) and (1, 0), somehow marked off on our plane.† Our problem is to decide which geometric figures can be constructed. Henceforth, when we refer to "construction" we will always mean "construction with straightedge and compass, given the above unit segment as initial data."

Let us now reduce our geometric problem to an algebraic problem. It is clear that each geometric figure which we can construct can be described by a finite number of points, line segments, and circular arcs. Let us describe a given geometrical figure by means of a family of complex numbers as follows: The finite number of points can be viewed as complex numbers as described above. A line segment is described by the complex numbers corresponding to its endpoints. A circular arc is described by four numbers: two corresponding to the endpoints of the arc, one corresponding to the center of the circle, and one equal to the radius of the circle. It is clear that the numbers described are both necessary and sufficient to construct the geometric figure. Thus, let us henceforth think of a geometric figure in terms of a collection of complex numbers describing the points, lines, and arcs in the figure. It is clear that in order to construct a given figure, it is both necessary and sufficient to be able to construct the line segments connecting (0, 0) to each of the complex numbers describing the figure. This leads us to make the following

DEFINITION 1: A complex number α is said to be constructible if the line segment connecting (0, 0) to α is constructible.

From our above discussion, we have

THEOREM 2: A geometrical figure is constructible if and only if each of the complex numbers describing the figure is constructible.

Thus, our original question concerning constructibility of geometric figures is reduced to one concerning constructibility of complex numbers. Let C denote the set of all constructible complex numbers. We will give a more-or-less complete description of C.

† This line segment of length 1 is just used as a comparison device. We could just as well begin our constructions by drawing a line segment of arbitrary length and using it for comparison.

Before proceeding further, let us recall some of the possible elementary constructions of Euclidean geometry. For the actual construction, we refer back to high school geometry. The reader will probably have little difficulty in supplying them. [*Hint:* (5), (6), (7) require some elementary facts about similar triangles.]

Basic Constructions

1. Bisect a given angle.
2. Construct a line perpendicular to a given line segment L at a given point P on L.
3. Construct a line through a given point P which is parallel to a given line L.
4. Given line segments of lengths l and l', construct a line segment of length ll'.
5. Given a line segment of length l, construct a line segment of length $1/l$.
6. Given a line segment of length l, construct a line segment of length $\sqrt{l}$.
7. Construct an angle equal to the sum of two given angles.

THEOREM 3: C is a subfield of $\mathbf{C}$.

Proof: Our assertion amounts to the following: If α and β are constructible, then so are $\alpha + \beta$, $-\beta$, $\alpha \cdot \beta$, $1/\beta$ ($\beta \neq 0$). Let us consider each case separately. Recall that addition of complex numbers can be accomplished geometrically via the parallelogram law. Therefore, let us construct $\overline{O\alpha}$ and $\overline{O\beta}$ as in Figure 7.1. Use basic construction 4 to construct line $\overline{\beta\gamma}$ parallel to $\overline{O\alpha}$. On $\overline{\beta\gamma}$ measure off a line segment of length $\overline{O\alpha}$, with one endpoint of the line segment at β. Then the other end of the measured segment is $\alpha + \beta$. Thus, $\alpha + \beta$ is constructible.

In order to construct $-\beta$, construct $\overline{O\beta}$ as in Figure 7.2. Continue the line segment $O\beta$ through the origin and on the continued portion measure off a segment of length $\overline{O\beta}$, having O as one endpoint. Then the other endpoint of the measured segment is $-\beta$.

Let us construct $\alpha \cdot \beta$. Without loss of generality, we may assume that $\alpha \neq 0$, $\beta \neq 0$. Then α and β can both be written in polar form

$$\alpha = r_1(\cos\theta_1 + i\sin\theta_1),$$
$$\beta = r_2(\cos\theta_2 + i\sin\theta_2)$$

By de Moivre's theorem,

$$\alpha \cdot \beta = r_1 r_2[\cos(\theta_1 + \theta_2) + i\sin(\theta_1 + \theta_2)]$$

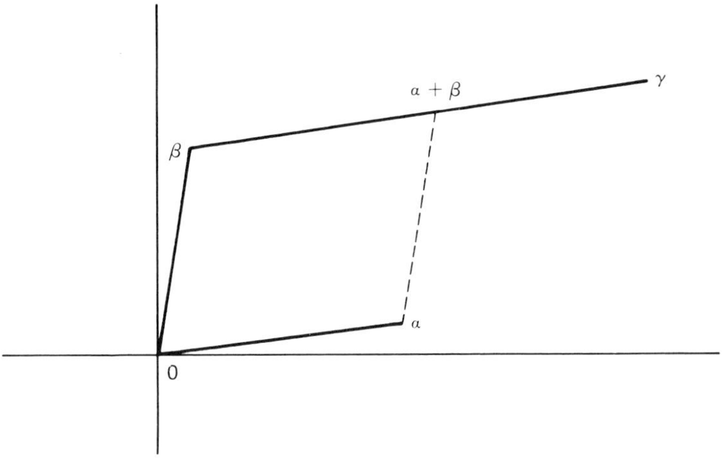

FIGURE 7-1: Construction of $\alpha + \beta$.

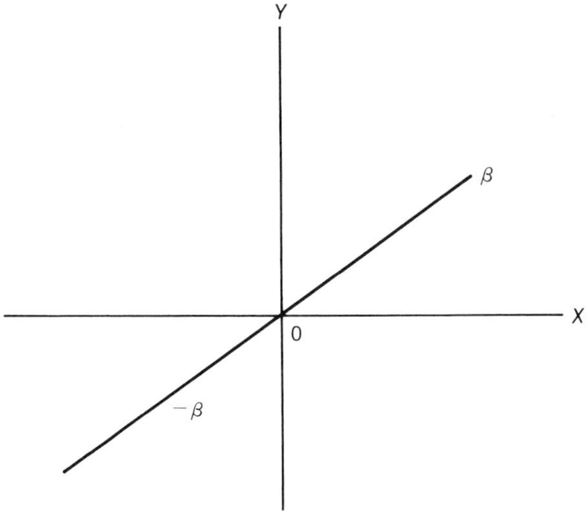

FIGURE 7-2: Construction of $-\beta$.

Thus, the line segment $O\alpha\cdot\beta$ has length r_1r_2 and makes an angle $\theta_1 + \theta_2$ with the positive X-axis. This is the clue to the construction of $\alpha\cdot\beta$. Refer to Figure 7.3, where α and β have been constructed. Construct the line segment $\overline{O\gamma}$ making an angle $\theta_1 + \theta_2$ with the positive X-axis. Measure off a segment $\overline{O\delta}$ of length r_1r_2 on $\overline{O\gamma}$. This is possible by basic construction 4. Then $\delta = \alpha\cdot\beta$.

Finally, assume that $\beta \neq 0$. Then, since $\sin^2\theta_2 + \cos^2\theta_2 = 1$ and $\cos(-\theta_2)$

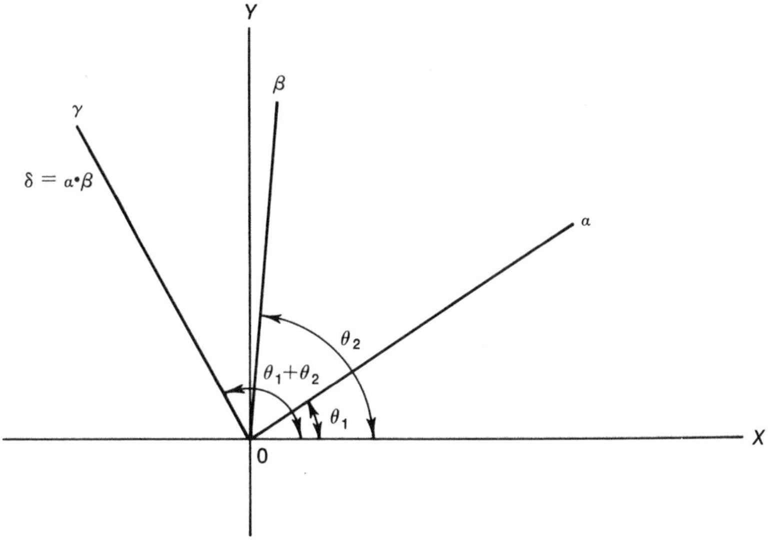

FIGURE 7-3: Construction of $\alpha \cdot \beta$.

$= \cos \theta_2$, $\sin(-\theta_2) = -\sin \theta_2$, we have

$$\begin{aligned}\frac{1}{\beta} &= r_2^{-1}(\cos \theta_2 + i \sin \theta_2)^{-1} \\ &= r_2^{-1} \frac{\cos \theta_2 - i \sin \theta_2}{\cos^2\theta_2 + \sin^2\theta_2} \\ &= r_2^{-1}[\cos(-\theta_2) + i \sin(-\theta_2)].\end{aligned}$$

Thus, $\overline{O\beta^{-1}}$ has length r_2^{-1} and makes an angle $-\theta_2$ with the positive X-axis. We leave it to the reader to apply basic construction 5 to construct β^{-1}.

COROLLARY 4: All rational numbers are constructible.

Proof: We have been given 1 as part of our initial data. Thus, $1 \in C$. But since C is a field, this implies that $\mathbf{Q} \subseteq C$.

PROPOSITION 5: Let $\alpha \in \mathbf{C}$ be constructible and let $\sqrt{\alpha}$ denote one of the square roots of α. Then $\sqrt{\alpha}$ is constructible.

Proof: Without loss of generality, assume $\alpha \neq 0$. By de Moivre's theorem,

$$\sqrt{\alpha} = \pm\sqrt{r_1}\left[\cos\left(\frac{\theta_1}{2}\right) + i \sin\left(\frac{\theta_1}{2}\right)\right].$$

Let us only consider the positive sign. The reasoning for the negative sign is similar. Then $\overline{O\sqrt{\alpha}}$ is a line segment of length $\sqrt{r_1}$ which makes an angle of $\theta_1/2$ with the positive X-axis. Therefore, $\sqrt{\alpha}$ can be constructed using basic constructions 1 and 6. We leave the details to the reader.

THEOREM 6: Let $\alpha_1, \ldots, \alpha_n$ be complex numbers such that

$$\alpha_1^2 \in \mathbf{Q},$$
$$\alpha_i^2 \in \mathbf{Q}(\alpha_1, \ldots, \alpha_{i-1}) \qquad (2 \leq i \leq n).$$

Then every element of $\mathbf{Q}(\alpha_1, \ldots, \alpha_n)$ is constructible.

Proof: By Corollary 7-7-4, every element of $\mathbf{Q}$ is constructible and by Proposition 5 and the assumption $\alpha_1^2 \in \mathbf{Q}$, we see that α_1 is constructible. Therefore, since the constructible numbers form a field, every element of $\mathbf{Q}(\sqrt{\alpha_1})$ is constructible. Thus the theorem is true for $n = 1$. Let us proceed by induction on n. Let $n > 1$ and assume the theorem for $n - 1$. Then every element of $\mathbf{Q}(\alpha_1, \ldots, \alpha_{n-1})$ is constructible. By hypothesis, $\alpha_n^2 \in \mathbf{Q}(\alpha_1, \ldots, \alpha_{n-1})$ so that α_n is a square root of a constructible number and hence is constructible by Proposition 5. But since the constructible numbers form a field, every element of $\mathbf{Q}(\alpha_1, \ldots, \alpha_{n-1})(\alpha_n) = \mathbf{Q}(\alpha_1, \ldots, \alpha_n)$ is constructible.

The amazing fact is that the converse of the last theorem also is true! We have

THEOREM 7: Let $\beta \in \mathbf{C}$ be constructible. Then there exists a set of complex numbers $\{\alpha_1, \ldots, \alpha_n\}$ such that

$$\alpha_1^2 \in \mathbf{Q},$$
$$\alpha_i^2 \in \mathbf{Q}(\alpha_1, \ldots, \alpha_{i-1}) \qquad (2 \leq i \leq n)$$

and such that $\beta \in \mathbf{Q}(\alpha_1, \ldots, \alpha_n)$. Thus, if β is constructible, $\deg(\mathbf{Q}(\beta)/\mathbf{Q}) = 2^k$ for some k.

Before proceeding with the proof of Theorem 7, let us do some preliminary work. Assume that the points $\beta_1, \beta_2, \ldots, \beta_{m-1}$ have been constructed. What points can we construct using $\beta_1, \ldots, \beta_{m-1}$? There are two elementary constructions which we can perform: (a) We can draw a line L connecting β_i to β_j $(i \neq j)$. (b) We can draw a circle C with center at β_i which passes through β_j $(i \neq j)$. Thus, if β_m is constructible from $\beta_1, \ldots, \beta_{m-1}$, then β_m is either the intersection of two lines L_1 and L_2, the intersection of a line L and a circle C, or the intersection of two circles C_1 and C_2. Let

$$\beta_j = \gamma_j + i\delta_j \qquad (\gamma_j, \delta_j \in \mathbf{R}, 1 \leq j \leq m), \tag{1}$$
$$F_j = \mathbf{Q}(i, \gamma_1, \delta_1, \ldots, \gamma_j, \delta_j) \qquad (1 \leq j \leq m). \tag{2}$$

We will consider separately three cases:

Case 1: β_m is the intersection of the lines L_1 and L_2.

The lines L_1 and L_2 have equations

$$\begin{aligned} L_1: &\quad a_1x + b_1y + c_1 = 0, \\ L_2: &\quad a_2x + b_2y + c_2 = 0, \end{aligned} \tag{3}$$

where $a_1, b_1, c_1, a_2, b_2, c_2 \in F_{m-1}$. The system of equations (3) has the solution $x = \gamma_m$, $y = \delta_m$, by assumption. But the solution of the system (3) can be calculated rationally in terms of $a_1, b_1, c_1, a_2, b_2, c_2$ and therefore lies in F_{m-1}. Thus, we see that

$$\begin{aligned} &\gamma_m \in F_{m-1}, \qquad \delta_m \in F_{m-1}. \\ &\Longrightarrow F_m = F_{m-1}, \end{aligned} \tag{4}$$

since $F_m = F_{m-1}(\gamma_m, \delta_m)$ if $m \geq 2$.

Case 2: β_m is the intersection of the line L and the circle C.

The line L and the circle C have equations

$$\begin{aligned} L: &\quad ax + by + c = 0, \\ C: &\quad dx^2 + ey^2 + fx + gy + h = 0, \end{aligned} \tag{5}$$

where $a, b, c, d, e, f, g, h \in F_{m-1}$. One of the solutions of the system of equations (5) is $x = \gamma_m$, $y = \delta_m$, by assumption. On the other hand, the system (5) can be solved by substituting the linear relation into the quadratic equation and then solving for, say, x, from which the corresponding value of y can be computed. Therefore, the solutions of the system (5) can be computed in terms of the square root of an element η, where η is computable rationally in terms of a, b, c, d, e, f, g, h. In particular, $\eta \in F_{m-1}$ and $\gamma_m \in F_{m-1}(\sqrt{\eta})$, $\delta_m \in F_{m-1}(\sqrt{\eta})$. Further,

$$F_m \subseteq F_{m-1}(\sqrt{\eta}), \tag{6}$$

since $F_m = F_{m-1}(\gamma_m, \delta_m)$ if $m \geq 2$.

Case 3: β_m is the intersection of the circles C_1 and C_2.

The equations of C_1 and C_2 are given by

$$\begin{aligned} C_1: &\quad x^2 + y^2 + a_1x + b_1y + c_1 = 0, \\ C_2: &\quad x^2 + y^2 + a_2x + b_2y + c_2 = 0, \end{aligned}$$

where $a_1, b_1, c_1, a_2, b_2, c_2 \in F_{m-1}$. The points of intersection of C_1 with C_2 are the same as the points of intersection of C_1 with the line whose equation is

$$(a_1 - a_2)x + (b_1 - b_2)y + (c_1 - c_2) = 0.$$

Therefore, by case 2, there exists η in F_{m-1} such that

$$F_m \subseteq F_{m-1}(\sqrt{\eta}). \tag{7}$$

By comparing the results of cases 1–3, we have

LEMMA 8: Suppose that the points $\beta_1, \ldots, \beta_{m-1}$ $(m \geq 2)$ have been constructed and that β_m is constructible from $\beta_1, \ldots, \beta_{m-1}$. Further, suppose that $\beta_j = \gamma_j + i\delta_j$, where γ_j and δ_j are real, and let

$$F_j = \mathbf{Q}(i, \gamma_1, \delta_1, \ldots, \gamma_j, \delta_j) \qquad (1 \leq j \leq m).$$

Then there exists $\eta \in F_{m-1}$ such that $F_m \subseteq F_{m-1}(\sqrt{\eta})$.

Proof: By (6) and (7) we are done in cases 2 and 3. In case 1 we may set $\eta = 1$ by (4).

Let us now prove Theorem 7. Suppose that β can be constructed by successively constructing $\beta_1, \beta_2, \ldots, \beta_n = \beta$, where β_1 is constructed from $\beta_0 = 1$. For $m = 1, \ldots, n$ let F_m be as in the Lemma, and set $F_0 = \mathbf{Q}(i)$. Let us apply Lemma 8 to each of the sets of numbers $\beta_0, \beta_1, \ldots, \beta_m$ $(1 \leq m \leq n)$ to get that there exists $\eta_m \in F_{m-1}$ such that

$$F_m \subseteq F_{m-1}(\sqrt{\eta_m}) \qquad (1 \leq m \leq n). \tag{8}$$

Define

$$\alpha_0 = i, \quad \alpha_j = \sqrt{\eta_j} \qquad (1 \leq j \leq n).$$

We assert that

$$F_m \subseteq \mathbf{Q}(\alpha_0, \alpha_1, \ldots, \alpha_m) \qquad (1 \leq m \leq n). \tag{9}$$

This follows trivially by induction from (8). It is clearly true for $m = 1$ by (8). Assume that $m > 1$ and that (9) is true for $m - 1$. Then, by the induction hypothesis and (8),

$$\begin{aligned} F_m &\subseteq F_{m-1}(\alpha_m) \\ &\subseteq \mathbf{Q}(\alpha_0, \ldots, \alpha_{m-1})(\alpha_m) \\ &= \mathbf{Q}(\alpha_0, \ldots, \alpha_m). \end{aligned}$$

This completes the induction and hence (9) is proved. It is clear that $\alpha_0^2 = -1 \in \mathbf{Q}$. Also, by (9), for $1 \leq m \leq n$, we have

$$\alpha_m^2 = \eta_m \in F_{m-1} \subseteq \mathbf{Q}(\alpha_0, \ldots, \alpha_{m-1}).$$

Finally, by (9) for $m = n$, and the fact that $\beta_n = \beta$, we have

$$\beta = \gamma_n + i\delta_n \in F_n \subseteq \mathbf{Q}(\alpha_0, \ldots, \alpha_n).$$

This completes the proof of the first assertion of Theorem 7. In order to show that $\deg(\mathbf{Q}(\beta)/\mathbf{Q}) = 2^k$ for some k, it suffices to show that

$$\deg(\mathbf{Q}(\alpha_0, \ldots, \alpha_n)/\mathbf{Q}) = 2^r$$

for some r. But, if we set $\tilde{F}_i = \mathbf{Q}(\alpha_0, \ldots, \alpha_i)$ $(0 \leq i \leq n)$, we have† $\deg(\tilde{F}_{i+1}/\tilde{F}_i) = 1$ or 2 since $\alpha_{i+1}^2 \in \tilde{F}_i$ and $\tilde{F}_{i+1} = \tilde{F}_i(\alpha_{i+1})$. Moreover,

$$\begin{aligned} \deg(\mathbf{Q}(\alpha_0, \ldots, \alpha_n)/\mathbf{Q}) &= \deg(\tilde{F}_n/\tilde{F}_{n-1}) \cdot \deg(\tilde{F}_{n-1}/\tilde{F}_{n-2}) \ldots \deg(\tilde{F}_1/\tilde{F}_0) \\ &= 2^r \end{aligned}$$

for some r.

† See Exercise 3.

Let us now illustrate Theorem 7 by applying it to the problem of trisecting an angle. Not only will we show that there is no general procedure for trisecting an angle with compass and straightedge, but there are particular angles which cannot be trisected! For example, let us prove:

THEOREM 9: A $60°$ angle cannot be trisected using only a compass and a straightedge.

Proof: It is well known that a $60°$ angle can be constructed using only a compass and straightedge. (Exercise.) Therefore, a $60°$ angle can be trisected if and only if it is possible to construct a $20°$ angle using only a compass and a straightedge. But let us see what it means for an angle θ to be constructible. If θ is constructible, then we may place one side of θ on the X-axis and construct the point where the other side of θ intersects the circle of radius 1 with center at the origin (see Figure 7.4). But by elementary trigonometry, this point is $\cos\theta + i\sin\theta$. Therefore, if an angle θ is constructible, the number $\cos\theta + i\sin\theta$ is constructible. Conversely, if $\cos\theta + i\sin\theta$ is constructible, then an angle θ is constructible (see Figure 7.4). Therefore, a $20°$ angle is constructible if and only if

$$\cos(20°) + i\sin(20°) = \zeta$$

is constructible.

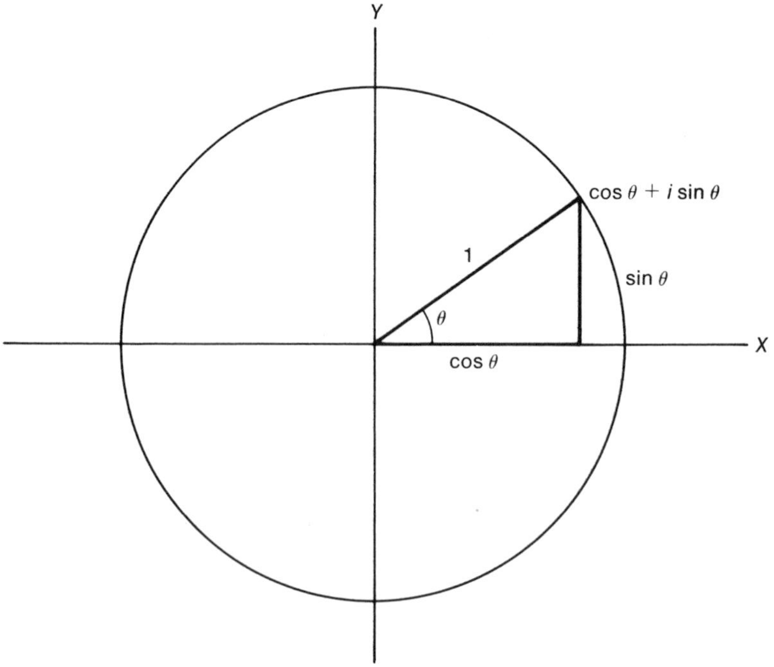

FIGURE 7-4: Construction of the Angle θ.

Let us assume that ζ is constructible. By Theorem 7, $\deg(\mathbf{Q}(\zeta)/\mathbf{Q}) = 2^r$ for some r. However, ζ is a primitive eighteenth root of 1, so that by Theorem 1 of Section 7.6,

$$\begin{aligned}\deg(\mathbf{Q}(\zeta)/\mathbf{Q}) &= \phi(18)\\ &= 6.\end{aligned}$$

Thus, a contradiction is reached and ζ is not constructible. Therefore, an angle of 60° cannot be trisected using only straightedge and compass.

Let us next turn to the problem of duplicating a cube. For simplicity's sake let us consider a cube C of side 1. We wish to construct a cube of volume 2. This is equivalent to constructing a line segment of length $\sqrt[3]{2}$. Let us show that the number $\sqrt[3]{2}$ is not constructible. Note that

$$\deg(\mathbf{Q}(\sqrt[3]{2})/\mathbf{Q}) = 3.$$

Therefore, by Theorem 7, we see that $\sqrt[3]{2}$ is not constructible. Thus, we have

THEOREM 10: It is impossible to duplicate the cube of side 1 using only a straightedge and compass.

Let us now turn to the problem of squaring the circle. For the sake of simplicity, let us consider the circle of radius 1. It has area equal to π. Thus, we are asked to construct a square of area π. This is equivalent to constructing a line segment of length $\sqrt{\pi}$. Thus, the question of squaring the circle boils down to: Is $\sqrt{\pi}$ constructible? We cannot give a complete proof of the fact that the answer is no. However, let us at least give an indication of the idea involved. If $\sqrt{\pi}$ is constructible, then $\deg(\mathbf{Q}(\sqrt{\pi})/\mathbf{Q})$ is a power of 2. In particular, $\sqrt{\pi}$ is algebraic over $\mathbf{Q}$ by Theorem 5 of Section 7.2. Actually, $\sqrt{\pi}$ is transcendental over $\mathbf{Q}$. However, the proof of this last statement is very deep and will not be included in this book.

Let us now take up the problem of the construction of a regular polygon of n sides. Constructing such a polygon is equivalent to constructing an angle of $360/n$. But, as we saw in the proof of Theorem 9, the angle θ can be constructed if and only if $\cos\theta + i\sin\theta$ is constructible. Therefore, a regular polygon of n sides can be constructed if and only if

$$\cos\left(\frac{360}{n}\right) + i\sin\left(\frac{360}{n}\right) = \zeta_n$$

is constructible. But ζ_n is a primitive nth root of 1 and $\deg(\mathbf{Q}(\zeta_n)/\mathbf{Q}) = \phi(n)$. Therefore, by Theorem 7, if ζ_n is constructible, $\phi(n) = 2^r$ for some r. Thus, if a regular polygon of n sides is constructible, then $\phi(n) = 2^r$. Actually, the converse is also true, but we will leave the proof of this fact to the exercises. Let us investigate the values of n for which $\phi(n)$ is a power of 2. By consulting Table 7.1, we see that this is not always true. For example $\phi(7) = 6$, so that it is impossible to construct a regular 7-gon.

TABLE 7.1
Values of $\phi(\mathbf{n})$

n	$\phi(n)$	n	$\phi(n)$
2	1	12	4
3	2	13	12
4	2	14	6
5	4	15	8
6	2	16	8
7	6	17	16
8	4	18	6
9	6	19	18
10	4	20	8
11	10	21	12

Let

$$n = p_1^{r_1} \cdot p_2^{r_2} \cdots p_t^{r_t} \qquad (r_i > 0)$$

be the decomposition of n into a product of powers of distinct primes $p_1, \ldots, p_t$. Then

$$\phi(n) = \phi(p_1^{r_1}) \cdots \phi(p_t^{r_t}), \tag{10}$$

so that $\phi(n)$ is a power of 2 if and only if $\phi(p_j^{r_j})$ is a power of 2 $(1 \leq j \leq t)$. Note that

$$\phi(p_j^{r_j}) = p_j^{r_j-1}(p_j - 1).$$

Therefore, if $p_j = 2$, $\phi(p_j^{r_j})$ is automatically a power of 2. Thus, we are reduced to the following question: Let p be an odd prime, r a positive integer. When is $p^{r-1}(p - 1)$ a power of 2? It is clearly necessary and sufficient that $r = 1$ and $p - 1 = 2^k$ for some k. However, if $p - 1 = 2^k$, then k must be a power of 2. For if k is divisible by an odd prime q, we must have $k = qu$ for some positive integer u. But then

$$\begin{aligned} p &= 2^k + 1 \\ &= 2^{qu} + 1 \\ &= (2^u + 1)(2^{(q-1)u} - 2^{(q-2)u} + \cdots + 1), \end{aligned}$$

which contradicts the fact that p is prime. Thus, we have shown that if $p^{r-1}(p - 1)$ is a power of 2, then $r = 1$ and $p = 2^{2^\nu} + 1$. And conversely, if $p = 2^{2^\nu} + 1$, then $\phi(p)$ is a power of 2. A prime p of the form

$$2^{2^\nu} + 1$$

is called a *Fermat prime*. The first few Fermat primes after 3 are

$$\begin{aligned} 2^{2^1} + 1 &= 5, \\ 2^{2^2} + 1 &= 17, \\ 2^{2^3} + 1 &= 257, \\ 2^{2^4} + 1 &= 65{,}537. \end{aligned}$$

It is not true that every integer of the form $2^{2^\nu} + 1$ is prime. For as Euler showed, $2^{2^5} + 1$ is divisible by 641. The net result of our discussion above is

THEOREM 11: In order for a regular polygon of n sides to be constructible using only straightedge and compass, it is necessary and sufficient for n to be of the form

$$n = 2^r p_1 p_2 \ldots p_w,$$

where $p_1, p_2, \ldots, p_w$ are distinct Fermat primes.

7.7 Exercises

1. Carry out each of the basic constructions.
2. Construct β^{-1} ($\beta \in \mathbf{C}$, $\beta \neq 0$).
3. Let F be a field, $E = F(\alpha)$ with $\alpha^2 \in F$. Show that $\deg(E/F) = 1$ or 2.
4. Let a be a positive integer.
   (a) If $360/a$ can be trisected, then $\phi(3a)$ is a power of 2, and conversely.
   (b) If a is a Fermat prime > 3, then $360/a$ can be trisected.
   [*Remark*: In Chapter 9 we will prove that ζ_n is constructible if and only if $\deg(\mathbf{Q}(\zeta_n)/\mathbf{Q})$ is a power of 2. One direction is clear from the results of this section. The reader may assume the converse in the proof of this problem.]
5. Let ζ be the primitive fifth root of unity $\zeta = \cos(360/5) + i\sin(360/5)$. Show that

$$(\zeta + \zeta^{-1})^2 + (\zeta + \zeta^{-1}) - 1 = 0,$$

   and thus

$$\zeta + \zeta^{-1} = \frac{-1 + \sqrt{5}}{2}.$$

   Deduce that $\cos(360/5) = (-1 + \sqrt{5})/4$, $\sin(360/5) = \frac{1}{4}\sqrt{10 + 2\sqrt{5}}$, and thus

$$\zeta = \frac{-1 + \sqrt{5}}{4} + i\frac{1}{4}\sqrt{10 + 2\sqrt{5}}$$

   Use these formulas to construct a regular pentagon. (This construction was known to the Greeks.)

*6. Prove that $\zeta = \cos(2\pi/17) + i\sin(2\pi/17)$ is constructible and hence a regular 17-gon is constructible. (This construction is due to the eighteenth-century Swiss mathematician Leonhard Euler.)

7. (a) Show that $\cos(3x) = 4\cos^3(x)\ 3 - \cos(x)$.
   (b) Show that an angle x is constructible $\Leftrightarrow \cos(x)$ is constructible.
   (c) Suppose that an angle x is given. Show that x is trisectable $\Leftrightarrow$ the "trisection equation" $4x^3 - 3x - \cos x$ is reducible in $\mathbf{Q}(\cos(x))[X]$.

7.8 Algebraically Closed Fields, I

In the beginning of the nineteenth century, Gauss proved the following remarkable property of the field **C** of complex numbers:

THEOREM 1: Let $f(X) \in \mathbf{C}[X]$ be nonconstant. Then, in $\mathbf{C}[X]$, $f(X)$ splits into a product of linear polynomials.

This fact, while not surprising to a student who has spent some time trying to solve polynomial equations, is rather deep. One of the shortest proofs of Gauss's theorem relies on complex analysis and uses Liouville's theorem to the effect that a bounded, entire function is constant. In Chapter 10 we will give a proof, due to Artin, which is almost purely algebraic in character and which uses Galois theory. Artin's proof makes use of the fact that every polynomial of odd degree and having real coefficients has a real zero. However, for the rest of this chapter, let us assume Gauss's theorem. This will not result in any circular reasoning since we will use the theorem only for purposes of illustration.

Gauss's theorem asserts that the field of complex numbers has a rather special property which we now formalize in a definition:

DEFINITION 2: A field F is said to be algebraically closed if every nonconstant polynomial in $F[X]$ splits into a product of linear factors in $F[X]$.

EXAMPLE 1: According to Gauss's theorem, **C** is algebraically closed.

EXAMPLE 2: Let

$$\bar{\mathbf{Q}} = \{\alpha \in \mathbf{C} \mid \alpha \text{ is algebraic over } \mathbf{Q}\}.$$

We will prove that $\bar{\mathbf{Q}}$ is algebraically closed.

PROPOSITION 3: The following statements are equivalent:

(1) F is algebraically closed.
(2) If E is an algebraic extension of F, then $E = F$.

Proof: (1) $\Longrightarrow$ (2). Let E be an algebraic extension of F and let $\alpha \in E$. Then α is algebraic over F. Moreover, $f = \operatorname{Irr}_F(\alpha, X)$ is a nonconstant, monic polynomial in $F[X]$. Therefore, since F is algebraically closed,

$$f = \prod_{i=1}^{n} (X - \alpha_i), \qquad \alpha_i \in F.$$

But $\alpha \in E$ is a zero of f, so that $X - \alpha$ divides f (in $E[X]$). Therefore, $X -$

$\alpha = X - \alpha_i$ for some i, and thus $\alpha = \alpha_i$ for some i, so that $\alpha \in F$. Therefore, $E \subseteq F$. But by assumption $E \supseteq F$, so we have proved $E = F$.

(2) $\Rightarrow$ (1). Let f be a nonconstant polynomial in $F[X]$. We must show that f splits into a product of linear factors in $F[X]$. Without loss of generality, we may restrict ourselves to the case of f monic and irreducible in $F[X]$. In this special case, we will prove that $\deg(f) = 1$ so that f is linear. Let $n = \deg(f)$. By adjoining a root α of f to F, we get an extension $F(\alpha)$ of F of degree n (Theorem 3 of Section 7.2). But by Proposition 1 of Section 7.3, $F(\alpha)$ is an algebraic extension of F. Therefore, by our hypothesis (2), $F(\alpha) = F$, so that $n = \deg(F(\alpha)/F) = 1$. Thus, f is linear.

In Section 7.5 we showed that it is possible to construct an extension E of a given field F in which any given finite collection of nonconstant polynomials in $F[X]$ splits into linear factors. At this point we would like to strengthen this result by proving that F is contained in an algebraically closed field. This result will prove to be of great utility to us in studying the properties of F, since we can fix, once and for all, an extension field F, which is so large as to accommodate the roots of all polynomial equations over F within it. Our main result is

THEOREM 4: Let F be a field. Then there exists an algebraically closed extension E of F.

The proof of Theorem 4 is rather difficult and will be deferred until Section 7.9. It is probably best for the beginner to omit the proof of Theorem 4 entirely and concentrate his efforts on understanding the many significant ramifications it has.

If F is a field and E is an algebraically closed field containing E, then E will usually be "too big." For example, E will usually contain elements which are transcendental over F. Since it is convenient to work only with algebraic elements, let us make the following definition:

DEFINITION 5: An *algebraic closure* of a field F is an extension $\bar{F}$ of F such that (1) $\bar{F}$ is algebraically closed, and (2) $\bar{F}$ is an algebraic extension of F.

The main result of this section is

THEOREM 6: Let F be a field. Then F has an algebraic closure.

Proof: Let E be an algebraically closed extension of F. E exists by Theorem 4. Set

$$\bar{F} = \{\alpha \in E \mid \alpha \text{ is algebraic over } F\}.$$

We assert that $\bar{F}$ is an algebraic closure of F. By Corollary 6 of Section 7.3, $\bar{F}$ is a field. It is clear that $\bar{F}$ is an algebraic extension of F. So we must show that $\bar{F}$ is algebraically closed. Let $f \in \bar{F}[X]$, $\deg(f) \geq 1$. In order to show that $\bar{F}$ is algebraically closed, we must show that f splits into a product of linear factors in $\bar{F}[X]$. Without loss of generality, assume that f is monic. Since E is algebraically closed,

$$f = \prod_{i=1}^{n} (X - \alpha_i), \qquad \alpha_i \in E. \tag{1}$$

Let $f = X^n + a_1 X^{n-1} + \cdots + a_n$, $a_i \in \bar{F}$. Then α_i is algebraic over $F(a_1, \ldots, a_n)$ which implies that

$$\deg(F(\alpha_i, a_1, \ldots, a_n)/F(a_1, \ldots, a_n)) < \infty$$

by Theorem 3 of Section 7.2. But $a_j \in \bar{F}\,(1 \leq j \leq n) \Rightarrow a_j$ is algebraic over $F \Rightarrow F(a_1, \ldots, a_n)$ is of finite degree over F (Corollary 5 of Section 7.3). Therefore, by Theorem 3 of Section 7.3

$$\deg(F(\alpha_i, a_1, \ldots, a_n)/F) < \infty$$

$\Longrightarrow \alpha_i$ is algebraic over F (Proposition 1 of Section 7.3)

$\Longrightarrow \alpha_i \in \bar{F}$.

Therefore, by (1), f splits into linear factors in $\bar{F}[X]$.

EXAMPLE 3: **C** is an algebraic closure of **R**.

EXAMPLE 4: Let

$$\bar{\mathbf{Q}} = \{\alpha \in \mathbf{C} \mid \alpha \text{ is algebraic over } \mathbf{Q}\}.$$

Then, by the proof of Theorem 6, $\bar{\mathbf{Q}}$ is an algebraic closure of **Q**. The field **Q** is called the *field of algebraic numbers*.

Note: $\bar{\mathbf{Q}} \neq \mathbf{C}$. For example, $\pi \in \mathbf{C} - \bar{\mathbf{Q}}$ by Lindemann's theorem.

7.8 Exercises

1. Let $F = \mathbf{Z}_p$ and let $\bar{F}$ be an algebraic closure of F. Show that $\bar{F}$ is infinite.
2. Exhibit an algebraically closed field containing **C** properly.
3. Let F be a field, E an algebraically closed field containing F. Show that E contains a unique algebraic closure of F.

7.9 Algebraically Closed Fields, II

In this section we will give a proof of Theorem 4 of Section 7.8. The results of Section 7.5 form the basis for the present proof. The principal difficulty in the proof is to pass from finite collections of polynomials to (possibly) infinite ones. This is usually accomplished via Zorn's lemma. Indeed, the present proof, which is due to Artin, makes use of Zorn's lemma, but in a rather disguised form. We use the fact that if I is a proper ideal of a commutative ring R with unity, then I is contained in a maximal ideal. And it is this fact which requires Zorn's lemma.

The bulk of the proof of Theorem 4 of Section 7.8 resides in the following lemma.

LEMMA 1: Let K be a field. Then there exists an extension P_K of K such that every nonconstant polynomial $f \in K[X]$ has a zero in P_K.

Proof: Let

$$A = \{f \in K[X] \mid f \text{ is nonconstant and irreducible}\}.$$

For each $f \in A$, let Y_f denote an indeterminate. Let us consider the ring $R = K[Y_f]_{f \in A}$, gotten by adjoining to K all the indeterminates Y_f. A typical element of R is a polynomial in finitely many of the indeterminates Y_f, having coefficients in K. Let I denote the ideal of R generated by the polynomials

$$f(Y_f) \qquad (f \in A).$$

We assert that $I \neq R$. For if $I = R$, then $1 \in I$, so that

$$1 = \sum_{i=1}^{m} g_i f_i(Y_{f_i}), \qquad g_i \in R, \tag{1}$$

where we may assume, without loss of generality, that

$$g_i = g_i(Y_{f_1}, \ldots, Y_{f_m}).$$

By Corollary 6 of Section 7.5, there exists an extension E of K in which each of the polynomials f_i has a zero α_i $(1 \leq i \leq m)$. The relation (1) only involves polynomials in $K[Y_{f_1}, \ldots, Y_{f_m}]$. Let us replace Y_{f_i} in (1) by α_i. Then, since α_i is a zero of f_i, equation (1) yields $1 = 0$, a contradiction. Therefore, $I \neq R$. Thus, I is contained in a maximal ideal M of R. Now R/M is a field by Theorem 8 of Section 5.9. Moreover, the mapping

$$x \longrightarrow x + M \qquad (x \in K)$$

is an isomorphism of K into R/M. Let us identify K with its image in R/M. Then R/M becomes a field extension of K. Set $P_K = R/M$. Let $f \in K[X]$ be nonconstant and irreducible. Then

$$f(Y_f + M) = f(Y_f) + M = 0 + M$$

since $f(Y_f) \in M$. Therefore, $Y_f + M$ is a zero of f in R/M. We have therefore shown that every irreducible, nonconstant polynomial in $K[X]$ has a zero in P_K, which completes the proof of Lemma 1.

Proof of Theorem 4 of Section 7.8: Let us construct a sequence of fields

$$E_0 \subseteq E_1 \subseteq E_2 \subseteq \cdots$$

as follows: Set $E_0 = F$ and for $i \geq 0$, set

$$E_{i+1} = P_{E_i},$$

where P_{E_i} is the field described in the lemma. Let

$$E = \bigcup_{i=0}^{\infty} E_i.$$

If $\alpha, \beta \in E$, then for all sufficiently large i, $\alpha, \beta \in E_i$. Therefore, $\alpha \pm \beta$, $\alpha \cdot \beta$, α/β $(\beta \neq 0)$ are defined as elements of E_i for all i sufficiently large and are independent of the choice of i. Thus, we can define the field operations on E and E becomes a field. We contend that E is algebraically closed. For let $f \in E[X]$, $n = \deg(f) \geq 1$. We must show that f splits into a product of linear factors in $E[X]$. This assertion is trivial for $n = 1$. Let us proceed by induction on n. Assume that $n > 1$ and assume the assertion for polynomials of degree $< n$. Since $f \in E[X]$, $f \in E_i[X]$ for some i. Therefore, by the construction of the E_i, f has a zero α in E_{i+1}, so that $f = (X - \alpha)g$ in $E_{i+1}[X]$, where $\deg(g) \geq 1$. Therefore, $f = (X - \alpha)g$ in $E[X]$, where $1 \leq \deg(g) < n$. Therefore, by induction, g is a product of linear factors in $E[X]$, so that f is a product of linear factors in $E[X]$.

Appendix *A*

The Irreducibility of the Cyclotomic Polynomials

Let m be a positive integer, ζ a primitive mth root of 1. In this appendix we will prove the following

THEOREM 1: Let

$$\Phi_m(X) = \prod_{\substack{0<a<m \\ (a,m)=1}} (X - \zeta^a)$$

be the mth cyclotomic polynomial. Then $\Phi_m(X)$ is irreducible in $\mathbf{Q}[X]$.

Let $r(X)$ be an irreducible, primitive polynomial having ζ as a zero. Our proof of Theorem 1 will be based on

THEOREM 2: Let p be a prime, $p \nmid m$. If $\eta \in \mathbf{C}$ is such that $r(\eta) = 0$, then $r(\eta^p) = 0$.

Let us first show that Theorem 2 implies Theorem 1. Since $r(X)$ is irreducible in $\mathbf{Q}[X]$, it suffices to show that

$$r(X) = \Phi_m(X). \tag{1}$$

If we show that

$$\Phi_m(X) \mid r(X), \tag{2}$$

then, since $\Phi_m(X) \in \mathbf{Q}[X]$, $\Phi_m(\zeta) = 0$ and since any polynomial in $\mathbf{Q}[X]$ which has ζ as a zero is divisible by $r(X)$, we conclude that (1) holds. Thus, let us show that Theorem 2 implies (2). Let $0 \leq a \leq m - 1$, $(a, m) = 1$. Then $a = p_1 \ldots, p_t$, p_i prime, $p_i \nmid m$. Then by repeated application of Theorem 2, we deduce that

$$\begin{aligned} r(\zeta) = 0 &\Longrightarrow r(\zeta^{p_1}) = 0 \\ &\Longrightarrow r(\zeta^{p_1 p_2}) = 0 \\ &\quad \vdots \\ &\Longrightarrow r(\zeta^a) = 0 \\ &\Longrightarrow X - \zeta^a \mid r(X) \\ &\Longrightarrow \Phi_m(X) \mid r(X). \end{aligned}$$

Let us now prove Theorem 2. Let $s(X)$ be an irreducible, primitive polynomial having η^p as a zero and let us assume that $r(\eta^p) \neq 0$. Then, since $r(X)$ and $s(X)$ are both irreducible, we see that $(r, s) = 1$. Therefore, since $s(X) \mid X^m - 1$, $r(X) \mid X^m - 1$, we conclude that

$$s(X)r(X) \mid X^m - 1.$$

Therefore, there exists $t(X) \in \mathbf{Q}[X]$ such that

$$X^m - 1 = s(X)r(X)t(X). \tag{3}$$

We assert that $t(X) \in \mathbf{Z}[X]$. Assume not. Let a denote the smallest positive integer such that $at(X) \in \mathbf{Z}[X]$. Then $a > 1$ and $at(X)$ is primitive. [For if the content of $at(X)$ is $b > 1$, we can replace a by a/b, contradicting the way in which a was chosen.] But then

$$a(X^m - 1) = s(X) \cdot r(X) \cdot at(X). \tag{4}$$

Since $s(X)$ and $r(X)$ are monic, they are both primitive. Therefore, by Gauss's lemma (Theorem 7 of Section 5.7), the right-hand side of (4) is primitive, so that its content is 1. But the content of $a(X^m - 1)$ is $a > 1$. Thus, we reach a contradiction and $t \in \mathbf{Z}[X]$.

Note that $s(X^p)$ has η as a zero, so that

$$s(X^p) = r(X) \cdot u(X) \tag{5}$$

for some $u(X) \in \mathbf{Q}[X]$. As above, we may prove that $u(X) \in \mathbf{Z}[X]$.

Let us interrupt our proof to derive some facts about $\mathbf{Z}_p$ and $\mathbf{Z}_p[X]$. Since $\mathbf{Z}_p$ is a field, $\mathbf{Z}_p[X]$ is a unique factorization domain by Corollary 3 of Section 5.7.

LEMMA 3: Let $\bar{a} \in \mathbf{Z}_p$. Then $\bar{a}^p = \bar{a}$.

Proof: Fermat's little theorem asserts that if $a \in \mathbf{Z}$, $p \nmid a$, then

$$a^{p-1} \equiv 1 \pmod{p}.$$

Therefore, if $p \nmid a$, we have

$$a^p \equiv a \pmod p. \tag{5}$$

However, (5) also holds if $p|a$. Therefore, (5) holds for all $a \in \mathbf{Z}$, so that $\bar{a}^p = \bar{a}$ for all $\bar{a} \in \mathbf{Z}_p$.

LEMMA 4: The mapping $f: \mathbf{Z}_p[X] \longrightarrow \mathbf{Z}_p[X]$, defined by

$$f(a) = a^p \qquad (a \in \mathbf{Z}_p[X])$$

is a ring homomorphism. In other words, if $a, b \in \mathbf{Z}_p[X]$, then

$$(a + b)^p = a^p + b^p, \tag{6}$$

$$(ab)^p = a^p b^p. \tag{7}$$

Proof: The proof of (7) is obvious. Let us verify (6). By the binomial theorem in $\mathbf{Z}_p[X]$ (Exercise 6 of Section 4.1),

$$(a + b)^p = a^p + \binom{p}{1}\cdot a^{p-1}b + \binom{p}{2}\cdot a^{p-2}b^2 + \cdots + \binom{p}{p-1}\cdot ab^{p-1} + b^p,$$

where $\binom{p}{j}$ is the integer defined by

$$\binom{p}{j} = \frac{p\cdot(p-1)\cdots(p-j+1)}{1\cdot 2\cdots j} \qquad (j = 1,\cdots, p-1).$$

Note that $1\cdot 2\cdots j$ is not divisible by p for $j = 1, 2,\cdots, p-1$ since p is prime. Therefore, $\binom{p}{j}$ is divisible by p for $j = 1, 2,\cdots, p-1$. But for x in $\mathbf{Z}_p[X]$, $p\cdot x = 0$. Therefore, $\binom{p}{j} a^{p-j}b^j = 0$ for $j = 1, 2,\cdots, p-1$, and (6) holds.

If $f(X) = a_0 + a_1 X + \cdots + a_n X^n \in \mathbf{Z}[X]$, let us define the *reduction of $f(X)$ modulo p*, denoted $\overline{f(X)}$, by

$$\overline{f(X)} = \bar{a}_0 + \bar{a}_1 X + \cdots + \bar{a}_n X^n \in \mathbf{Z}_p[X].$$

It is easy to verify that the mapping

$$\mathbf{Z}[X] \longrightarrow \mathbf{Z}_p[X],$$
$$f \longrightarrow \bar{f}$$

is a ring homomorphism.

LEMMA 5: Let $f = a_0 + a_1 X + \cdots + a_n X^n \in \mathbf{Z}[X]$. Then

$$\overline{f(X^p)} = \overline{f(X)}^p.$$

Proof: By Lemma 3,

$$\begin{aligned}\overline{f(X^p)} &= \bar{a}_0 + \bar{a}_1 X^p + \cdots + \bar{a}_n X^{np}\\ &= \bar{a}_0^p + \bar{a}_1^p X^p + \cdots + \bar{a}_n^p X^{np}\\ &= \bar{a}_0^p + (\bar{a}_1 X)^p + \cdots + (\bar{a}_n X^n)^p.\end{aligned}$$

Therefore, by Lemma 4,

$$\begin{aligned}\overline{f(X^p)} &= (\bar{a}_0 + \bar{a}_1 X + \cdots + \bar{a}_n X^n)^p \\ &= \overline{f(X)}^p.\end{aligned}$$

Let us now return to our proof of Theorem 2.

Proof of Theorem 2: By Lemma 5 and (5), we deduce that

$$\overline{s(X)}^p = \overline{r(X)} \cdot \overline{u(X)}. \tag{8}$$

Let $\overline{v(X)}$ be an irreducible factor of $\overline{r(X)}$ in $\mathbf{Z}_p[X]$. Then by (8), $\overline{v(X)}$ divides $\overline{s(X)}$. By (3), however,

$$X^m - \bar{1} = \overline{X^m - 1} = \overline{s(X)} \cdot \overline{r(X)} \cdot \overline{t(X)},$$

so that $\overline{v(X)}^2$ divides $X^m - \bar{1}$. Thus,

$$X^m - \bar{1} = \overline{v(X)}^2 \cdot \overline{w(X)}. \tag{9}$$

By taking the formal derivatives of both sides, we see that $\overline{v(X)}$ divides $\bar{m}X^{m-1}$. But $p \nmid m$, so that $\bar{m} \neq \bar{0}$. Therefore, $\overline{v(X)} = \bar{a} \cdot X^b$ for some $b \geq 1$. But since $\overline{v(X)}$ is irreducible, $\overline{v(X)} = \bar{a} \cdot X$, which contradicts (9). Thus, $r(\eta^p) = 0$.

Appendix B

Dirichlet's Theorem on Primes in Arithmetic Progressions

In this appendix we will give an interesting and simple application of the properties of the cyclotomic polynomials to the problem of the existence of primes in an arithmetic progression. Let k and n be positive integers. The question we seek to answer is the following: Are there infinitely many primes in the arithmetic progression

$$k, k + n, k + 2n, \cdots ?$$

If d denotes the g.c.d. of k and n, then every element of the arithmetic progression is divisible by d; and if $d > 1$ and $k + rn > d$, then we see immediately that $k + rn$ is not prime, since $k + rn$ is of the form ds for some integer s greater than 1. Therefore, if $d > 1$, there are only finitely many primes in the arithmetic progression. Thus, without loss of generality, let us assume that k and n are relatively prime. In this case, the answer was provided by Lejeune Dirichlet in 1837:

THEOREM 1 (DIRICHLET): Let k and n be relatively prime positive integers. Then there exist infinitely many primes in the arithmetic progression

$$k, k + n, k + 2n, \cdots .$$

Many proofs of Dirichlet's theorem are known and all are fairly complicated. However, in the special case $k = 1$, we can use the theory of cyclotomic polynomials to give a proof. Thus, in this section, our goal is the proof of the following result:

THEOREM 2: Let n be a positive integer. Then there are infinitely many primes p such that p is congruent to 1 modulo n.

Let $\Phi_n(X)$ denote the nth cyclotomic polynomial. Then $\Phi_n(X)$ has integer coefficients. (Exercise 5, Section 7.6.) Our proof of Theorem 2 will rest on the following lemma.

LEMMA 3: Let k be a positive integer. If p does not divide n and p divides $\Phi_n(k)$, then p is congruent to 1 modulo n.

Proof: We know that

$$X^n - 1 = \prod_{d|n} \Phi_d(X). \tag{1}$$

Since $p \mid \Phi_n(k)$, this implies that $p \mid k^n - 1$. In particular, k is not divisible by p. Let $\bar{x}$ denote the residue class of x modulo p. Then $\bar{k} \in \mathbf{Z}_p^\times$ since $p \nmid k$. Moreover, $\bar{k}^n = \bar{1}$ since $p \mid k^n - 1$. Thus, if d_0 denotes the order of $\bar{k}$ in the group $\mathbf{Z}_p^\times$, then n is a multiple of d_0, say $n = md_0$. Let us prove that $m = 1$. If $m \neq 1$, then $1 \leq d_0 < n$. Since d_0 is the order of $\bar{k}$ in $\mathbf{Z}_p^\times$, we know that $\bar{k}^{d_0} = \bar{1}$, in other words that $k^{d_0} - 1$ is divisible by p. However,

$$X^{d_0} - 1 = \prod_{e|d_0} \Phi_e(X). \tag{2}$$

Thus, by equations (1) and (2), we have

$$\begin{aligned} X^n - 1 &= \Phi_n(X) \prod_{\substack{d|n \\ 1 \leq d < n}} \Phi_d(X) \\ &= \Phi_n(X) \prod_{e|d_0} \Phi_e(X) h(X) \\ &\qquad\qquad \text{for some } h(X) \in \mathbf{Z}[X] \\ &= \Phi_n(X)(X^{d_0} - 1)h(X). \end{aligned}$$

As in the preceding Appendix, if $f \in \mathbf{Z}[X]$, let $\bar{f}$ denote the reduction of f modulo p. Then

$$X^n - \bar{1} = \overline{\Phi_n(X)}(X^{d_0} - \bar{1})\bar{h}(X). \tag{3}$$

However, $p \mid \Phi_n(k)$, so that $\bar{\Phi}_n(\bar{k}) = 0$, and thus $\bar{\Phi}_n(X)$ is divisible by $X - \bar{k}$ in $\mathbf{Z}_p[X]$. Moreover, since $p \mid k^{d_0} - 1$, we see that $X^{d_0} - \bar{1}$ has $\bar{k}$ as a zero, and therefore is divisible by $X - \bar{k}$. Thus, by Equation (3), $X^n - \bar{1}$ is divisible by $(X - \bar{k})^2$ in $\mathbf{Z}_p[X]$—that is,

$$X^n - \bar{1} = (X - \bar{k})^2 \cdot \bar{r}(X), \qquad \bar{r}(X) \in \mathbf{Z}_p[X]. \tag{4}$$

Let us take formal derivatives of both sides of Equation (4):

$$\bar{n}X^{n-1} = \bar{2}(X - \bar{k})\bar{r}(X) + (X - \bar{k})^2 \cdot D\bar{r}(X), \tag{5}$$

where $D\bar{r}(X)$ denotes the formal derivative of $\bar{r}(X)$. Since p does not divide n, we see that $\bar{n} \neq \bar{0}$. Since k is not divisible by p, $\bar{k} \neq \bar{0}$. Thus,

$$\bar{n}\bar{k}^{n-1} \neq \bar{0}.$$

On the other hand, we see that for $X = \bar{k}$, the right-hand side of (5) is $\bar{0}$. Thus, a contradiction is reached and we see that $m = 1$. In other words, $n = d_0$ and the order of $\bar{k}$ in $\mathbf{Z}_p^\times$ is n. However, the order of $\mathbf{Z}_p^\times$ is $p - 1$, so the order of $\bar{k}$ divides $p - 1$; therefore, $n \mid p - 1$ and $p \equiv 1 \pmod{n}$.

Proof of Theorem 2: First let us observe that if m is any positive integer, then

$$(\Phi_m(km), m) = 1 \qquad \text{for all } k \geq 1. \tag{6}$$

Indeed, suppose that q is a prime dividing both $\Phi_m(km)$ and m. Then if $\bar{x} \equiv x \pmod{q}$, then $\overline{km}$ is a zero of $\overline{\Phi_m(X)} \in \mathbf{Z}_q[X]$. However, since q divides m, we see that $\overline{km} = \bar{0}$, so that $\overline{\Phi_m(X)}$ is divisible by X in $\mathbf{Z}_q[X]$. However,

$$X^m - \bar{1} = \prod_{d \mid m} \overline{\Phi_d(X)},$$

which implies that $X^m - \bar{1}$ is divisible by X, which is nonsense. Thus, (6) is proved. Next note that if $m > 1$, then

$$|\Phi_m(km)| > 1. \tag{7}$$

for all sufficiently large positive integers k. For if $|\Phi_m(km)| \leq 1$ for arbitrarily large k, then one of the polynomials $\Phi_m(mX)$, $\Phi_m(mX) - 1$, $\Phi_m(mX) + 1$ has infinitely many zeros, and hence is the zero polynomial, which is a contradiction. Let us now prove Theorem 2. Let us choose k_1, a positive integer so large that $|\Phi_n(k_1 n)| > 1$. This is possible by (7). Then there exists a prime p_1 dividing $\Phi_n(k_1 n)$ and by (6), we have $p_1 \nmid n$. By Lemma 3, $p_1 \equiv 1 \pmod{n}$. Next, let us choose k_2 so large that $|\Phi_{np_1}(k_2 np_1)| > 1$, and let p_2 be a prime dividing $\Phi_{np_1}(k_2 np_1)$. Then $p_2 \nmid np_1$ by (5). In particular, p_2 and p_1 are different. Moreover, by Lemma 3, $p_2 \equiv 1 \pmod{np_1}$, so that $p_2 \equiv 1 \pmod{n}$. Proceeding in this way, we can construct a sequence

$$p_1, p_2, p_3, \ldots$$

of distinct primes congruent to 1 modulo n.

8 THE THEORY OF GROUPS, II

In Chapter 3 we introduced the notion of a group and proved a number of very elementary results. In the present chapter let us return to the theory of groups and let us make a somewhat deeper analysis of the structure of groups, especially finite groups, than was presented in Chapter 3. In this chapter we will show, among other things, how to classify all finite abelian groups. We will expand our theory of nonabelian finite groups by proving the Sylow theorems. Finally, we will introduce solvable groups, to pave the way for our study of Galois theory in Chapter 9.

8.1 The Isomorphism Theorems of Group Theory

In Chapter 3 we proved the *first isomorphism theorem* of group theory. Let us begin this chapter by complementing our previous result with two additional isomorphism theorems which actually follow from the first. Throughout this chapter the following notation will be in effect: G is a group; if H is a subgroup, then we will write $H \leq G$ (or $G \geq H$); if H is a normal subgroup of G, then we will write $H \lhd G$ (or $G \rhd H$).

Let $K \lhd G$. The elements of the factor group G/K are the cosets gK ($g \in G$). Let us begin by describing the subgroups of G/K.

Let H be a subgroup of G such that

$$G \geq H \geq K. \tag{1}$$

Since $K \lhd G$, we see that $K \lhd H$, so that the factor group H/K makes sense.

The elements of H/K are the cosets hK ($h \in H$), so that $H/K \subseteq G/K$. If $hK, h'K \in H/K$, then

$$(hK)(h'K)^{-1} = hh'^{-1}K \in H/K,$$

so that H/K is a subgroup of G/K. Thus, we have described a procedure for constructing subgroups of G/K.

PROPOSITION 1: Let $\bar{H} \leq G/K$. Then there exists a subgroup H of G, satisfying (1), such that $\bar{H} = H/K$.

Proof: Set

$$H = \{g \in G \mid gK \in \bar{H}\}.$$

If $h,h' \in H$, then $hK, h'K \in \bar{H}$, so that

$$hh'^{-1}K = (hK)(h'K)^{-1} \in \bar{H},$$

since $\bar{H}$ is a subgroup of G/K. Therefore, $hh'^{-1} \in H$ and H is a subgroup of G. Moreover, if $k \in K$, then $kK = K \in \bar{H}$, so that $K \leq H$, and (1) is satisfied. It is clear from the definition of H that $H/K = \bar{H}$.

PROPOSTION 2: Let $H \lhd G$, $K \lhd G$, $K \leq H$. Then $H/K \lhd G/K$.

Proof: Exercise.

THEOREM 3 (SECOND ISOMORPHISM THEOREM): Let $H \lhd G$, $K \lhd G$, $K \leq H$. Then

$$(G/K)/(H/K) \approx G/H. \tag{2}$$

Proof: By Proposition 2, $H/K \lhd G/K$, so that the left-hand side of (2) makes sense. Let i and j denote the natural homomorphisms defined by

$$G \xrightarrow{i} G/K$$

$$G/K \xrightarrow{j} (G/K)/(H/K).$$

Since i and j are surjective, the homomorphism

$$ji\colon G \longrightarrow (G/K)/(H/K) = M$$

is surjective. Moreover,

$$\ker(j) = H/K. \tag{3}$$

Note that if $g \in G$, then

$$\begin{aligned} g \in \ker(ji) &\longleftrightarrow (ji)(g) = 1_M \\ &\longleftrightarrow j(i(g)) = 1_M \\ &\longleftrightarrow i(g) \in \ker(j) = H/K \qquad \text{[by (3)]} \\ &\longleftrightarrow g \in i^{-1}(H/K) = H. \end{aligned}$$

Therefore, $\ker(ji) = H$, so that (2) follows from the first isomorphism theorem.

EXAMPLE 1: Let $G = \mathbf{Z}$, $H = 2\mathbf{Z}$, $K = 6\mathbf{Z}$. Then G is abelian, so that $H \lhd G$, $K \lhd G$. Thus, the hypotheses of Theorem 3 are satisfied. Now $G/H = \mathbf{Z}/2\mathbf{Z} = \mathbf{Z}_2$, $G/K = \mathbf{Z}/6\mathbf{Z} = \mathbf{Z}_6$, $H/K = 2\mathbf{Z}/6\mathbf{Z}$. Therefore, by Theorem 3,

$$\mathbf{Z}_6/(2\mathbf{Z}/6\mathbf{Z}) \approx \mathbf{Z}_2,$$

a fact which is readily verified directly.

Throughout the remainder of this section. let H and K be subgroups of G. Recall that $HK = \{hk \mid h \in H, k \in K\}$. Let us first determine when HK is a group.

PROPOSITION 4: HK is a group if and only if $HK = KH$.

Proof: $\Rightarrow$ Assume that HK is a group and let $h \in H$, $k \in K$. Then $h^{-1}k^{-1} \in HK$, and, since HK is a group, $kh = (h^{-1}k^{-1})^{-1} \in HK$. Therefore, $KH \subseteq HK$. Similarly, $HK \subseteq KH$, so that $HK = KH$.
$\Leftarrow$ Assume that $HK = KH$, and let $hk, h'k' \in HK$. Then, since $HK = KH$, there exist $h'' \in H$, $k'' \in K$ such that $(kk'^{-1})\cdot h'^{-1} = h''k''$. But then

$$\begin{aligned}(hk)(h'k')^{-1} &= h(kk'^{-1}h'^{-1})\\ &= hh''k'' \in HK.\end{aligned}$$

Therefore, HK is a subgroup of G.

COROLLARY 5: If one of H, K is a normal subgroup of G, then HK is a subgroup of G.

Proof: Assume that $H \lhd G$. If $h \in H$, $k \in K$, then $k^{-1}hk \in H$, say $k^{-1}hk = h'$. Thus, $hk = kh' \in KH$. Thus, $HK \subseteq KH$. A similar argument shows that $KH \subseteq HK$, so that $HK = KH$ and the result follows from Proposition 4.

Let us now suppose that $H \lhd G$ and $K \leq G$. By Corollary 5, $HK \leq G$. Moreover, since $1 \in K$, we see that $H \leq HK$. In fact, $H \lhd HK$, since $H \lhd G$. Similarly, $K \leq HK$. Further, we see that $H \cap K$ is a subgroup of K and of H. Actually, if $k \in K$, $h \in H \cap K$, then $khk^{-1} \in K$, since $h \in K$ and K is a group; moreover, since $k \in G$ and $H \lhd G$, we have $khk^{-1} \in H \cap K$. Thus, we have shown that $G \cap K \lhd K$. We may summarize the situation in the following diagram, where single lines denote subgroups and double lines denote normal subgroups:

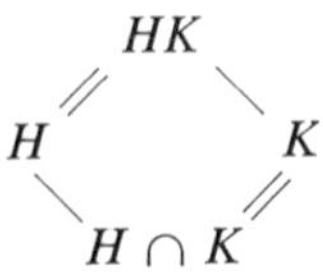

THEOREM 6 (THIRD ISOMORPHISM THEOREM): Assume that $H \lhd G$ and $K \leq G$.

(1) $HK \leq G$.

(2) $H \lhd HK$; $H \cap K \lhd K$.

(3) $HK/H \approx K/H \cap K$.

Proof: We have already proved (1) and (2). In order to prove (3), let us consider the homomorphisms

$$i: K \longrightarrow HK,$$
$$j: HK \longrightarrow HK/H,$$

defined by

$$i(k) = k \qquad (k \in K),$$
$$j = \text{the natural homomorphism.}$$

Note that i is an injection, so that $\ker(i) = \{1\}$, while $\ker(j) = H$. Let us now consider the homomorphism

$$ji: K \longrightarrow HK/H.$$

Then it is clear that $k \in \ker(ji)$ if and only if $k \in \ker(j)$ if and only if $k \in H$. Therefore, $\ker(ji) = H \cap K$. Therefore, by the first isomorphism theorem, it suffices to prove that ji is surjective. Let $hk \in HK$. We claim that

$$(ji)(k) = hkH.$$

Indeed, since $H \lhd G$, we see that $k^{-1}hk \in H$. Therefore,

$$hkH = k(k^{-1}hk)H = kH.$$

But $(ji)(k) = kH$. Thus, our claim is proved and hkH is contained in the image of ji. Therefore, ji is surjective and our result is proved.

EXAMPLE 2: Let $G = \mathbf{Z}$ with respect to addition, and let $H = 2\mathbf{Z}$, $K = 3\mathbf{Z}$. Then $HK = 2\mathbf{Z} + 3\mathbf{Z} = \mathbf{Z}$, $H \cap K = 2\mathbf{Z} \cap 3\mathbf{Z} = 6\mathbf{Z}$. The third isomorphism theorem then asserts that

$$\mathbf{Z}/2\mathbf{Z} \approx 3\mathbf{Z}/6\mathbf{Z},$$

a result which can be trivially checked.

Let H and K be subgroups of a finite group G. As our last result in this section, let us compute the number of elements in the set HK. If $H \lhd G$, then Theorem 6 already provides us with the answer. For HK/H has $|HK|/|H|$ elements, and $K/(H \cap K)$ has $|K|/|H \cap K|$ elements. Therefore, by Theorem 6, we have

$$|HK| = \frac{|H| \cdot |K|}{|H \cap K|}. \tag{4}$$

If H is not a normal subgroup of G, then HK is not necessarily a group. However, we will prove that (4) still holds in this case.

Even if HK is not a group, it still makes sense to consider the set of right cosets $H\backslash HK$. A typical coset belonging to $H\backslash HK$ is Hhk, where $h \in H$, $k \in K$. Each such coset contains $|H|$ elements. Moreover, we have already seen that two cosets intersect if and only if they are equal. Therefore, the elements of HK are distributed into nonoverlapping cosets, each having $|H|$ elements, so that $|HK|$ is a multiple of $|H|$ and the number of cosets in $H\backslash HK$ is given by

$$|H\backslash HK| = \frac{|HK|}{|H|}. \tag{5}$$

Similarly,

$$|H \cap K\backslash K| = \frac{|K|}{|H \cap K|}. \tag{6}$$

Thus, to prove (4), it suffices to prove that

$$|H\backslash HK| = |H \cap K\backslash K|. \tag{7}$$

Let us define a function

$$\psi\colon H\backslash HK \longrightarrow H \cap K\backslash K$$

by

$$\psi(Hhk) = (H \cap K)k, \qquad h \in H, \quad k \in K.$$

ψ is well defined, since if $Hhk = Hh'k'$, then $Hk = Hk' \Rightarrow k = hk'$ for some $h \in H$. But since $h = kk'^{-1}$, $h \in K$ and thus $h \in H \cap K$. Therefore,

$$\begin{aligned}(H \cap K)k = (H \cap K)hk' &= (H \cap K)k',\\ \Longrightarrow \psi(Hhk) &= \psi(Hh'k'),\\ \Longrightarrow \psi &\text{ is well defined.}\end{aligned}$$

If $(H \cap K)k = (H \cap K)k'$, then $k = k^*k'$ ($k^* \in H \cap K$). Therefore, if $\psi(Hhk) = \psi(Hh'k')$, then $Hhk = Hk = Hk^*k' = Hk' = Hh'k'$. And therefore ψ is injective. It is clear that ψ is surjective. Therefore, ψ is a bijection and (7) is proved.

8.1 *Exercises*

1. Prove Proposition 2.
2. Let $G = D_k$, $H = [R]$, $K = [F]$. Verify that the third isomorphism theorem holds in this special case.
3. Let G be a group, H_1, H_2, H_3 subgroups. How many elements are contained in $H_1H_2H_3$?

*4. Let G be a group, $H_1, H_2, \ldots, H_n$ subgroups. How many elements are contained in $H_1H_2 \ldots H_n$?

5. Let G be a group, $a \in G$. Define the *normalizer of a in G*, denoted $N_G(a)$, by
$$N_G(a) = \{x \in G \mid xax^{-1} = a\}.$$
   (a) Show that $N_G(a)$ is a subgroup of G.
   (b) Show that $N_G(a)$ is the largest subgroup of G in which $[a]$ is normal.
   (c) Let C_a denote the conjugacy class of G which contains a—that is, $C_a = \{xax^{-1} \mid x \in G\}$. Show that
$$|C_a| = \frac{|G|}{|N_G(a)|}.$$
   (d) Show that
$$|G| = \sum_a \frac{|G|}{|N_G(a)|}.$$
   where a runs over a set of representatives of the conjugacy classes of G. This relation is called the *class equation*.

*6. Using Exercise 5, show that if $|G| = p^n$, p prime, then the center of G is nontrivial—that is, $\{g \in G \mid gx = xg \text{ for all } x \in G\} \neq \{1_G\}$.

7. Show that any group of order p^2, p prime, is abelian.

8.2 Decomposition with Respect to Two Subgroups

In Chapter 2 we discovered that we could decompose a group G into (right or left) cosets with respect to a subgroup H. In this section we will generalize this decomposition to a decomposition of G with respect to two subgroups, H and K. The decomposition derived in this section was first discovered by Frobenius at the end of the nineteenth century and will be very useful in our discussion of the Sylow theorems, later in this chapter. Throughout this section, let G be a group and let H and K be subgroups of G.

Let us define an equivalence relation $\sim$ on G as follows: Say that $g \sim g'$ if and only if $g = hg'k$ for some $h \in H, k \in K$. Let us first prove that $\sim$ is an equivalence relation:

$\sim$ is reflexive: $g = 1 \cdot g \cdot 1 \Longrightarrow g \sim g$.

$\sim$ is symmetric: $g \sim g' \Longrightarrow g = hg'k$
$$(h \in H, k \in K) \Longrightarrow g' = h^{-1}gk^{-1} \Longrightarrow g' \sim g.$$

$\sim$ is transitive: $g \sim g', g' \sim g'' \Longrightarrow g = hg'k, g' = h'g''k'$
$$(h, h' \in H; k, k' \in K) \Longrightarrow g = (hh')g''(k'k) \Longrightarrow g \sim g''.$$

Thus, $\sim$ is an equivalence relation on G and therefore decomposes G into equivalence classes C_i $(i \in I)$, where I is some set which we use to label the equivalence classes. If $g_i \in C_i$, then C_i consists of those elements of G which are equivalent to g_i, so that
$$C_i = Hg_iK.$$

Thus,

$$G = \bigcup_{i \in I} Hg_iK. \tag{1}$$

Moreover, since $C_i \cap C_j = \varnothing$ if $i \neq j$, we have

$$Hg_iK \cap Hg_jK = \varnothing \qquad (i \neq j). \tag{2}$$

We will be interested in the case where G is a finite group of order n. Suppose that there are t equivalence classes $C_1, \ldots, C_t$. Then (1) may be written

$$G = Hg_1K \cup \cdots \cup Hg_tK. \tag{3}$$

Therefore, by (2) and (3),

$$|G| = |Hg_1K| + \cdots + |Hg_tK|. \tag{4}$$

Now it is clear that

$$|Hg_iK| = |g_i^{-1}Hg_iK|, \tag{5}$$

since if $Hg_iK = \{x_1, \ldots, x_a\}$, then $g_i^{-1}Hg_iK = \{g_i^{-1}x_1, \ldots, g_i^{-1}x_a\}$. Let $H_i = g_i^{-1}Hg_i$. Then $H_i \leq G$ and $|H_i| = |H|$. Moreover, by (5),

$$|Hg_iK| = |H_iK|. \tag{6}$$

By Equation (4) of Section 8.1 we see that

$$|H_iK| = \frac{|H_i||K|}{|H_i \cap K|} = \frac{|H||K|}{|H_i \cap K|}. \tag{7}$$

Therefore, by (6) and (7), we have

$$|Hg_iK| = \frac{|H||K|}{|H_i \cap K|}. \tag{8}$$

By combining (4) and (8), we may summarize our results as follows:

THEOREM 1 (FROBENIUS): Let G be a finite group of order n, let H and K be subgroups of order p and q, respectively, and let $g_1, \ldots, g_t$ be representatives of the equivalence classes with respect to $\sim$. Then

$$n = \sum_{i=1}^{t} \frac{pq}{u_i},$$

where $u_i = |g_i^{-1}Hg_i \cap K|$.

If in Theorem 1, we set $H = \{1\}$, then we see that $p = 1$ and $u_i = 1$ for all i, so that $n = qt$, which is Lagrange's theorem, since the equivalence classes C_i are just cosets g_iK.

8.3 Direct Products of Groups

Let G and H be groups. The Cartesian product $G \times H$ consists of the ordered pairs (g, h), for $g \in G$, $h \in H$. Let us put a group structure on $G \times H$.

We define multiplication in $G \times H$ by

$$(g, h)(g', h') = (gg', hh'). \tag{1}$$

It is clear that (1) defines a binary operation on $G \times H$. We leave the proof of associativity as an exercise. Since

$$(g, h)(1_G, 1_H) = (g \cdot 1_G, h \cdot 1_H) = (g, h),$$
$$(1_G, 1_H)(g, h) = (1_G \cdot g, 1_H \cdot h) = (g, h),$$

we see that $(1_G, 1_H)$ is an identity element. Moreover,

$$(g, h)(g^{-1}, h^{-1}) = (g^{-1}, h^{-1})(g, h) = (1_G, 1_H),$$

so that every element of $G \times H$ has an inverse with respect to the multiplication (1). Thus, with respect to the multiplication (1), $G \times H$ is a group.

DEFINITION 2: $G \times H$ is called the *direct product* of G and H.

EXAMPLE 1: Let $G = H = \mathbf{Z}_2$. Then

$$\begin{aligned} \mathbf{Z}_2 \times \mathbf{Z}_2 &= \{(a, b) \mid a, b \in \mathbf{Z}_2\} \\ &= \{(\bar{0}, \bar{0}), (\bar{0}, \bar{1}), (\bar{1}, \bar{0}), (\bar{1}, \bar{1})\}. \end{aligned}$$

The identity element is $(\bar{0}, \bar{0})$, and if we set $I = (\bar{0}, \bar{0})$, $A = (\bar{0}, \bar{1})$, $B = (\bar{1}, \bar{0})$, $C = (\bar{1}, \bar{1})$, then we easily see that

$$AB = BA, \qquad A^2 = B^2 = I, \qquad C = AB.$$

Thus, $\mathbf{Z}_2 \times \mathbf{Z}_2$ is isomorphic to the Klein 4-group, which we discussed in Section 3.4. (Another way of establishing this fact is to observe that $\mathbf{Z}_2 \times \mathbf{Z}_2$ is a group of order 4 which is not cyclic. There is only one such group—the Klein 4-group.) Note that this example shows that a direct product of cyclic groups need not be cyclic.

EXAMPLE 2: Let $G = \mathbf{Z}_2$, $H = \mathbf{Z}_3$. Then

$$\begin{aligned} G \times H &= \{(a, b) \mid a \in \mathbf{Z}_2, b \in \mathbf{Z}_3\} \\ &= \{(\bar{0}, \bar{0}), (\bar{0}, \bar{1}), (\bar{0}, \bar{2}), (\bar{1}, \bar{0}), (\bar{1}, \bar{1}), (\bar{1}, \bar{2})\}. \end{aligned}$$

Note that $(\bar{0}, \bar{0})$ is the identity and if $\alpha = (\bar{1}, \bar{1})$, then

$$\begin{aligned} \alpha^2 &= (\bar{0}, \bar{2}), \\ \alpha^3 &= (\bar{1}, \bar{0}), \\ \alpha^4 &= (\bar{0}, \bar{1}), \\ \alpha^5 &= (\bar{1}, \bar{2}), \\ \alpha^6 &= (\bar{0}, \bar{0}), \end{aligned}$$

so that α has order 6. Thus, $\mathbf{Z}_2 \times \mathbf{Z}_3$ is a cyclic group of order 6, with α as generator.

Let $G_1, \ldots, G_n$ be a finite collection of groups. The direct product $G_1 \times$

$\ldots \times G_n$ is the group whose elements are the ordered n-tuples $(g_1, \ldots, g_n)$, where $g_i \in G_i$ $(1 \leq i \leq n)$, and whose multiplication is defined by

$$(g_1, \ldots, g_n)(g'_1, \ldots, g'_n) = (g_1 g'_1, \ldots, g_n g'_n).$$

The verification that $G_1 \times \cdots \times G_n$ is a group is similar to the proof given above for the special case $n = 2$.

If G_1, G_2, and G_3 are groups, we can form the direct product $(G_1 \times G_2) \times G_3$, whose elements are of the form $((g_1, g_2), g_3)$ $(g_i \in G_i, i = 1, 2, 3)$. It is trivial to see that the mapping

$$\psi: (G_1 \times G_2) \times G_3 \longrightarrow G_1 \times G_2 \times G_3,$$
$$\psi(((g_1, g_2), g_3)) = (g_1, g_2, g_3),$$

is a surjective isomorphism, so that

$$(G_1 \times G_2) \times G_3 \approx G_1 \times G_2 \times G_3.$$

Similarly,

$$G_1 \times (G_2 \times G_3) \approx G_1 \times G_2 \times G_3.$$

Subsequently, we will always identify the groups $(G_1 \times G_2) \times G_3$, $G_1 \times (G_2 \times G_3)$, $G_1 \times G_2 \times G_3$. Similar comments apply to direct products of more than three groups. For example, $G_1 \times (G_2 \times G_3) \times G_4$ will be identified with $G_1 \times G_2 \times G_3 \times G_4$, and so on.

The next result is a convenient one for decomposing a given group into a direct product of two subgroups.

THEOREM 3: Let G be a group. $H \leq G$, $K \leq G$. Assume that H and K satisfy the following:

(1) $H \cap K = \{1\}$.
(2) If $h \in H$, $k \in K$, then $hk = kh$.
(3) If $g \in G$, then there exist $h \in H$, $k \in K$ such that $g = hk$.

Then $G \approx H \times K$.

Proof: First let us prove that if $g \in G$, then g can be *uniquely* written in the form $g = hk$ $(h \in H, k \in K)$. Indeed, if $g = hk = h'k'$, then

$$h'^{-1}h = k'k^{-1} \in H \cap K = \{1\},$$
$$\Longrightarrow h'^{-1}h = k'k^{-1} = 1,$$
$$\Longrightarrow h = h', k = k'.$$

Let $\psi: G \longrightarrow H \times K$ be the mapping defined by $\psi(g) = (h, k)$, where $g = hk$. This mapping is well defined by (3) and the fact that the representation of g is unique. It is clear that ψ is surjective since $\psi(hk) = (h, k)$ for any $h \in H$, $k \in K$. It is also clear that ψ is injective. Thus, it suffices to show that ψ is a homomorphism for then $G \approx H \times K$. Let $g = hk$, $g' = h'k'$. Then, by (2), $gg' = (hk)(h'k') = (hh')(kk')$, so that

$$\begin{aligned}\psi(gg') &= (hh', kk') \\ &= (h, k)(h', k') \\ &= \psi(g)\psi(g').\end{aligned}$$

Thus, ψ is a homomorphism.

A simple, but important application of Theorem 3 is the following.

THEOREM 4: Let m and n be positive integers such that $(m, n) = 1$. Then if G is a cyclic group of order mn, we have

$$G \approx \mathbf{Z}_m \times \mathbf{Z}_n.$$

Proof: Let a be a generator of G. By Theorem 16 of Section 3.3, the order of a^m is n and the order of a^n is m. Let $H = [a^n]$, $K = \langle a^m \rangle$. Then H and K are cyclic groups of orders m and n, respectively. Therefore, by Theorem 3 of Section 3.4 we see that

$$H \approx \mathbf{Z}_m, \qquad K \approx \mathbf{Z}_n.$$

Therefore, it suffices to prove that $G \approx H \times K$. Let us apply Theorem 3. Let $g \in H \cap K$. Then the order of g divides the orders of both H and K, so that the order of g is a common divisor of m and n. But since $(m, n) = 1$, we see that g is of order 1—that is, $g = 1_G$ and $H \cap K = \{1_G\}$. Every element of H commutes with every element of K since G is abelian. Finally, if $g \in K$, then $g = a^r$ for some integer r. Since $(m, n) = 1$, we see that there exist integers c and d such that $mc + nd = 1$. Therefore, $r = mcr + ndr$, and

$$\begin{aligned}g = a^r &= a^{mcr+ndr} \\ &= (a^m)^{cr} \cdot (a^n)^{dr}.\end{aligned}$$

Setting $h = (a^n)^{dr}$, $k = (a^m)^{cr}$, we see that $h \in H$, $k \in K$ and

$$g = hk.$$

Therefore, (1)–(3) of Theorem 3 hold, and $G \approx H \times K$.

COROLLARY 5: Let G be a cyclic group of order $p_1^{a_1} p_2^{a_2} \ldots p_t^{a_t}$, where $p_1, \ldots, p_t$ are distinct primes and $a_1, \ldots, a_t$ are positive integers. Then

$$G \approx \mathbf{Z}_{p_1{}^{a_1}} \times \cdots \times \mathbf{Z}_{p_t{}^{a_t}}.$$

Proof: Let us proceed by induction on t. The assertion is true for $t = 1$ by Theorem 3.3.4. Assume $t > 1$ and the assertion $t - 1$. Set $m = p_1^{a_1}$, $n = p_2^{a_2} \ldots p_t^{a_t}$. Then by Theorem 4 above,

$$G \approx \mathbf{Z}_{p_1{}^{a_1}} \times \mathbf{Z}_n. \tag{2}$$

But $\mathbf{Z}_n$ is a cyclic group of order $p_2^{a_2} \ldots p_t^{a_t}$. Therefore, by the assertion for $t - 1$, we see that

$$\mathbf{Z}_n \approx \mathbf{Z}_{p_2^{a_2}} \times \cdots \times \mathbf{Z}_{p_t^{a_t}}. \tag{3}$$

From (2) and (3), we conclude the assertion for t and the induction step is proved.

8.3 *Exercises*

1. Write down all the elements and multiplication tables for the following groups:
   (a) $\mathbf{Z}_2 \times S_3$. (b) $\mathbf{Z}_3 \times \mathbf{Z}_3$. (c) $\mathbf{Z}_2 \times D_3$.
2. Let $m > 1$ be an integer. Prove that $\mathbf{Z}_m \times \mathbf{Z}_m$ is not isomorphic to $\mathbf{Z}_{m^2}$.
3. Let G_1 and G_2 be groups, and let
$$p_i\colon G_1 \times G_2 \longrightarrow G_i \qquad (i = 1, 2),$$
$$p_i((g_1, g_2)) = g_i.$$
   (a) Show that p_i $(i = 1, 2)$ are surjective homomorphisms. These homomorphisms are called *projection homomorphisms.*
   (b) Suppose that a group H and a pair of homomorphisms $q_i\colon H \longrightarrow G_i$ $(i = 1, 2)$ are given. Show that there exists a unique homomorphism $\phi\colon G_1 \times G_2 \longrightarrow H$ such that $q_i\phi = p_i$ $(i = 1, 2)$.

*4. Show that $\mathbf{Z}_{p^n}$, p a prime cannot be written as the direct product of two non-trivial subgroups.

8.4 *The Fundamental Theorem of Abelian Groups*

We have already mentioned in Chapter 3 that the broad goal of the theory of finite groups is to arrive at a classification of all finite groups. In this section we will achieve a much more modest goal: We will classify all finite abelian groups. Our main result will be the *fundamental theorem of abelian groups* (FT), which asserts that a finite abelian group is isomorphic to a direct product of cyclic groups. We will even be able to pin down the structure of a finite abelian group further by specifying to some extent the cyclic groups which occur.

THEOREM 1 (FT): Let G be a finite abelian group. Then G is isomorphic to a direct product of cyclic subgroups.

Proof: Suppose that G has order n. The theorem is clear if $n = 1$. Thus, we may assume that $n > 1$, and we may proceed by induction on n. Let x_0

be an element of G of highest order, and let $C_0 = [x_0]$. Since G is abelian, $C_0 \lhd G$. Moreover, G/C_0 has order less than n, so the induction hypothesis implies that there exist cyclic subgroups $H_1, H_2, \ldots, H_s$ of G/C_0 such that

$$G/C_0 \approx H_1 \times H_2 \times \cdots \times H_s. \tag{1}$$

Each H_i is of the form C_i/C_0, where C_i is a subgroup of G which contains C_0. Since H_i is cyclic, $H_i = [x_iC_0]$ for some $x_i \in G$. Let us prove that

$$G \approx [x_0] \times [x_1, \ldots, x_s]. \tag{2}$$

for some choice of x_i such that $H_i = [x_iC_0]$. Let us apply Theorem 3 of Section 8.3. Since G is abelian, every element of $[x_0]$ commutes with every element of $[x_1, \ldots, x_s]$. If $g \in G$, then $gC_0 \in G/C_0$, so that by (1), $gC_0 = (x_1C_0)^{\alpha_1} \ldots (x_sC_0)^{\alpha_s} = x_1^{\alpha_1} \ldots x_s^{\alpha_s}C_0$. Thus,

$$g = c_0x_1^{\alpha_1} \ldots x_s^{\alpha_s}$$

for some $c_0 \in C_0$. But $c_0 = x_0^{\alpha_0}$ since $C_0 = [x_0]$. Therefore,

$$g = x_0^{\alpha_0}(x_1^{\alpha_1} \ldots x_s^{\alpha_s}),$$

and g is the product of an element of $[x_0]$ and an element of $[x_1, \ldots, x_s]$. In order to apply Theorem 3 of Section 8.3, we must prove that

$$[x_0] \cap [x_1, \ldots, x_s] = \{1\}. \tag{3}$$

This is the heart of the proof. Let a_i denote the order of x_i. Let us first show that the order of x_iC_0 (in G/C_0) is also a_i. Let b_i denote the order of x_iC_0. Since $(x_iC_0)^{a_i} = x_i^{a_i}C_0 = C_0$, we see that

$$b_i \mid a_i. \tag{4}$$

Since $(x_iC_0)^{b_i} = C_0$, we see that $x_i^{b_i} \in C_0$, so that $x_i^{b_i} = x_0^c$ for some integer $c \geq 0$. Since the order of x_iC_0 is the same as the order of $x_ix_0^dC_0$, and since $x_ix_0^dC_0$ is a generator of H_i, we may replace x_i by $x_ix_0^d$ for any integer d. In doing so, we replace c by $c + b_id$. By choosing d appropriately, we can guarantee that $0 \leq c + b_id < b_i$. Thus, without loss of generality, we may assume that

$$x_i^{b_i} = x_0^c, \qquad 0 \leq c < b_i. \tag{5}$$

The order of $x_i^{b_i}$ is $a_i/(b_i, a_i) = a_i/b_i$ [by (4)]. The order of x_0^c is $r/(r, c)$, where $r =$ the order of x_0. Therefore, by (5),

$$a_i = \frac{b_ir}{(r, c)}.$$

On the other hand, since x_0 has maximal order in G, $a_i \leq r$. Therefore,

$$\frac{b_ir}{(r, c)} \leq r \text{ which is equivalent to } b_i \leq (r, c).$$

If $c > 0$, then $(r, c) \leq c$ and thus we have

$$b_i \leq c,$$

which contradicts (5). Therefore, $c = 0$ and $x_i^{b_i} = x_0^0 = 1$. Thus, $a_i | b_i$, which together with (4) implies that $a_i = b_i$. We have thus proved that the order of $x_i C_0$ in G/C_0 is a_i. Every element in G/C_0 is of the form

$$(x_1 C_0)^{\alpha_1} \cdots (x_s C_0)^{\alpha_0} = x_1^{\alpha_1} \cdots x_s^{\alpha_s} C_0, \qquad 0 \leq \alpha_i < a_i. \tag{6}$$

By what we have proved above, $H_i = C_i/C_0$ has order a_i, so that by (1), G/C_0 has order $a_1 a_2 \cdots a_s$. Therefore, all the elements (6) are distinct. In particular, if $x_1^{\alpha_1} \cdots x_s^{\alpha_s} \in C_0$ $(0 \leq \alpha_i < a_i)$, then $\alpha_1 = \cdots = \alpha_s = 0$. Let us finally prove (3). If $g \in [x_0] \cap [x_1, \ldots, x_s]$, then $g = x_1^{\alpha_1} \ldots x_s^{\alpha_s}$ $(0 \leq \alpha_i < a_i)$ and $g \in C_0$. Therefore, as we have shown above, $\alpha_1 = \cdots = \alpha_s = 0$, and $g = 1$. Thus, (3) is proved and with it (2). However, by induction,

$$[x_1, \ldots, x_s] \approx [y_1] \times \cdots \times [y_t], \tag{7}$$

so that by (2),

$$G \approx [x_1] \times [y_1] \times \cdots \times [y_t],$$

and G is a direct product of cyclic groups.

Corollary 2: Let G be a finite abelian group. Then there exist positive integers $n_1, \ldots, n_t$ such that

$$G \approx \mathbf{Z}_{n_1} \times \cdots \times \mathbf{Z}_{n_t}.$$

Proof: Every cyclic group is isomorphic to $\mathbf{Z}_n$ for some positive integer n.

Corollary 3: Let G be a finite abelian group. Then there exists a set of prime powers $p_1^{a_1}, p_2^{a_2}, \ldots, p_s^{a_s}$ (not necessarily distinct) such that

$$G \approx \mathbf{Z}_{p_1^{a_1}} \times \mathbf{Z}_{p_2^{a_2}} \times \cdots \times \mathbf{Z}_{p_s^{a_s}}.$$

Proof: Let $n_1, \ldots, n_t$ be as in Corollary 2, and let

$$n_i = p_{i1}^{a_{i1}} \cdots p_{ij_i}^{a_{ij_i}} \qquad (1 \leq i \leq t).$$

Then by Corollary 5 of Section 8.3,

$$G \approx \mathbf{Z}_{p_{11}^{a_{11}}} \times \mathbf{Z}_{p_{12}^{a_{12}}} \times \cdots \times \mathbf{Z}_{p_{t1}^{a_{t1}}} \times \cdots \times \mathbf{Z}_{p_{tj_t}^{a_{tj_t}}}.$$

Note that $H \times K \approx K \times H$ with respect to the mapping $\psi\colon H \times K \to K \times H$ defined by $\psi(h, k) = (k, h)$ $(h \in H, K \in K)$. Therefore, we may rearrange factors in a direct product and still get a group isomorphic to the original one. Using this fact, we can normalize the decomposition of Corollary 3 as follows: Let $q_1, \ldots, q_v$ be the distinct primes among $\{p_1, \ldots, p_s\}$, arranged so that

$$q_1 < q_2 < \cdots < q_v.$$

Let

$$q_i^{a_{i1}}, q_i^{a_{i2}}, \ldots, q_i^{a_{ij(i)}} \tag{8}$$

be the powers of the prime q_i (not necessarily distinct) appearing in the decomposition of Corollary 3, arranged so that

$$0 < a_{i1} \leq a_{i2} \leq \cdots \leq a_{ij(i)}.$$

Define

$$G(q_i) = \mathbf{Z}_{q_i^{a_{i1}}} \times \mathbf{Z}_{q_i^{a_{i2}}} \times \cdots.$$

Then Corollary 3 and the above remark imply that

$$G \approx G(q_1) \times G(q_2) \times \cdots \times G(q_v). \tag{9}$$

It is clear that every element of $G(q_i)$ has order a power of q_i. Moreover, from (9), we see that the set of all elements of $G(q_1) \times \cdots \times G(q_v)$ whose order is a power of q_i is $\{1\} \times \{1\} \times \cdots \times G(q_i) \times \cdots \times \{1\} \approx G(q_i)$. Thus, we have

COROLLARY 4: Let $H(q_i)$ denote the subgroup of G consisting of all elements whose order is a power of q_i. Then

$$H(q_i) \approx \mathbf{Z}_{q_i^{a_{i1}}} \times \cdots \times \mathbf{Z}_{q_i^{a_{ij(i)}}}$$

and

$$G \approx H(q_1) \times \cdots \times H(q_v).$$

If q is a prime, then $H(q)$ is called the *q-primary part* of G. Let us consider an example. Let $G = \mathbf{Z}_{10} \times \mathbf{Z}_{25}$. Then

$$\mathbf{Z}_{10} \approx \mathbf{Z}_2 \times \mathbf{Z}_5, \qquad \mathbf{Z}_{25} = \mathbf{Z}_{5^2}.$$

Therefore,

$$G \approx \mathbf{Z}_2 \times \mathbf{Z}_5 \times \mathbf{Z}_{5^2}.$$

The 2-primary part of G is $\approx \mathbf{Z}_2$, while the 5-primary part of G is $\approx \mathbf{Z}_5 \times \mathbf{Z}_{5^2}$.

DEFINITION 5: The prime powers

$$q_i^{a_{ij}} \qquad [1 \leq j \leq j(i),\ 1 \leq i \leq v]$$

are called the *elementary divisors* of G.

It is clear that if we are given the elementary divisors of G, then it is possible to determine G up to isomorphism using (9).

THEOREM 6: The elementary divisors of G are uniquely determined by G. In other words, the decomposition (9) of G is unique.

Proof: The primes $q_1, \ldots, q_v$ are uniquely determined as the distinct primes dividing the order of G. Let $q = q_i$ for some i and let $H =$ the q-primary part of G. Then H depends only on q and G. Suppose that we are given a decomposition of H of the form

$$H \approx \underbrace{\mathbf{Z}_q \times \cdots \times \mathbf{Z}_q}_{\alpha(1)} \times \underbrace{\mathbf{Z}_{q^2} \times \cdots \times \mathbf{Z}_{q^2}}_{\alpha(2)} \times \cdots \times \underbrace{\mathbf{Z}_{q^a} \times \cdots \times \mathbf{Z}_{q^a}}_{\alpha(a)}.$$

It suffices to show that $\alpha(1), \alpha(2), \ldots \alpha(a)$ depend only on G and q. If b is a positive integer, then set $H^b = \{h^b \mid h \in H\}$. Note that $q^c\, \mathbf{Z}_{q^b} = \{0\}$ if $c \geq b$, $\approx \mathbf{Z}_{q^{b-c}}$ if $c < b$. Thus,

$$H^{q^{a-1}} \approx \underbrace{\mathbf{Z}_q \times \cdots \times \mathbf{Z}_q}_{\alpha(a)},$$

so that the order of $H^{q^{a-1}}$ is $q^{\alpha(a)}$. Thus, $\alpha(a)$ is determined only by G, q, and a. Next, note that

$$H^{q^{a-2}} \approx \underbrace{\mathbf{Z}_q \times \cdots \times \mathbf{Z}_q}_{\alpha(a-1)} \times \underbrace{\mathbf{Z}_{q^2} \times \cdots \times \mathbf{Z}_{q^2}}_{\alpha(a)}.$$

Thus, the order of $H^{q^{a-2}}$ is

$$q^{\alpha(a-1)+2\alpha(a)}.$$

Thus, $\alpha(a-1)$ depends only on $G, q, a-1$. Proceeding in this way, we see that $\alpha(1), \ldots, \alpha(a)$ are determined uniquely by G and q.

EXAMPLE 1: Let us determine up to isomorphism all abelian groups of order 16. Since $16 = 2^4$, the possible sets of elementary divisors are

(a) 2^4.
(b) $2, 2^3$.
(c) $2, 2, 2^2$.
(d) $2, 2, 2, 2$.
(e) $2^2, 2^2$.

The corresponding abelian groups are

(a) $\mathbf{Z}_{16}$.
(b) $\mathbf{Z}_2 \times \mathbf{Z}_8$.
(c) $\mathbf{Z}_2 \times \mathbf{Z}_2 \times \mathbf{Z}_4$.
(d) $\mathbf{Z}_2 \times \mathbf{Z}_2 \times \mathbf{Z}_2 \times \mathbf{Z}_2$.
(e) $\mathbf{Z}_4 \times \mathbf{Z}_4$.

None of the groups (a)–(e) are isomorphic to one another by Theorem 6. This example suggests the following generalization. Let p be a prime, m a positive integer. Let us determine the abelian groups of order p^m. We know that every such group is uniquely specified by its elementary divisors

$$p^{a_1}, p^{a_2}, \ldots, p^{a_s},$$

where

$$a_1 \leq a_2 \leq \cdots \leq a_s, \qquad a_1 + a_2 + \cdots + a_s = m.$$

The last property comes from the fact that

$$\mathbf{Z}_{p^{a_1}} \times \mathbf{Z}_{p^{a_2}} \times \cdots \times \mathbf{Z}_{p^{a_s}} \tag{10}$$

has order $p^{a_1} \cdot p^{a_2} \cdots p^{a_s}$. Thus, to each abelian group of order p^m is associated set $\{a_1, \ldots, a_s\}$ of positive integers such that $a_1 + \cdots + a_s = m$, $a_1 \leq a_2 \leq \cdots \leq a_s$. Such a set is called a *partition* of m. Conversely, to each partition $\{a_1, \ldots, a_s\}$ of m there corresponds an abelian group of order p^m—the group (10). Moreover, by Theorem 6, distinct partitions of m correspond to nonisomorphic groups. Let $p(m)$ denote the number of distinct partitions of m. Then, we see that the number of nonisomorphic abelian groups of order p^m is $p(m)$.

In the above example, $p = 2$, $m = 4$. The partitions of 4 are

$$\{4\}, \quad \{1, 3\}, \quad \{1, 1, 2\}, \quad \{1, 1, 1, 1\}, \quad \{2, 2\}.$$

Thus, there are five nonisomorphic abelian groups of order 2^4, as we discovered.

We have proved in Chapter 2 that the number of nonisomorphic groups of order n is at most n^{n^2}. If $n = p^m$, there are at most $p^{mp^{2m}}$. How does this number compare with the number of nonisomorphic abelian groups of order p^m? In other words, how big is $p(m)$?

The problem of determining the law of growth of the partition function $p(m)$ has been solved completely, but only in the twentieth century, and belongs to a stunning chapter of contemporary mathematics called *additive number theory*. Although many results about partitions had been discovered as early as the eighteenth century by Euler and Lagrange, it was not until the invention of the *circle method* by Hardy and Ramanujan in 1918 that any real progress was made in determining the order of magnitude of $p(m)$. Hardy and Ramanujan proved that

$$\lim_{m\to\infty} \frac{p(m)}{(1/(4\sqrt{3}\,m))\exp(\pi\sqrt{2m/3})} = 1. \tag{11}$$

In other words, for large m, $p(m)$ is approximately equal to

$$\frac{1}{4\sqrt{3}\,m}\exp\left(\pi\sqrt{\frac{2m}{3}}\right).$$

We leave it as an exercise to show that this quantity is much smaller than $p^{mp^{2m}}$, in the sense that

$$\lim_{m\to\infty} \frac{(1/(4\sqrt{3}\,m))\exp(\pi\sqrt{2m/3})}{p^{mp^{2m}}} = 0.$$

The proof of (11) is very difficult and relies on analysis. Much more precise results than (11) are known. In fact, in 1936 Rademacher found a beautiful exact formula for $p(m)$ in terms of an infinite series.

Let us close this section with a very simple, but useful application of the fundamental theorem.

PROPOSITION 7: Let G be an abelian group whose order is divisible by a prime p. Then G contains an element of order p.

Proof: Without loss of generality assume G to be of the form

$$\mathbf{Z}_{p^{a_1}} \times \mathbf{Z}_{p^{a_2}} \times \cdots \times \mathbf{Z}_{p^{a_t}} \times \mathbf{Z}_{q^{b_1}} \times \cdots.$$

Then

$$(p^{a_1-1}, p^{a_2-2}, \ldots, p^{a_t-1}, 0, 0, \ldots, 0)$$

has order p.

8.4 Exercises

1. Write the following groups as direct products of cyclic groups of prime power order:
   (a) $\mathbf{Z}_{12}$. (b) $\mathbf{Z}_5 \times \mathbf{Z}_{15}$. (c) $\mathbf{Z}_{125}$.
   (d) $\mathbf{Z}_{169}$. (e) $\mathbf{Z}_{60} \times \mathbf{Z}_{20}$.
2. Find, up to isomorphism, all abelian groups of the following orders:
   (a) 32. (b) 25. (c) 30. (d) 100.
3. Let p be a prime, $n \geq 1$ an integer, $f(p, n)$ = the number of nonisomorphic abelian groups of order p^n.
   (a) Prove that $f(p, n)$ does not depend on p.
   (b) Prove that $f(p, 1) = 1$.
   (c) Compute $f(p, n)$ for $n \leq 20$.

8.5 Sylow's Theorems

We now come to the three theorems of Sylow, which are among the most important results in the theory of finite groups. Recall that in Chapter 2 we proved Lagrange's theorem, which asserts that if G is a finite group and H is a subgroup of G, then the order of H divides the order of G. Note, however, that the converse of Lagrange's theorem is false. That is, if G has order m and if $m \mid n$, then there does not necessarily exist a subgroup of G of order m. For example, A_4 is a group of order 12 which has no subgroups of order 6 (Exercise 12 of Section 3.8). Sylow's theorem comes about as close as one can come to providing a converse to Lagrange's theorem.

Throughout this section, let G be a group of order n, let p be a prime dividing n, and let $n = p^a n_0$, $(n_0, p) = 1$.

DEFINITION 1: A subgroup of G of order p^a is called a *p-Sylow subgroup.*

Our main results are the following results of the Norwegian mathematician Sylow, denoted SI, SII, and SIII.

THEOREM 2 (SI): G contains a p-Sylow subgroup.

THEOREM 3 (SII): If P and Q are any two p-Sylow subgroups of G, then there exists $g \in G$ such that $P = gQg^{-1}$.

THEOREM 4 (SIII): Suppose that G has r p-Sylow subgroups. The $r \mid n$ and $r \equiv 1 \pmod{p}$.

In order to prove these remarkable results, a number of preliminaries must be introduced.

If G is any group, then the *center* of G, denoted Z_G, is defined as

$$Z_G = \{g \in G \mid gx = xg \text{ for all } x \in G\}.$$

Then $Z_G \lhd G$. (Easy exercise.) Moreover, G is abelian if and only if $Z_G = G$.

If S is any subset of G, then the *normalizer* of S, denoted $N_G(S)$, is defined by

$$N_G(S) = \{g \in G \mid gSg^{-1} \subseteq S\}.$$

Then $N_G(S) \leq G$. (Easy exercise.) Moreover, if S is a subgroup of G, then $S \lhd N_G(S)$. (Obvious!)

Let $g, g' \in G$. We say that g is *conjugate* to g' if there exists $x \in G$ such that $g = xg'x^{-1}$. It is immediate that the relation of conjugacy is an equivalence relation on G. The equivalence classes of G with respect to conjugacy are called *conjugacy classes*. If C is a conjugacy class of G, and $g \in C$, then

$$C = \{xgx^{-1} \mid x \in G\}.$$

Moreover, if $C_1, \ldots, C_h$ are the conjugacy classes of G, then

$$G = C_1 \cup \cdots \cup C_h$$

and $C_i \cap C_j = \varnothing$ if $i \neq j$.

PROPOSITION 5: Let $g \in G$, $S = \{g\}$, $h_g =$ the number of elements in the conjugacy class containing g. Then

$$|N_G(S)| = n/h_g.$$

Proof: Let C denote the conjugacy class containing g, $h = h_g$, $C = \{g_1 g g_1^{-1}, \ldots, g_h g g_h^{-1}\}$. Let us show that

$$G = g_1 N_G(S) \cup \cdots \cup g_h N_G(S) \tag{1}$$

and

$$g_i N_G(S) \cap g_j N_G(S) = \varnothing \qquad (i \neq j). \tag{2}$$

This will suffice to prove the proposition. For $g_iN_G(S)$ contains $|N_G(S)|$ elements for $i = 1, \ldots, h$ and by (2), none of these sets overlap. Therefore, by (1), $n = h_g \cdot |N_G(S)|$. If $x \in g_iN_G(S) \cap g_jN_G(S)$ $(i \neq j)$, then there exist $m,m' \in N_G(S)$ such that

$$x = g_im = g_jm'. \tag{3}$$

However,

$$\begin{aligned} N_G(S) &= \{x \in G \mid xgx^{-1} = g\} \\ &= \{x \in G \mid xg = gx\}. \end{aligned}$$

But, by (3), $g_j^{-1}g_i = m'm^{-1} \in N_G(S)$, so that

$$g_j^{-1}g_ig = gg_j^{-1}g_i \Longrightarrow g_igg_i^{-1} = g_jgg_j^{-1},$$

which contradicts the choice of the g_i's. Thus (2) is proved. Let $x \in G$. Then $xgx^{-1} = g_igg_i^{-1}$ for some i, so that

$$\begin{aligned} g_i^{-1}xgx^{-1}g_i = g &\Longrightarrow (g_i^{-1}x)g(g_i^{-1}x)^{-1} = g \\ &\Longrightarrow g_i^{-1}x \in N_G(S) \\ &\Longrightarrow x \in g_iN_G(S). \end{aligned}$$

Thus (1) is proved.

Let us now give proofs of the Sylow theorems.

Proof of SI: The result is clearly true if $n = 2$, so let us proceed by induction on n. Let us consider three cases:

Case A: $p \mid |Z_G|$. In this case, by Proposition 7 of Section 8.4, Z_G contains an element x of order p. Let $P = [x]$. Then $P \lhd G$ since $x \in Z_G$ and thus G/P is defined and is clearly of order $p^{a-1}n_0$. By induction G/P has a subgroup W of order p^{a-1}. But W is of the form H/P for some subgroup H of G. Then H is a subgroup of G of order p^a.

Case B: $p \nmid |Z_G|$ and $1 < |Z_G| < n_0$. In this case, G/Z_G is of order p^an' for some $n' < n_0$, $(n', p) = 1$. By induction, there exists a subgroup W of G/Z_G of order p^a. But W is of the form H/Z_G for some subgroup H of G. But then H is of order $p^an' < n$, so that by induction, H has a subgroup H_0 of order p^a.

Case C: $|Z_G| = n_0$ or 1. Let $C_1, \ldots, C_t$ denote the conjugacy classes of G. Note that $C_i = \{g_i\}$ if and only if $g_i \in Z_G$. Thus, let $Z_G = \{g_1, \ldots, g_s\}$ and number the C_i so that $C_i = \{g_i\}$ $(1 \leq i \leq s)$. Then, for $i > s$, C_i contains more than one element. Assume that $p \mid |C_i|$ for all $i > s$. Then

$$\begin{aligned} n &= |Z_G| + |C_{s+1}| + \cdots + |C_t| \\ &\Longrightarrow n \equiv n_0 \text{ or } 1 \pmod{p} \\ &\Longrightarrow n_0 \text{ or } 1 \equiv 0 \pmod{p} \qquad (\text{since } p \mid n). \end{aligned}$$

But $(n_0, p) = 1$, so that $p \nmid |C_i|$ for some $i > s$. Let c be any element belonging to C_i. Then, if $S = \{c\}$, Proposition 5 implies that $|N_G(S)| = n/|C_i| = p^a n'$ $[n' < n_0, (n', p) = 1]$, since $|C_i| > 1$ and $p \nmid |C_i|$. Therefore, by induction, $N_G(S)$ has a subgroup of order p^a.

Proof of SII: Let S_1 and S_2 be two p-Sylow subgroups of G, and let

$$G = S_1 g_1 S_2 \cup \cdots \cup S_1 g_t S_2$$

be the decomposition of G relative to the two subgroups S_1 and S_2. Then, since $|S_1| = |S_2| = p^a$, Theorem 1 of Section 8.2 asserts that

$$n = \sum_{i=1}^{t} \frac{p^a \cdot p^a}{u_i},$$

where $u_i =$ the order of $U_i = g_i^{-1} S_1 g_i \cap S_2$. But since $n = p^a n_0$, we see that

$$n_0 = \sum_i \frac{p^a}{u_i}.$$

Since $U_i \leq S_2$, $u_i | p^a$, and thus $p^a/u_i = p^b$ for some $b > 0$ or $p^a/u_i = 1$. Since $p \nmid n_0$, there exists i for which $p^a/u_i = 1$. But then $p^a = u_i$ and thus, $g_i^{-1} S_1 g_i \cap S_2 = S_2$, and therefore

$$g_i^{-1} S_1 g_i \supseteq S_2.$$

But both $g_i^{-1} S_1 g_i$ and S_2 have order p^a, so that we conclude that

$$g_i^{-1} S_1 g_i = S_2.$$

COROLLARY 6: Let S be a p-Sylow subgroup of G. Then S is the only p-Sylow subgroup of G if and only if $S \lhd G$.

Proof: $\Rightarrow$ Let $g \in G$. Then gSg^{-1} is a p-Sylow subgroup of G, so that $gSg^{-1} = S$. Thus, $S \lhd G$.
$\Leftarrow$ Let S' be any p-Sylow subgroup of G. Then by SII, $S' = gSg^{-1}$ for some $g \in G$. But since $S \lhd G$, $gSg^{-1} = S$, so that $S' = S$.

PROPOSITION 7: Let S be a p-Sylow subgroup of G, $N = N_G(S)$. Then the number of p-Sylow subgroups of G is $[G: N]$.

Proof: Let

$$G = g_1 N \cup g_2 N \cup \cdots \cup g_t N, \qquad t = [G: N]$$

be a coset decomposition. Then

$$g_1 S g_1^{-1}, g_2 S g_2^{-1}, \ldots, g_t S g_t^{-1} \tag{4}$$

are p-Sylow subgroups of G. These are distinct, since if

$$g_i S g_i^{-1} = g_j S g_j^{-1},$$

we see that $(g_j^{-1} g_i) S (g_j^{-1} g_i)^{-1} = S$, so that $g_j^{-1} g_i \in N$. Thus, $g_i \in g_j N$ and

$g_iN = g_jN$, so that $i = j$. Thus, the p-Sylow subgroups (4) are distinct. Let S' be any p-Sylow subgroup. Then $S' = gSg^{-1}$ for some $g \in G$. Now $g = g_in$ for some $n \in N$. Therefore, $S' = g_i(nSn^{-1})g_i^{-1} = g_iSg_i^{-1}$ since $n \in N$. Thus, S' is contained in the list (4), which must therefore include all p-Sylow subgroups of G.

Proof of SIII: Let t be the number of p-Sylow subgroups of G. Then by Proposition 7, $t = [G:N]$, where $N = N_G(S)$ and S is a fixed p-Sylow subgroup. In particular, $n = t\,|N|$, so that $t\,|\,n$. Let

$$G = Sg_1N \cup Sg_2N \cup \cdots \cup Sg_vN$$

be the decomposition of G with respect to the two subgroups S and N. Without loss of generality, assume that $g_1 = 1$, so that $Sg_1N = N$. Let $|N| = r = p^am_0$, $(m_0, p) = 1$. By Theorem 1 of Section 8.2, we have

$$n = p^an_0 = \sum_{i=1}^{v} \frac{p^ar}{c_i}$$

where $c_i = |g_i^{-1}Sg_i \cap N|$. Note that since $g_1 = 1$, $c_1 = |S \cap N| = |S| = p^a$. Therefore,

$$\begin{aligned} n_0 = r\sum_{i=1}^{v} \frac{1}{c_i} &= m_0 \sum_{i=1}^{v} \frac{p^a}{c_i} \\ &= m_0\left(1 + \sum_{i=2}^{v} \frac{p^a}{c_i}\right) \end{aligned} \tag{5}$$

$$\Longrightarrow \frac{n_0}{m_0} = 1 + \sum_{i=2}^{v} \frac{p^a}{c_i}$$

$$\Longrightarrow t = \frac{n}{r} = 1 + \sum_{i=2}^{v} \frac{p^a}{c_i}$$

Note that since $g_i^{-1}Sg_i \cap N$ is a subgroup of $g_i^{-1}Sg_i$, which has order p^a, we see that $c_i = p^b$ for some b $(0 \leq b \leq a)$. We assert that $c_i < p^a$ for $i = 2, \ldots, v$. This fact, in conjunction with (4), suffices to complete the proof, since then $\sum_{i=2}^{v} p^a/c_i \equiv 0 \pmod{p}$ and $t \equiv 1 \pmod{p}$. Assume, on the contrary, that $i \geq 2$ and $c_i = p^a$. Then $g_i^{-1}Sg_i \cap N$ is a p-Sylow subgroup of N. But S is a p-Sylow subgroup of N and $S \lhd N$, since $N = N_G(S)$. Therefore, by Corollary 6, $g_i^{-1}Sg_i \cap N = S$ and thus $g_i^{-1}Sg_i \supseteq S$. However, both $g_i^{-1}Sg_i$ and S contain p^a elements, so that $g_i^{-1}Sg_i = S$. Thus, $g_i \in N$ and $Sg_iN = SN = N$. However, since $Sg_iN \cap Sg_jN = \varnothing$ if $i \neq j$ and since $Sg_1N = N$, we see that $i = 1$, which is a contradiction to the choice of i. Thus, $c_i < p^a$.

Let us now give a few examples of applications of the Sylow theorems.

PROPOSITION 8: Let p and q be distinct primes, $p \not\equiv 1 \pmod{q}$, $q \not\equiv 1 \pmod{p}$, and let G be any abelian group of order pq. Then $G \approx \mathbf{Z}_p \times \mathbf{Z}_q$.

Proof: By the fundamental theorem of abelian groups, it suffices to show that G is abelian. Let P and Q, be respectively, a p-Sylow and a q-Sylow subgroup of G. Since every element of P except the identity has order p and every element of Q except the identity has order q, we see that $P \cap Q = \{1\}$. The number of p-Sylow subgroups divides pq and is therefore 1, p, q, pq. Moreover, this number is congruent to 1 (mod p). Since $q \not\equiv 1 \pmod{p}$, and $p, pq \equiv 0 \pmod{p}$, we see that P is the only p-Sylow subgroup of G and that $P \lhd G$ by Corollary 6. Similarly, $Q \lhd G$. Therefore by Theorem 6 of Section 8.1, PQ is a subgroup of G of order $pq/|P \cap Q| = pq$. Thus, every element of G can be written in the form xy ($x \in P$, $y \in Q$). Finally, if $x \in P$, $y \in Q$, then $(xyx^{-1})y^{-1} \in Q$, since $Q \lhd G$. Moreover, $xyx^{-1}y^{-1} = x(yx^{-1}y^{-1}) \in P$ since $P \lhd G$, so that $xyx^{-1}y^{-1} \in P \cap Q = \{1\}$ and thus $xy = yx$. By Theorem 3 of Section 8.3, we see that $G \approx P \times Q$. Thus, G is abelian, since P and Q are cyclic.

Thus, for example, since $5 \not\equiv 1 \pmod 3$ and $3 \not\equiv 1 \pmod 5$, we see that the only group of order 15 is $\mathbf{Z}_{15} \approx \mathbf{Z}_5 \times \mathbf{Z}_3$.

EXAMPLE: No group of order 20 is simple.

Let G be a group of order 20. The number of 5-Sylow subgroups is one of 1, 2, 4, 5, 10, 20. However, the number of 5-Sylow subgroups is congruent to 1 (mod 5), so that this number is 1. Thus, if P is the unique 5-Sylow subgroup, then $P \lhd G$ and G is not simple.

8.5 *Exercises*

1. Find the conjugacy classes of S_3; of D_4.
2. Prove that any group of order 35 is cyclic.
3. Prove that no group of order 56 is simple.
4. Show that if S is any subset of a group G, then $N_G(S) \leq G$.
5. Show that if S is a subgroup of G, then $S \lhd N_G(S)$.
6. Show that if S is a subgroup of G, then $N_G(S)$ is the largest subgroup of G which contains S as a normal subgroup.

8.6 *Solvable Groups*

Let us now introduce the class of groups known as *solvable groups*, which play a crucial role in the theory of the solution of equations in radicals. In some sense, one could say that the solvable groups are one order of magnitude more complicated than abelian groups. Before introducing the concept of a solvable group, we must define several preliminary notions. Throughout this section let G be a group of order n.

DEFINITION 1: A *normal series* for G is a chain of subgroups of G of the form

$$G = G_0 \triangleright G_1 \triangleright G_2 \triangleright \cdots \triangleright G_r = \{1\}, \qquad G_i \neq G_{i+1}.$$

If two normal series for G are given

$$G = G_0 \triangleright G_1 \triangleright \cdots \triangleright G_r = \{1\},$$
$$G = H_0 \triangleright H_1 \triangleright \cdots \triangleright H_s = \{1\},$$

then the second is said to be a *refinement* of the first if every H_i is a G_j for some j. The second is said to be a *proper refinement* of the first if it is a refinement and $s > r$. A normal series without any proper refinements is called a *composition series.*

It is clear that by refining any normal series for G, we can eventually arrive at a composition series for G.

EXAMPLE 1: Let $G = V_4 = \mathbf{Z}_2 \times \mathbf{Z}_2$. Then a normal series for G is given by

$$\mathbf{Z}_2 \times \mathbf{Z}_2 \triangleright \mathbf{Z}_2 \times \{\bar{1}\} \triangleright \{\bar{1}\} \times \{\bar{1}\}.$$

EXAMPLE 2: Let $G = S_3$. A normal series for G is given by

$$S_3 \triangleright A_3 \triangleright \{1\}.$$

We will show below that the normal series in Examples 1 and 2 are composition series.

DEFINITION 2: Let

$$G = G_0 \triangleright G_1 \triangleright G_2 \triangleright \cdots \triangleright G_r = \{1\}$$

be a normal series for G. The groups

$$G_0/G_1,\ G_1/G_2, \ldots, G_{r-1}/G_r \tag{1}$$

are called *composition factors of the normal series.* If the series is a composition series, then the composition factors (1) are said to be *composition factors* for G.

An important result about composition factors, which we will state but not prove is the theorem of Jordan and Hölder:

THEOREM 3 (JORDAN–HÖLDER): Let

$$G = G_0 \triangleright G_1 \triangleright \cdots \triangleright G_r = \{1\},$$
$$G = H_0 \triangleright H_1 \triangleright \cdots \triangleright H_s = \{1\}$$

be two composition series for G. Then $r = s$ and the composition factors

$$\bar{G}_j = G_{j-1}/G_j \qquad (j = 1, \ldots, r)$$

can be rearranged, so that the ith composition factor is isomorphic to

$$\bar{H}_i = H_{i-1}/H_i \qquad (i = 1, \ldots, r).$$

Thus, the composition factors of G are uniquely determined up to isomorphism and rearrangement, independent of the choice of composition series.

Recall that a simple group is a group with no nontrivial normal subgroups. Then the following result is very useful:

PROPOSITION 4: Let

$$G = G_0 \rhd G_1 \rhd \cdots \rhd G_r = \{1\} \tag{2}$$

be a normal series for G. This normal series is a composition series if and only if all the composition factors are simple groups.

Proof: Note that (2) is a composition series if and only if for every i $(1 \leq i \leq r)$, there does not exist a subgroup H of G such that $G_{i-1} \rhd H \rhd G_i$, $H \neq G_{i-1}$, $H \neq G_i$. The condition $G_{i-1} \rhd H \rhd G_i$ is equivalent to $G_{i-1}/G_i \rhd H/G_i$, and the conditions $H \neq G_{i-1}$, G_i are equivalent, respectively, to $H/G_i \neq G_{i-1}/G_i, \{1\}$. Thus, (2) is a composition series for $G \Longleftrightarrow$ for every i $(1 \leq i \leq r)$, the composition factor G_{i-1}/G_i has no nontrivial normal subgroups.

Let us consider the normal series of Examples 1 and 2. Since

$$\mathbf{Z}_2 \times \mathbf{Z}_2/\mathbf{Z}_2 \times \{1\} \approx \mathbf{Z}_2, \qquad \mathbf{Z}_2 \times \{1\}/\{1\} \times \{1\} \approx \mathbf{Z}_2,$$
$$S_3/A_3 \approx \mathbf{Z}_2, \qquad A_3/\{1\} \approx \mathbf{Z}_3,$$

and since $\mathbf{Z}_2$ and $\mathbf{Z}_3$ are simple groups, we see that the normal series of examples 1 and 2 are composition series. Thus, the composition factors for V_4 are $\mathbf{Z}_2$ and $\mathbf{Z}_2$, and the composition factors for S_3 are $\mathbf{Z}_2$ and $\mathbf{Z}_3$.

DEFINITION 5: G is said to be *solvable* if it has a composition series whose composition factors are cyclic of prime order.

PROPOSITION 6: G is solvable if and only if the composition factors of G are abelian.

Proof: $\Rightarrow$ Obvious.
$\Leftarrow$ Let C be a composition factor of G. Then C is a simple, abelian group. Let p be a prime dividing $|C|$. Then C contains an element g of order p. Since C is simple and abelian, $[g] \lhd C$, so that $C = [g]$. Thus, C is cyclic of prime order.

EXAMPLE 3: V_4 and S_3 are solvable.

EXAMPLE 4: Let G be abelian. Then the composition factors of G are all abelian, so that G is solvable.

EXAMPLE 5: Let $n \geq 5$. Then the composition factors of the normal series

$$S_n \rhd A_n \rhd \{1\}$$

are

$$S_n/A_n \approx \mathbf{Z}_2, \qquad A_n/\{1\} \approx A_n.$$

Now A_n is simple by Abel's theorem, so that (3) is a composition series for S_n. However, for $n \geq 5$, A_n is not abelian. Thus, S_n is not solvable.

EXAMPLE 6: Every group of prime power order is solvable. We will sketch a proof of this fact in the exercises.

PROPOSITION 7: Let G be a solvable group, H a subgroup of G. Then H is solvable.

Proof: Let $G = G_0 \rhd G_1 \rhd \cdots \rhd G_r = \{1\}$ be a composition series whose composition factors are cyclic of prime order. Define

$$H_i = G_i \cap H \qquad (0 \leq i \leq r).$$

Then $H = H_0 \rhd H_1 \rhd \cdots \rhd H_r = \{1\}$. Moreover, the mapping

$$\phi\colon G_{i-1} \cap H \longrightarrow G_{i-1}/G_i,$$
$$\phi(g) = gG_i \qquad (g \in G_{i-1})$$

is a homomorphism with kernel $G_i \cap H$. Therefore, H_{i-1}/H_i is isomorphic to a subgroup of G_{i-1}/G_i and therefore is either trivial or is a cyclic group of prime order. The former case occurs if and only if $H_{i-1} = H_i$. Let us omit any duplicates from among $H_0, \ldots, H_r$. Then the resulting chain of subgroups of H is a composition series whose composition factors are cyclic of prime order.

PROPOSITION 8: Let G be a solvable group, H any group, $f\colon G \to H$ a homomorphism. Then $f(G)$ is a solvable group.

Proof: Let $G = G_0 \rhd G_1 \rhd \cdots \rhd G_r = \{1\}$ be a composition series with composition factors cyclic of prime order. Then

$$f(G) = f(G_0) \rhd f(G_1) \rhd \cdots \rhd f(G_r) \rhd \{1\}.$$

Moreover, for $i = 1, \ldots, r$, $f(G_{i-1})/f(G_i)$ is either trivial or is cyclic of prime order. [Exercise. *Hint*: $f(G_{i-1})/f(G_i) \approx G_{i-1}/G_i \cdot \ker(f)$.] Thus, as in Proposition 8, we deduce that $f(G)$ is solvable.

COROLLARY 9: Let G be a solvable group, $H \lhd G$. Then G/H is solvable.

Proof: Let $f\colon G \longrightarrow G/H$ be the natural homomorphism. Then $f(G) = G/H$. The corollary now follows from Proposition 8.

8.6 *Exercises*

1. Find a composition series and the corresponding composition factors for the following groups:
   (a) $\mathbf{Z}_9$. (b) $\mathbf{Z}_9 \times \mathbf{Z}_{27}$. (c) $\mathbf{Z}_{p^n}$, p prime.
   (d) S_4. (e) D_5.
2. Let G be a group and let $H \lhd G$. Suppose that both H and G/H are solvable. Show that G is solvable.
3. Let G be a group of order p^n, where p is a prime.
   (a) Show that $Z(G) \neq \{1_G\}$ if $n \geq 1$. (*Hint*: Use the class equation.)
   (b) Use induction on n and (a) to show that G is solvable. [*Hint*: $Z(G)$ is solvable and $G/Z(G)$ is solvable by induction, because of (a). Then apply Exercise 2.]
4. Complete the proof of Proposition 8.

9 GALOIS THEORY

9.1 A Restrictive Assumption

Let F be a field, $\bar{F}$ its algebraic closure, $f \in F[X]$ a monic, nonconstant polynomial. In $\bar{F}[X]$, we may factor f into linear factors:

$$f = \prod_{i=1}^{n} (X - \alpha_i), \qquad \alpha_i \in \bar{F}.$$

Suppose that m of the α_i are distinct. Let us renumber $\alpha_1, \ldots, \alpha_n$ so that $\alpha_1, \ldots, \alpha_m$ are all different. Then

$$f = \prod_{i=1}^{m} (X - \alpha_i)^{\nu_i}, \qquad \nu_i \geq 1.$$

The positive integer ν_i is called the *multiplicity of the zero* α_i. If $\nu_i = 1$, then α_i is called a *simple zero* of f; if $\nu_i > 1$, then α_i is called a *multiple zero of* f. A zero of multiplicity > 1 is called a *multiple zero.*

An irreducible polynomial $f \in F[X]$ can have multiple zeros. For example, let $F = \mathbf{Z}_2(Y)$, where Y is transcendental over $\mathbf{Z}_2$, and let $f = X^2 - Y \in F[X]$. Then f is irreducible in $F[X]$. Otherwise, f factors into linear factors in $F[X]$ and there exists $Z \in F$ such that $Z^2 = Y$. But Z is the quotient of two polynomials in Y, so that we get an algebraic equation for Y over $\mathbf{Z}_2$, a contradiction to the fact that Y is transcendental over $\mathbf{Z}_2$. Thus, f is irreducible in $F[X]$. In $\bar{F}[X]$,

$$f = (X - \sqrt{Y})(X + \sqrt{Y}),$$

where $\sqrt{Y}$ denotes a fixed square root of Y in $\bar{F}$. But $1 + 1 = 0$ in F, so that

$(1 + 1)\sqrt{Y} = 0 \Rightarrow \sqrt{Y} = -\sqrt{Y}$. Thus,

$$f = (X - \sqrt{Y})^2$$

and $\sqrt{Y}$ is a multiple zero of f.

In order to avoid this and other pathologies, let us restrict our fields somewhat. Henceforth, unless explicit mention to the contrary is made, we will assume that *all fields are extensions* of $\mathbf{Q}$. For example, we rule out $\mathbf{Z}_2(Y)$ since if F is an extension of $\mathbf{Q}$, $1 + 1 \neq 0$ in F. It is possible to develop field theory somewhat more generally than we are going to develop it. However, this can only be done by introducing a number of new concepts as well as by using more complicated arguments. Therefore, we have opted, in this introductory text, to develop a somewhat specialized Galois theory, for extensions of $\mathbf{Q}$. In the remainder of this section, let us show what this restriction does for us. First we will prove that the phenomenon exhibited above cannot occur.

THEOREM 1: Let $f \in F[X]$ be irreducible and of degree ≥ 1. Then all zeros of f in $\bar{F}$ are simple.

Before proving Theorem 1, let us introduce a few tools. Let $n \in N$, $a \in \bar{F}$. In Chapter 4 we defined the product $n \cdot a$ as follows:

$$n \cdot a = 0 \qquad \text{if } n = 0,$$
$$= \underbrace{a + a + \cdots + a}_{n \text{ times}} \qquad \text{if } n > 0.$$

If $h = a_0 + a_1X + \cdots + a_rX^r \in \bar{F}[X]$, let us define the formal *derivative* Dh of h by

$$Dh = 1 \cdot a_1 + 2 \cdot a_2X + \cdots + r \cdot a_rX^{r-1}.$$

Then $Dh \in \bar{F}[X]$. It is trivial to verify the following properties of the formal derivative:

$$D(ag + bh) = aDg + bDh \quad (a, b \in \bar{F};\ g, h \in \bar{F}[X]), \tag{1}$$
$$D(gh) = hDg + gDh \qquad (g, h \in \bar{F}[X]), \tag{2}$$
$$D(x - a)^\nu = \nu(x - a)^{\nu-1} \qquad (a \in \bar{F},\ \nu \geq 1). \tag{3}$$
$$\text{If } \deg(f) = n, \text{ then } \deg(Df) = n - 1 \qquad (n \geq 1, f \in \bar{F}[X]). \tag{4}$$

The proofs of (1)–(3) are given as exercises in Chapter 4.

LEMMA 2: Suppose that f has a multiple zero. Then f and Df have a nonconstant common factor in $F[X]$.

Proof: If α is a multiple zero of f, then $f = (x - \alpha)^\nu g$, $g \in \bar{F}[X]$, $\nu > 1$. By (1)–(3), we have

$$Df = \nu(X - \alpha)^{\nu-1}g + (X - \alpha)^\nu Dg.$$

But since $\nu > 1$, α is a zero of Df. If f and Df have no nonconstant common factor in $F[X]$, then there exist $\gamma, \beta \in F[X]$ such that $1 = \gamma f + \beta Df$. Therefore, if we replace X by α in this last equation, we get $1 = 0$, which is a contradiction. Thus, $(f, Df) \neq 1$.

Proof of Theorem 1: Without loss of generality, assume that f is monic and assume that f has a multiple zero. Then $\deg(f) \geq 2$, and if $\deg(f) = n$, then $\deg(Df) = n - 1 \geq 1$. Moreover by Lemma 2, $(f, Df) \neq 1$, so that f and Df have a nontrivial factor in common. Since f is irreducible, this implies that Df is divisible by f. But this is impossible since $1 \leq \deg(Df) < \deg(f)$.

The next result is another extremely useful consequence of our restrictive assumption.

THEOREM 3 (PRIMITIVE ELEMENT THEOREM): Let $E = F(\alpha, \beta)$ be an algebraic extension of F. Then there exists $\gamma \in E$ such that $E = F(\gamma)$. Thus, E is a simple extension of F.

Proof: Let $\bar{F}$ be the algebraic closure of F and let $f = \operatorname{Irr}_F(\alpha, X)$, $g = \operatorname{Irr}_F(\beta, X)$. Then in $\bar{F}$, we have

$$f = (X - \alpha_1)\cdots(X - \alpha_n), \qquad \alpha_i \in \bar{F}, \quad \alpha_1 = \alpha,$$
$$g = (X - \beta_1)\cdots(X - \beta_m), \qquad \beta_j \in \bar{F}, \quad \beta_1 = \beta.$$

Consider the following set of elements of $\bar{F}$:

$$\frac{\alpha_i - \alpha_1}{\beta_1 - \beta_j} \qquad (1 \leq i \leq n,\ 2 \leq j \leq m). \tag{5}$$

This set is finite and, since $F \supseteq \mathbf{Q}$, we can choose $t \in F$ distinct from all the elements (5). We will prove that $E = F(\gamma)$ with $\gamma = \alpha + t\beta$. Since $\alpha + t\beta \in E$, it is clear that $E \supseteq F(\gamma)$. Let us prove the reverse inclusion. Consider the polynomial

$$h(X) = f(\gamma - tX) \in F(\gamma)[X].$$

Then $h(\beta) = f(\gamma - t\beta) = f(\alpha) = 0$. Therefore, in $\bar{F}[X]$, $h(X)$ is divisible by $X - \beta_1$. If $h(X)$ is divisible by $X - \beta_j$ for some $j > 1$, then $h(\beta_j) = 0$, which implies that $f(\gamma - t\beta_j) = 0$. Therefore, $\gamma - t\beta_j$ is a zero of f, and thus $\gamma - t\beta_j = \alpha_i$ for some i. But since $\gamma = \alpha_1 + t\beta_1$, this implies that

$$t = \frac{\alpha_i - \alpha_1}{\beta_1 - \beta_j},$$

which contradicts the choice of t. Thus, $h(X)$ is divisible by $X - \beta_1$, but not by $X - \beta_j$ for $j > 1$. Therefore, in $\bar{F}[X]$, the g.c.d. of $h(X)$ and $g(X)$ is $X - \beta_1$, and consequently, in $F(\gamma)[X]$, the g.c.d. of $h(X)$ and $g(X)$ is either 1 or $X - \beta_1$. If it were 1, then there exists $a(X), b(X) \in F(\gamma)[X]$ such that $a(X)h(X) + b(X)g(X) = 1$. Setting $X = \beta_1$, we then get $0 = 1$, a contradiction. Therefore, the g.c.d. of $h(X)$ and $g(X)$ in $F(\gamma)[X]$ is $X - \beta_1$. In particular, $\beta_1 \in F(\gamma)$. But then $\gamma - \beta_1 t = \alpha \in F(\gamma)$, so that $F(\alpha, \beta) \subseteq F(\gamma)$. This completes the proof of the theorem.

COROLLARY 4: Let $F(\alpha_1, \alpha_2, \ldots, \alpha_n)$ be an algebraic extension of F. Then there exists $\gamma \in F(\alpha_1, \ldots, \alpha_n)$ such that $F(\alpha_1, \ldots, \alpha_n) = F(\gamma)$.

Proof: Induction on n.

9.1 *Exercises*

1. For each of the following extensions E of $\mathbf{Q}$, determine $\beta \in E$ such that $E = \mathbf{Q}(\beta)$:
   (a) $E = \mathbf{Q}(\sqrt{2}, \sqrt{3})$.
   (b) $E = \mathbf{Q}(\sqrt{5}, \omega)$, $\omega = (-1 + \sqrt{3}i)/2$.
   (c) $E = \mathbf{Q}(\sqrt{2}, \sqrt{3}, \sqrt{5})$.
   (d) $E = \mathbf{Q}(\sqrt[3]{2}, i)$.
   (e) $E =$ the splitting field over $\mathbf{Q}$ of $X^3 - 3$.
2. Determine whether any of the following polynomials have multiple zeros.
   (a) $X^2 - X + 1$.
   (b) $X^3 - 6X^2 + 12X - 8$.
   (c) $X^5 - 5X^3 + 10X^2 + 5$.
   (d) $X^6 - 6X^3 + 9$.
   (e) $\Phi_m(X)$, $m \geq 1$.
3. Determine the multiplicities of each of the zeros of the polynomials (a)–(e) of Exercise 2.
4. Let $F = \mathbf{C}$, $E = \mathbf{C}(X, Y)$, where X and Y are indeterminates over $\mathbf{C}$. Does there exist $Z \in E$ such that $E = \mathbf{C}(Z)$?
5. Suppose that $f \in F[X]$ is a nonconstant, monic polynomial and suppose that $(f, Df) \neq 1$, where $(f, Df) =$ the g.c.d. of f and Df in $F[X]$. Show that f has a multiple zero.
6. Let E be an extension of the field F, X an indeterminate over E. Further, let f and g belong to $F[X]$. Show that a g.c.d. of f and g in $F[X]$ is also a g.c.d. of f and g in $E[X]$. (*Hint:* If h is a g.c.d. of f and g in $F[X]$, then $hF[X] = fF[X] + gF[X]$.)
7. Prove Corollary 4.

9.2 *Conjugates and Normal Extensions*

Let F be a field, $\bar{F}$ its algebraic closure, $\alpha \in \bar{F}$. Then α is algebraic over F. Let $f = \mathrm{Irr}_F(\alpha, X)$. Then

$$f = (X - \alpha_1)(X - \alpha_2)\cdots(X - \alpha_n), \qquad \alpha_i \in \bar{F}, \quad \alpha_1 = \alpha.$$

DEFINITION 1: The elements $\alpha_1, \ldots, \alpha_n$ are called the *conjugates* of α over F.

EXAMPLE 1: If $\alpha \in F$, then $\text{Irr}_F(\alpha, X) = X - \alpha$ and the only conjugate of α over F is α itself.

EXAMPLE 2: Let $F = \mathbf{Q}$, $\alpha = \sqrt{5}$. Then $\text{Irr}_{\mathbf{Q}}(\alpha, X) = X^2 - 5$, and the conjugates of α over $\mathbf{Q}$ are $\sqrt{5}$ and $-\sqrt{5}$.

EXAMPLE 3: Let $F = \mathbf{Q}$, $\alpha = (-1 + \sqrt{3}i)/2$. Then $\text{Irr}_{\mathbf{Q}}(\alpha, X) = X^2 + X + 1$, so that the conjugates of α over $\mathbf{Q}$ are $(-1+\sqrt{3}i)/2$, $(-1-\sqrt{3}i)/2$.

EXAMPLE 4: Let $F = \mathbf{Q}$, $\alpha = \zeta$, a primitive nth root of 1. Then $\text{Irr}_{\mathbf{Q}}(\zeta, X) = \Phi_n(X)$, the nth cyclotomic polynomial. Therefore, the conjugates of ζ over $\mathbf{Q}$ are $\zeta^a [0 \leq a \leq n-1, (a, n) = 1]$.

EXAMPLE 5: Let $F = \mathbf{Q}$, $\alpha = \sqrt[3]{2}$. Then $\text{Irr}_{\mathbf{Q}}(\alpha, X) = X^3 - 2$. Therefore, the conjugates of $\sqrt[3]{2}$ over $\mathbf{Q}$ are $\sqrt[3]{2}$, $\sqrt[3]{2}\omega$, $\sqrt[3]{2}\omega^2$, where $\omega = (-1 + \sqrt{3}i)/2$ is a primitive cube root of 1.

Let E be a finite extension of F. Let us use the notion of conjugate elements to determine all the F-isomorphisms of E into $\bar{F}$. Since E is a finite extension of F, E is an algebraic extension of F and there exist $\alpha_1, \ldots, \alpha_n \in E$ such that $E = F(\alpha_1, \ldots, \alpha_n)$. By Corollary 4 of Section 9.1, there exists $\beta \in E$ such that $E = F(\beta)$. Moreover, by Theorem 3 of Section 7.2, every element x of E can be written uniquely in the form

$$x = a_0 + a_1\beta + \cdots + a_{n-1}\beta^{n-1}, \qquad a_i \in F, \quad n = \deg(E/F).$$

Therefore, if σ is an F-isomorphism of E into $\bar{F}$, then

$$\begin{aligned} \sigma(x) &= \sigma(a_0) + \sigma(a_1)\sigma(\beta) + \cdots + \sigma(a_{n-1})\sigma(\beta)^{n-1} \\ &= a_0 + a_1\sigma(\beta) + \cdots + a_{n-1}\sigma(\beta)^{n-1}. \end{aligned} \tag{1}$$

Thus, σ is completely determined once $\sigma(\beta)$ is specified. We assert that $\sigma(\beta)$ is a conjugate of β over F. Let $f = b_0 + b_1X + \cdots + b_{n-1}X^{n-1} + X^n = \text{Irr}_F(\beta, X)$. Then

$$\begin{aligned} 0 &= b_0 + b_1\beta + \cdots + b_{n-1}\beta^{n-1} + \beta^n \\ \Longrightarrow 0 = \sigma(0) &= \sigma(b_0 + b_1\beta + \cdots + b_{n-1}\beta^{n-1} + \beta^n) \\ &= \sigma(b_0) + \sigma(b_1)\sigma(\beta) + \cdots + \sigma(b_{n-1})\sigma(\beta)^{n-1} + \sigma(\beta)^n \\ &= b_0 + b_1\sigma(\beta) + \cdots + b_{n-1}\sigma(\beta)^{n-1} + \sigma(\beta)^n \\ &= f(\sigma(\beta)). \end{aligned}$$

Therefore, $\sigma(\beta)$ is a zero of f, and hence is a conjugate of β over F. We may summarize our findings in the following proposition.

PROPOSITION 2: Let E be an extension of F of degree n and let $\beta \in E$ be such that $E = F(\beta)$. Further, let σ be an F-isomorphism of E into $\bar{F}$. Then σ has the form (1), where $\sigma(\beta)$ is a conjugate of β over F. In particular, there are at most n F-isomorphisms of E into $\bar{F}$.

We will refine Proposition 2 by determining precisely all the F-isomorphisms of E into $\bar{F}$. As a first step in this direction, let us establish the following general result.

THEOREM 3: Let F be a field, $F(\alpha)$ an algebraic extension of F of degree n, K an algebraically closed field, $f = \mathrm{Irr}_F(\alpha, X)$. Suppose that an isomorphism $\sigma: F \longrightarrow K$ is given. If

$$f = a_0 + a_1 X + \cdots + a_{n-1}X^{n-1} + X^n \in F[X],$$

let us define

$$f^\sigma = \sigma(a_0) + \sigma(a_1)X + \cdots + \sigma(a_{n-1})X^{n-1} + X^n \in K[X].$$

Let β be any zero of f^σ in K. Then there exists an isomorphism $\eta: F(\alpha) \longrightarrow K$ such that (1) the restriction of η to F coincides with σ and (2) $\eta(\alpha) = \beta$. In other words, η is an extension of σ to $F(\alpha)$ which maps α into β.

Proof: Every element x of $F(\alpha)$ can be uniquely written in the form

$$x = a_0 + a_1\alpha + \cdots + a_{n-1}\alpha^{n-1}, \qquad a_i \in F. \tag{2}$$

Therefore, let us define $\eta: F(\alpha) \longrightarrow K$ by

$$\eta(x) = \sigma(a_0) + \sigma(a_1)\beta + \cdots + \sigma(a_{n-1})\beta^{n-1}. \tag{3}$$

It is clear that η restricted to F equals σ and that $\eta(\alpha) = \beta$. Thus, it remains to prove that η is an isomorphism. Let x be defined by (2) and let

$$y = b_1 + b_1\alpha + \cdots + b_{n-1}\alpha^{n-1}, \qquad b_i \in F.$$

It is straightforward to verify that $\eta(x + y) = \eta(x) + \eta(y)$. Let us show that $\eta(x \cdot y) = \eta(x) \cdot \eta(y)$. Set $g = \sum_{i=0}^{n-1} a_i X^i$, $h = \sum_{j=0}^{n-1} b_j X^j$. By the division algorithm in $F[X]$, there exist polynomials $q, r \in F[X]$ such that

$$g \cdot h = q \cdot f + r, \qquad r = d_0 + d_1 X + \cdots + d_{n-1}X^{n-1}. \tag{4}$$

Since $f(\alpha) = 0$, (4) implies that

$$x \cdot y = g(\alpha)h(\alpha) = r(\alpha).$$

For any polynomial $k = e_0 + e_1 X + \cdots + e_s X^s \in F[X]$, set

$$k^\sigma = \sigma(e_0) + \sigma(e_1)X + \cdots + \sigma(e_s)X^s \in K[X].$$

Then by the definition of η and the fact that $f^\sigma(\beta) = 0$, we have

$$\begin{aligned} \eta(x \cdot y) &= r^\sigma(\beta) \\ &= f^\sigma(\beta) \cdot q^\sigma(\beta) + r^\sigma(\beta) \\ &= (f \cdot q + r)^\sigma(\beta) \\ &= (g \cdot h)^\sigma(\beta) \\ &= g^\sigma(\beta) \cdot h^\sigma(\beta) \\ &= \eta(x) \cdot \eta(y). \end{aligned}$$

Thus, η is a homomorphism. But since $\eta(1) = 1 \neq 0$, Proposition 2 of Section 7.1 implies that η is an isomorphism.

It is now easy to determine all F-isomorphisms of E into $\bar{F}$.

THEOREM 4: Let E be an extension of F of degree n, and let $\beta \in E$ be such that $E = F(\beta)$. Further, let $\beta_1, \ldots, \beta_n$ be the conjugates of β over F. Then

(1) $\beta_1, \ldots, \beta_n$ are distinct.

(2) For each $i (1 \leq i \leq n)$, there exists an F-isomorphism σ_i of E into $\bar{F}$ such that $\sigma_i(\beta) = \beta_i$. Moreover, σ_i is unique and is given by

$$\sigma_i(b_0 + b_1\beta + \cdots + b_{n-1}\beta^{n-1}) = b_0 + b_1\beta_i + \cdots + b_{n-1}\beta_i^{n-1}, \qquad b_j \in F. \tag{5}$$

(3) There are exactly n F-isomorphisms of E into $\bar{F}$ and these are $\sigma_1, \ldots, \sigma_n$.

Proof: (1) $\beta_1, \ldots, \beta_n$ are the zeros of $\mathrm{Irr}_F(\beta, X)$. Therefore, $\beta_1, \ldots, \beta_n$ are distinct by Theorem 1 of Section 9.1.

(2) Let $\sigma: F \to \bar{F}$ be the isomorphism defined by $\sigma(x) = x$ $(x \in F)$. Then $f^\sigma = f$ and Theorem 3 implies that there exists an F-isomorphism σ_i such that $\sigma_i(\beta) = \beta_i$. Moreover, by (1), σ_i is given by (5).

(3) By Proposition 2, the number of F-isomorphisms of E into $\bar{F}$ is at most n. Since the β_i are all distinct by part (1), the σ_i are all distinct. Therefore, we have exhibited n different F-isomorphisms of E into $\bar{F}$. This proves part (3).

The isomorphisms σ_i $(1 \leq i \leq n)$ are called *conjugation mappings* for the extension E of F. Let us denote the set of all conjugation mappings by $G(E/F)$. For example, if $F = \mathbf{Q}$, $E = \mathbf{Q}(\sqrt{2})$, then

$$G(E/F) = \{\sigma_1, \sigma_2\},$$

where $\sigma_1(\sqrt{2}) = \sqrt{2}$ and $\sigma_2(\sqrt{2}) = -\sqrt{2}$. If $F = \mathbf{Q}$, $E = \mathbf{Q}(\sqrt[3]{2})$, then

$$G(E/F) = \{\lambda_1, \lambda_2, \lambda_3\},$$

where $\lambda_1(\sqrt[3]{2}) = \sqrt[3]{2}$, $\lambda_2(\sqrt[3]{2}) = \sqrt[3]{2}\,\omega$, $\lambda_3(\sqrt[3]{2}) = \sqrt[3]{2}\,\omega^2$.

Note that the extension $F(\alpha)$ may not contain all the conjugates of α. For example, $\mathbf{Q}(\sqrt[3]{2}) \subseteq \mathbf{R}$, but $\sqrt[3]{2}\,\omega \notin \mathbf{R}$. The extensions which are obtained by adjoining to F *all* the conjugates of a set of elements of F are a very important class of extensions, called normal extensions. More precisely, let us make the following definition.

DEFINITION 5: Let F be a field. A *normal extension* of F is an extension of F obtained by adjoining all zeros of a finite set of polynomials $\{f_1, \ldots, f_m\}$, $f_j \in F[X]$.

For example, if $f \in F[X]$ is a nonconstant polynomial, then the splitting field E_f of f is a normal extension of F. Since a normal extension of F is gotten by adjoining a finite number of algebraic elements to F, a normal extension is finite and algebraic. Let us derive a number of equivalent conditions for an extension of F to be normal.

THEOREM 6: Let F be a field, E a finite, algebraic extension of F. Then the following conditions are equivalent:

(1) E is the splitting field of a polynomial $f \in F[X]$.
(2) E is a normal extension of F.
(3) If $\sigma \in G(E/F)$, then $\sigma(E) = E$.
(4) If $x \in E$, then all conjugates of x over F lie in E.

Proof: We will prove the theorem by proving the series of implications $(1) \Rightarrow (2) \Rightarrow (3) \Rightarrow (4) \Rightarrow (1)$.

$(1) \Rightarrow (2)$. We observed this fact above.

$(2) \Rightarrow (3)$. Suppose that E is a normal extension of F and that $\sigma \in G(E/F)$. Then E is obtained from F by adjoining all zeros $\{\alpha_1, \ldots, \alpha_s\}$ of a finite collection of polynomials $\{f_1, \ldots, f_m\}, f_j \in F[X]$. In particular, if $\alpha \in \{\alpha_1, \ldots, \alpha_s\}$, then all conjugates of α over F belong to $\{\alpha_1, \ldots, \alpha_s\}$, and hence to $E = F(\alpha_1, \ldots, \alpha_s)$. But $\sigma(\alpha_i)$ is one of the conjugates of α_i over F. Therefore, $\sigma(\alpha_i) \in E$. Hence, since σ is an F-isomorphism, we see that $\sigma(E) \subseteq E$. Let $\deg(E/F) = n$. Then, since σ is an F-isomorphism, $\deg(\sigma(E)/F) = n$ (an easy exercise). Therefore, since $F \subseteq \sigma(E) \subseteq E$, and since $\deg(E/F) = \deg(E/\sigma(E)) \cdot \deg(\sigma(E)/F)$, we see that $\deg(E/\sigma(E)) = 1$. Therefore, $E = \sigma(E)$ by Proposition 2 of Sec 7.3.

$(3) \Rightarrow (4)$. Suppose that $\sigma \in G(E/F)$ implies that $\sigma(E) = E$, and assume that $x \in E$. Let x' be a conjugate of x over F. We wish to show that $x' \in E$. By Theorem 3, there exists an F-isomorphism $\sigma: F(x) \longrightarrow \bar{F}$ such that $\sigma(x) = x'$. By Theorem 3 of Section 9.1, there exists $\beta \in E$ such that $E = F(x)(\beta)$. Then, by Theorem 3, σ can be extended to an F-isomorphism $\eta: E \longrightarrow \bar{F}$. [Map β onto any one of the zeros of $f^\sigma, f = \mathrm{Irr}_{F(x)}(\beta, X)$.] But then $\eta(x) = x'$, so that by our assumption, $x' \in \eta(E) = E$.

$(4) \Rightarrow (1)$. By Theorem 3 of Section 9.1, there exists $\beta \in E$ such that $E = F(\beta)$. Let $\beta_1, \ldots, \beta_n$ be the conjugates of β over F. By (4), $\beta_1, \ldots, \beta_n$ all belong to E, so that $E = F(\beta_1, \ldots, \beta_n)$. Thus, E can be gotten by adjoining to F all zeros of $\mathrm{Irr}_F(\beta, X)$, so that E is the splitting field of $\mathrm{Irr}_F(\beta, X)$ over F.

As an immediate consequence of the above result we deduce that $\mathbf{Q}(\sqrt[3]{2})$ is not a normal extension of $\mathbf{Q}$, since the conjugates $\sqrt[3]{2}\,\omega$ and $\sqrt[3]{2}\,\omega^2$ of $\sqrt[3]{2}$ over $\mathbf{Q}$ are not in $\mathbf{Q}(\sqrt[3]{2})$, which violates condition (4). Moreover, if ζ_m is a primitive mth root of 1, then $\mathbf{Q}(\zeta_m)$ is a normal extension of $\mathbf{Q}$, since it is the splitting field over $\mathbf{Q}$ of $\Phi_m(X)$. [The zeros of $\Phi_m(X)$ are ζ_m^a, $0 \le a \le m-1$, $(a, m) = 1$.]

9.2 Exercises

1. Find all conjugates over $\mathbf{Q}$ of the following:
   (a) $\sqrt{5}$. (b) $\sqrt[4]{3}$. (c) $\sqrt{2}+\sqrt{3}$.
   (d) $\sqrt{1+\sqrt{3}}$. (e) $\sqrt[3]{1+\sqrt{3}}$. (f) $\sqrt{1+\sqrt{1+\sqrt{2}}}$.
   [*Hint:* For each element θ, it is best to first determine $\mathrm{Irr}_{\mathbf{Q}}(\theta, X)$.]
2. For each of the elements θ of Exercise 1, describe $G(\mathbf{Q}(\theta)/\mathbf{Q})$.
3. For each of the elements θ of Exercise 1, determine whether or not $\mathbf{Q}(\theta)$ is a normal extension of $\mathbf{Q}$.
4. Let F be a field and let $\alpha, \beta \in F$ be such that α and β are not squares of elements of F. Let $E_1 = F(\sqrt{\alpha})$, $E_2 = F(\sqrt{\beta})$. Show that $E_1 = E_2 \Leftrightarrow \alpha = \gamma^2\beta$ for some $\gamma \in F$.
5. Give an example of fields D, E, F such that (1) $F \subseteq E \subseteq D$, (2) E/F and D/E are normal, (3) D/F is not normal. [*Hint*: Consider the field $\mathbf{Q}(\sqrt{3+\sqrt{2}})$, and use Exercise 4.]
6. Let D, E, and F be fields such that $F \subseteq E \subseteq D$ and D/F is finite. Show that if $\sigma \in G(E/F)$, then $\sigma = \eta|_E$ for some $\eta \in G(D/F)$.
7. Let $E = F(\beta)$ and let the conjugates of β over F be $\beta_1, \ldots, \beta_n$. Let $\alpha = a_0 + a_1\beta + \cdots + a_{n-1}\beta^{n-1} \in E$ and let $m = \deg(F(\alpha)/F)$.
   (a) Show that $m \mid n$.
   (b) Show that the images of α under $G(E/F)$ are $a_0 + a_1\beta_i + \cdots + a_{n-1}\beta_i^{n-1}$, where each conjugate is repeated n/m times.
   [*Hint*: Use Exercise 6.)
8. Let E/F be a normal extension and let $f \in F[X]$ be an irreducible polynomial. Assume that f has a zero in E. Prove that f splits into a product of linear factors in E.

*9. Does there exist a normal extension of $\mathbf{Q}$ of degree 3?

9.3 The Galois Group of an Extension

Let F be a field and let E be a finite extension of F. Our objective at this point is to associate a finite group to the extension E/F, called the *Galois group*. We will show, in the next section, how it is possible to utilize the Galois group to describe the arithmetic of the extension. In this section we will limit ourselves to defining the Galois group and computing some examples.

DEFINITION 1: An *F-automorphism* of E is an F-isomorphism of E onto itself. The set of all F-automorphisms of E will be denoted $\mathrm{Gal}(E/F)$.

Since an F-automorphism of E is an F-isomorphism of E into $\bar{F}$, we see that $\mathrm{Gal}(E/F) \subseteq G(E/F)$. Therefore, Theorem 4 of Section 9.2 implies the following result.

LEMMA 2: Suppose that $\deg(E/F) = n$. Then $\mathrm{Gal}(E/F)$ contains at most n elements.

The following lemma helps to pick out the elements of $\mathrm{Gal}(E/F)$ from $G(E/F)$.

LEMMA 3: Let $\sigma \in G(E/F)$. Then $\sigma \in \mathrm{Gal}(E/F)$ if and only if $\sigma(E) \subseteq E$.

Proof: $\Rightarrow$ Obvious.

$\Leftarrow$ It suffices to show that if $\sigma(E) \subseteq E$, then $\sigma(E) = E$. Suppose that $\deg(E/F) = n$, and let $\{\alpha_1, \ldots, \alpha_n\}$ be a basis of E over F. Since σ is an F-isomorphism, $\{\sigma(\alpha_1), \ldots, \sigma(\alpha_n)\}$ is a basis of $\sigma(E)$ over F. Thus, $\deg(\sigma(E)/F) = n$. However, since $\sigma(E) \subseteq E$,

$$\deg(E/F) = \deg(\sigma(E)/F) \cdot \deg(E/\sigma(E)),$$

and thus $\deg(E/\sigma(E)) = 1$, which implies that $E = \sigma(E)$.

By Lemma 3 and Theorem 4 of Section 9.2, we can deduce the following description of $\mathrm{Gal}(E/F)$:

THEOREM 4: Let E be an extension of F of degree n, and let $\beta \in E$ be such that $E = F(\beta)$. Further, let $\beta_1, \ldots, \beta_n$ be the conjugates of β over F and let $\sigma_i \in G(E/F)$ be such that $\sigma_i(\beta) = \beta_i$. Then $\sigma_i \in \mathrm{Gal}(E/F)$ if and only if $\beta_i \in E$. In particular, if E is a normal extension of F, then all β_i belong to E and $\mathrm{Gal}(E/F)$ contains n elements.

Let us define a group structure on $\mathrm{Gal}(E/F)$. If $\sigma, \eta \in \mathrm{Gal}(E/F)$, then the composite function $\eta\sigma$, defined by

$$(\eta\sigma)(x) = \eta(\sigma(x))$$

is an F-automorphism of E. Indeed, $\eta\sigma$ is a homomorphism of E into E such that $(\eta\sigma)(1) = 1$. Therefore, $\eta\sigma$ is an isomorphism by Proposition 2 of Section 7.1. And since η and σ are F-isomorphisms, if $x \in F$, we have $(\eta\sigma)(x) = \eta(\sigma(x)) = \eta(x) = x$. Therefore, $\eta\sigma$ is an F-isomorphism of E into E, so that by Lemma 3, $\eta\sigma$ is an F-automorphism of E. Let us define the product of η and σ to be the F-automorphism $\eta\sigma$. With respect to this law of multiplication, $\mathrm{Gal}(E/F)$ is a group. Indeed, the identity element in the identity automorphism i defined $i(x) = x$ for all $x \in E$. If $\eta \in \mathrm{Gal}(E/F)$, let η^{-1} denote the inverse function of η, when η is considered as a bijection from E onto E. Then η^{-1} is an F-automorphism of E (exercise), and $\eta\eta^{-1} = \eta^{-1}\eta = i$. We leave the proof of the associativity of the multiplication as an exercise.

DEFINITION 5: $\mathrm{Gal}(E/F)$ is called the *Galois group of the extension E* over F.

Let us compute $\mathrm{Gal}(E/F)$ in a few examples.

EXAMPLE 1: $F = \mathbf{Q}$, $E = \mathbf{Q}(\sqrt{2})$. Then E is a normal extension of F and $\mathrm{Gal}(E/F) = \{\sigma_0, \sigma_1\}$, where

$$\sigma_0(\sqrt{2}) = \sqrt{2}, \qquad \sigma_1(\sqrt{2}) = -\sqrt{2}.$$

The mapping $\psi: \mathrm{Gal}(E/F) \longrightarrow \mathbf{Z}_2$ defined by $\psi(\sigma_a) = a \pmod 2$ is a surjective isomorphism. Therefore, $\mathrm{Gal}(E/F) \approx \mathbf{Z}_2$.

EXAMPLE 2: $F = \mathbf{Q}$, $E = \mathbf{Q}(\zeta_m)$, where ζ_m is a primitive mth root of 1. Then E is a normal extension of F of degree $\phi(n)$. The conjugates of ζ_m are ζ_m^a $(0 \le a \le m-1, (a, m) = 1)$. Therefore, $\mathrm{Gal}(E/F)$ has order $\phi(m)$ and consists of the elements σ_a $(0 \le a \le m-1, (a, m) = 1)$ defined by

$$\sigma_a(\zeta_m) = \zeta_m^a.$$

Notice that

$$\sigma_a\sigma_b(\zeta_m) = \sigma_a(\zeta_m^b) = \sigma_a(\zeta_m)^b = \zeta_m^{ab}.$$

Therefore, if q and c are integers such that $ab = qm + c$, $0 \le c \le m - 1$, we have

$$\sigma_a\sigma_b(\zeta_m) = \zeta_m^{qm+c} = (\zeta_m^m)^q\zeta_m^c = \sigma_c(\zeta_m) \tag{1}$$

since $\zeta_m^m = 1$. Let $\psi: \mathrm{Gal}(E/F) \longrightarrow \mathbf{Z}_m^\times$ be the mapping defined by $\psi(\sigma_a) = a \pmod m$. Then formula (1) implies that

$$\psi(\sigma_a\sigma_b) = c \pmod m = ab \pmod m = \psi(\sigma_a)\psi(\sigma_b).$$

Therefore, ψ is a homomorphism. It is clear that ψ is a bijection, and thus

$$\mathrm{Gal}(E/F) \approx \mathbf{Z}_m^\times.$$

EXAMPLE 3: $F = \mathbf{Q}$, $E = \mathbf{Q}(\sqrt[3]{2})$, where $\sqrt[3]{2}$ denotes the real cube root of 2. The conjugates of $\sqrt[3]{2}$ over $\mathbf{Q}$ are $\sqrt[3]{2}, \sqrt[3]{2}\,\omega, \sqrt[3]{2}\,\omega^2$, where $\omega = (-1 + \sqrt{3}\,i)/2$. The only conjugate of $\sqrt[3]{2}$ which lies in E is $\sqrt[3]{2}$, so that

$$\mathrm{Gal}(E/F) = \{\sigma_0\},$$

where $\sigma_0(\sqrt[3]{2}) = \sqrt[3]{2}$. Note that $\deg(E/F) = 3$, but the order of $\mathrm{Gal}(E/F)$ is 1. This is not a contradiction to our above discussion since E is not a normal extension of F.

9.3 *Exercises*

1. Let ζ be a primitive cube root of 1 and let E denote the splitting field of $X^3 - 2$ over $F = \mathbf{Q}(\zeta)$.

(a) Show that $E = F(\sqrt[3]{2})$.
(b) Show that $\mathrm{Gal}(E/F) = \{\sigma_0, \sigma_1, \sigma_2\}$, where $\sigma_a(\sqrt[3]{2}) = \zeta^a \sqrt[3]{2}$.
(c) Show that $\mathrm{Gal}(E/F) \approx \mathbf{Z}_3$.

2. Let ζ be a primitive cube root of 1 and let E denote the splitting field of $X^3 - 2$ over $\mathbf{Q}$.
(a) Show that $E = \mathbf{Q}(\sqrt[3]{2}, \zeta)$.
(b) If $\sigma \in \mathrm{Gal}(E/F)$, show that $\sigma(\zeta) = \zeta^a$ [$a \equiv 1$ or $2 \pmod 3$], $\sigma(\sqrt[3]{2}) = \zeta^b \sqrt[3]{2}$ ($b \in \mathbf{Z}$). Show that such σ is completely determined by a and b. Let us den‸ ‘e this element of $\mathrm{Gal}(E/F)$ by $\sigma_{a,b}$.
(c) Show that $\sigma_{a,b} \cdot \sigma_{c,d} = \sigma_{ac,ad+b}$.
(d) Show that $\mathrm{Gal}(E/\mathbf{Q}) = \{\sigma_{a,b} \mid a = 1, 2; b = 0, 1, 2\}$.
(e) Show that $\mathrm{Gal}(E/\mathbf{Q}) \approx S_3$.

3. Let E denote the splitting field of $X^8 - 2$ over $\mathbf{Q}$.
(a) Show that $E = \mathbf{Q}(\sqrt[8]{2}, \zeta_8)$, where ζ_8 is a primitive eighth root of 1.
(b) Show that $\sqrt{2} \in \mathbf{Q}(\zeta_8)$.
(c) Show that $\mathrm{Irr}_{\mathbf{Q}(\zeta_8)}(\sqrt[8]{2}, X) = X^4 - \sqrt{2}$.
(d) Show that $\deg(E/\mathbf{Q}) = 16$.
*(e) Let $\sigma_{a,b}(\sqrt[8]{2}) = \zeta_8^b \sqrt[8]{2}$, $\sigma_{a,b}(\zeta_8) = \zeta_8^a$. Show that $\mathrm{Gal}(E/\mathbf{Q}) = \{\sigma_{a,b} \mid a = 1, 7; b = 0, 2, 4, 6\} \cup \{\sigma_{a,b} \mid a = 3, 5; b = 1, 3, 5, 7\}$. [*Hint*: $\sigma_{a,b} = (\sqrt{2})\,\zeta_8^{4b}$ on the one hand and $= \zeta_8^a + \zeta_8^{-a}$ on the other.]

4. Let p be an odd prime and let E denote the splitting field of $X^p - 2$ over $\mathbf{Q}$.
(a) Show that $E = \mathbf{Q}(\zeta_p, \sqrt[p]{2})$, where ζ_p is a primitive pth root of 1.
(b) Show that $\deg(E/\mathbf{Q}) = p(p - 1)$.
(c) Describe $\mathrm{Gal}(E/\mathbf{Q})$.
(d) Show that $\mathrm{Gal}(E/\mathbf{Q})$ is always nonabelian.

5. Let E/F be an extension of degree n. Show that $\mathrm{Gal}(E/F)$ has order n if and only if E/F is a normal extension.

6. Let E/F be a finite extension, let H be a subgroup of $\mathrm{Gal}(E/F)$, and let $D = \{x \in E \mid \sigma(x) = x \text{ for all } \sigma \in H\}$. Show that D is a subfield of E containing F.

9.4 The Fundamental Theorem of Galois Theory

Let us now determine what information about an extension E of F can be gleaned from the Galois group. Usually, very little can be said. For example, Example 3 of Section 9.3 shows that E can be of degree > 1, while the Galois group is of order 1. In this case one would not expect the Galois group to contain very much information about the extension. In some sense, the Galois group is "too small." However, we have shown that if E is a normal extension of F, then the order of $\mathrm{Gal}(E/F)$ equals $\deg(E/F)$. This situation looks more hopeful.

DEFINITION 1: If E is a normal extension of F, then we say that E/F is a *Galois extension.*

Let us fix a Galois extension E/F of degree n and let $G = \text{Gal}(E/F)$ By an *intermediate field* D we will mean a field such that

$$F \subseteq D \subseteq E.$$

The *fundamental theorem of Galois theory* allows a complete description of all intermediate fields D in terms of the Galois group G.

One way of manufacturing an intermediate field is as follows: Let H be a subgroup of G, and set

$$\mathfrak{F}(H) = \{x \in E \mid \sigma(x) = x \text{ for all } \sigma \in H\}. \tag{1}$$

Then we have

PROPOSITION 2: $\mathfrak{F}(H)$ is an intermediate field called the *fixed field* of H.

Proof: It is clear that $\mathfrak{F}(H) \subseteq E$. Moreover, since $\sigma \in H$ implies that σ is an F-automorphism of E, we see that $F \subseteq \mathfrak{F}(H)$. Therefore, it suffices to show that $\mathfrak{F}(H)$ is a field. If $x, y \in \mathfrak{F}(H)$, then

$$\begin{aligned} &\sigma(x) = x, \quad \sigma(y) = y && \text{for all } \sigma \in H \\ \Longrightarrow\ &\sigma(x \pm y) = \sigma(x) \pm \sigma(y) = x \pm y && \text{for all } \sigma \in H \\ \Longrightarrow\ &x \pm y \in \mathfrak{F}(H). \end{aligned}$$

Thus, $\mathfrak{F}(H)$ is an additive subgroup of E. If $x, y \in \mathfrak{F}(H) - \{0\}$, then

$$\begin{aligned} &\sigma(x) = x, \quad \sigma(y) = y \quad \text{for all } \sigma \in H \\ \Longrightarrow\ &\sigma(xy^{-1}) = \sigma(x)\sigma(y^{-1}) = \sigma(x)\sigma(y)^{-1} \\ &\qquad = xy^{-1} \quad \text{for all } \sigma \in H \\ \Longrightarrow\ &xy^{-1} \in \mathfrak{F}(H). \end{aligned}$$

Therefore, the nonzero elements of $\mathfrak{F}(H)$ form a group under multiplication and $\mathfrak{F}(H)$ is a field.

We have therefore shown how to go from a subgroup H of G to an intermediate field $\mathfrak{F}(H)$. One of the main results we will prove below states that every intermediate field is of the form $\mathfrak{F}(H)$ for some subgroup H of G. Thus, the subgroups of G "parametrize" the intermediate fields for the extension E/F.

PROPOSITION 3: Let E/F be a Galois extension, H a subgroup of $\text{Gal}(E/F)$. Then $E/\mathfrak{F}(H)$ is a Galois extension and

$$\text{Gal}(E/\mathfrak{F}(H)) = H.$$

Proof: Since E/F is finite and normal, the same is true of $E/\mathfrak{F}(H)$. Therefore, $E/\mathfrak{F}(H)$ is a Galois extension. Moreover, $\text{Gal}(E/\mathfrak{F}(H))$ consists of all

$\mathfrak{F}(H)$-automorphisms of E. However, if $\sigma \in H$, then by the definition of $\mathfrak{F}(H)$, σ is an $\mathfrak{F}(H)$-automorphism of E. Therefore, $\sigma \in \mathrm{Gal}(E/\mathfrak{F}(H))$ and

$$H \subseteq \mathrm{Gal}(E/\mathfrak{F}(H)). \tag{2}$$

Suppose that H has s elements and $\mathrm{Gal}(E/\mathfrak{F}(H))$ has t elements. Then, by (2),

$$s \leq t. \tag{3}$$

In order to show that $\mathrm{Gal}(E/\mathfrak{F}(H)) = H$, it therefore suffices to show that $s = t$. Let us assume that $s < t$ and reason by way of contradiction. Since $s + 1 \leq t$, we can find $\alpha_1, \ldots, \alpha_{s+1} \in E$ which are linearly independent over $\mathfrak{F}(H)$. The homogeneous system of equations

$$\sum_{i=1}^{s+1} \sigma_j(\alpha_i)X_i = 0 \qquad (1 \leq j \leq s). \tag{4}$$

has s equations in $s + 1$ unknowns. By Theorem 6 of Section 6.4, this system has a solution $(c_1, \ldots, c_{s+1})$, $c_i \in E$, c_i not all zero. Among all nonzero solutions of (4), let us assume that $(c_1, \ldots, c_{s+1})$ is one with as few nonzero entries as possible. By possibly renumbering the α_i, we may assume that this solution has the form

$$(c_1, c_2, \ldots, c_r, 0, 0, \ldots, 0), \quad c_i \neq 0 \qquad (1 \leq i \leq r) \tag{5}$$

Since the system (4) is homogeneous, we may normalize c_r to be 1 by replacing (5) by the solution $(c_1c_r^{-1}, c_2c_r^{-1}, \ldots, 1, 0, 0, \ldots, 0)$. If all c_i belong to $\mathfrak{F}(H)$, then

$$\sum_{i=1}^{s+1} \sigma_1(\alpha_i)c_i = 0 \Longrightarrow \sigma_1\left(\sum_{i=1}^{s+1} \alpha_i c_i\right) = 0 \qquad [\text{since } \sigma_1 \text{ is an } \mathfrak{F}(H)\text{-automorphism}]$$

$$\Longrightarrow \sum_{i=1}^{s+1} \alpha_i c_i = 0,$$

which is a contradiction to the assumed linear independence of $\alpha_1, \ldots, \alpha_{s+1}$ over $\mathfrak{F}(H)$. Therefore, not all c_i belong to $\mathfrak{F}(H)$. Without loss of generality, we may assume that $c_1 \notin \mathfrak{F}(H)$. Then there exists $\sigma \in H$ such that $\sigma(c_1) \neq c_1$. Apply σ to the system of equations (4) to get

$$\sum_{i=1}^{s+1} \sigma\sigma_j(\alpha_i)\sigma(c_i) = 0.$$

Therefore, since $\sigma\sigma_j$ runs over H as σ_j runs over H, we see that

$$(\sigma(c_1), \sigma(c_2), \ldots, \sigma(c_{r-1}), 1, 0, 0, \ldots, 0)$$

is a solution of the system (4). But the difference of two solutions of (4) is also a solution of (4), so we see that

$$(\sigma(c_1) - c_1, \sigma(c_2) - c_2, \ldots, \sigma(c_{r-1}) - c_{r-1}, 0, 0, \ldots, 0)$$

is a solution of (4). Since $\sigma(c_1) - c_1 \neq 0$, we see that this solution is nonzero, having at most $r - 1$ nonzero entries. But this is a contradiction to our original choice of the solution $(c_1, \ldots, c_r, 0, 0, \ldots, 0)$.

In our discussion above, we constructed the correspondence

$$H \longrightarrow \mathfrak{F}(H),$$

which associates to a subgroup H of Gal(E/F), the intermediate field $\mathfrak{F}(H)$. It is also possible to construct a correspondence which associates to an intermediate field D a subgroup $\mathfrak{G}(D)$ of Gal(E/F)—let us set

$$\mathfrak{G}(D) = \{\sigma \in G \mid \sigma(x) = x \text{ for all } x \in D\}.$$

It is clear that

$$\mathfrak{G}(D) = \text{Gal}(E/D). \tag{6}$$

Therefore, by Proposition 3, if H is any subgroup of Gal(E/F), then

$$\boxed{\mathfrak{G}(\mathfrak{F}(H)) = H.} \tag{7}$$

Thus, if D is any intermediate field,

$$\mathfrak{G}(\mathfrak{F}(\mathfrak{G}(D))) = \mathfrak{G}(D) \qquad \text{[Equation (7) with } H = \mathfrak{G}(D)] \tag{8}$$

$$\Longrightarrow \deg(E/\mathfrak{F}(\mathfrak{G}(D))) = |\mathfrak{G}(\mathfrak{F}(\mathfrak{G}(D)))| \begin{array}{l}\text{[Equation (6) with } D \text{ replaced by } \mathfrak{F}(\mathfrak{G}(D)),\\ \text{and Theorem 4 of Section 9.3.}\end{array}$$

$$= |\mathfrak{G}(D)| \qquad \text{[Equation (8)]} \tag{9}$$

$$= \deg(E/D) \text{ [Equation (6) and Theorem 4 of Section 9.3]}$$

But it is clear that

$$\mathfrak{F}(\mathfrak{G}(D)) \supseteq D,$$

so that, by Equation (9),

$$\deg(\mathfrak{F}(\mathfrak{G}(D))/D) = 1 \Longrightarrow \mathfrak{F}(\mathfrak{G}(D)) = D.$$

Thus, we see that if D is any intermediate field, then

$$\boxed{\mathfrak{F}(\mathfrak{G}(D)) = D.} \tag{10}$$

Note that this last equation implies that the intermediate field D is of the form $\mathfrak{F}(H)$, where $H = \mathfrak{G}(D)$.

Finally, we may state

THEOREM 4 (FUNDAMENTAL THEOREM OF GALOIS THEORY): Let E/F be a Galois extension with Galois group G. Then the correspondence

$$H \longrightarrow \mathfrak{F}(H)$$

is a one-to-one mapping of the set of subgroups of G onto the set of intermediate fields for the extension E/F. Moreover, if H corresponds to D under this correspondence, then E/D is a Galois extension and Gal$(E/D) = H$.

Proof: The fact that the correspondence $H \longrightarrow \mathfrak{F}(H)$ is one to one follows from Equation (7), since $\mathfrak{F}(H_1) = \mathfrak{F}(H_2) \Rightarrow \mathfrak{G}(\mathfrak{F}(H_1)) = \mathfrak{G}(\mathfrak{F}(H_2) \Rightarrow H_1 =$

H_2. We have already noted above that every intermediate field is of the form $\mathfrak{F}(H)$, and thus the correspondence is onto. The last assertion follows from Proposition 3.

COROLLARY 5: Let E/F be a Galois extension. There are only finitely many intermediate fields D such that $F \subseteq D \subseteq E$.

This result is somewhat surprising.

Proof: $\text{Gal}(E/F)$ is a finite group and therefore has only finitely many subgroups. Thus, by the fundamental theorem, there are only finitely many intermediate fields.

EXAMPLE 1: Let $F = \mathbf{Q}$, $E = \mathbf{Q}(\sqrt{2}, \sqrt{3})$. Then $\text{Gal}(E/F)$ is of order 4 and consists of the following elements:

$$\psi_1: \begin{matrix} \sqrt{2} \longrightarrow \sqrt{2} \\ \sqrt{3} \longrightarrow \sqrt{3} \end{matrix}, \qquad \psi_2: \begin{matrix} \sqrt{2} \longrightarrow -\sqrt{2} \\ \sqrt{3} \longrightarrow \sqrt{3} \end{matrix},$$

$$\psi_3: \begin{matrix} \sqrt{2} \longrightarrow \sqrt{2} \\ \sqrt{3} \longrightarrow -\sqrt{3} \end{matrix}, \qquad \psi_4: \begin{matrix} \sqrt{2} \longrightarrow -\sqrt{2} \\ \sqrt{3} \longrightarrow -\sqrt{3} \end{matrix}.$$

The subgroups of $\text{Gal}(E/F)$ are

$$H_1 = \{\psi_1\}, \quad H_2 = \{\psi_1, \psi_2\}, \quad H_3 = \{\psi_1, \psi_3\}, \quad H_4 = \{\psi_1, \psi_4\},$$
$$H_5 = \{\psi_1, \psi_2, \psi_3, \psi_4\}.$$

The corresponding fixed fields are

$$\begin{aligned} \mathfrak{F}(H_1) &= \mathbf{Q}(\sqrt{2}, \sqrt{3}), \\ \mathfrak{F}(H_2) &= \mathbf{Q}(\sqrt{3}), \\ \mathfrak{F}(H_3) &= \mathbf{Q}(\sqrt{2}), \\ \mathfrak{F}(H_4) &= \mathbf{Q}(\sqrt{6}), \\ \mathfrak{F}(H_5) &= \mathbf{Q}. \end{aligned}$$

[The only computation which deserves comment is the one leading to the assertion $\mathfrak{F}(H_4) = \mathbf{Q}(\sqrt{6})$. It is clear that $\sqrt{6} = \sqrt{2}\cdot\sqrt{3} \in \mathfrak{F}(H_4)$, so that $\mathbf{Q}(\sqrt{6}) \subseteq \mathfrak{F}(H_4)$. But H_4 has order 2, so that by the fundamental theorem and Proposition 4 of Section 9.3, we see that $\deg(\mathbf{Q}\sqrt{2},\sqrt{3})/\mathfrak{F}(H_4))$ $= 2$. However, $\deg(\mathbf{Q}(\sqrt{2}, \sqrt{3})/\mathbf{Q}(\sqrt{6})) = 2$, so that $\deg(\mathfrak{F}(H_4)/\mathbf{Q}(\sqrt{6}))$ $= 1 \Rightarrow \mathbf{Q}(\sqrt{6}) = \mathfrak{F}(H_4)$.]

As a consequence of the fundamental theorem, we see that $\mathbf{Q}(\sqrt{2}, \sqrt{3})$ has only five subfields: $\mathbf{Q}$, $\mathbf{Q}(\sqrt{6})$, $\mathbf{Q}(\sqrt{2})$, $\mathbf{Q}(\sqrt{3})$, and $\mathbf{Q}(\sqrt{2}, \sqrt{3})$.

We shall refer to the correspondence $H \longrightarrow \mathfrak{F}(H)$ as the *Galois correspondence*. Let us conclude this section by proving two very fundamental facts concerning the Galois correspondence.

THEOREM 6: Let E/F be a Galois extension with Galois group G. Let H_1 and H_2 be subgroups of G, and let D_1 and D_2, respectively, correspond to H_1 and H_2 under the Galois correspondence. Then

(1) $H_1 \subseteq H_2$ if and only if $D_1 \supseteq D_2$.

(2) The intermediate field corresponding to $H_1 \cap H_2$ is $D_1 D_2$, where by $D_1 D_2$ we mean the smallest subfield of E containing both D_1 and D_2.

(3) The intermediate field corresponding to $[H_1 \cup H_2]$ is $D_1 \cap D_2$.

Proof: By the fundamental theorem,

$$D_1 = \mathfrak{F}(H_1), \qquad D_2 = \mathfrak{F}(H_2).$$

(1) It is clear that $H_1 \subseteq H_2$ implies that $\mathfrak{F}(H_1) \supseteq \mathfrak{F}(H_2)$. Conversely, if $\mathfrak{F}(H_1) \supseteq \mathfrak{F}(H_2)$, then $\mathcal{G}(\mathfrak{F}(H_1)) \subseteq \mathcal{G}(\mathfrak{F}(H_2))$, so that $H_1 \subseteq H_2$ by Equation (7).

(2) $H_1 \cap H_2$ is the largest subgroup of G contained in both H_1 and H_2. Therefore, by (1) and the fundamental Theorem, $\mathfrak{F}(H_1 \cap H_2)$ is the smallest subfield of E containing $\mathfrak{F}(H_1)$ and $\mathfrak{F}(H_2)$.

(3) $[H_1 \cup H_2]$ is the smallest subgroup of G containing both H_1 and H_2. Therefore, by (1) and the fundamental theorem, $\mathfrak{F}([H_1 \cup H_2])$ is the largest subfield contained in both $\mathfrak{F}(H_1)$ and $\mathfrak{F}(H_2)$. That is, $\mathfrak{F}([H_1 \cup H_2]) = \mathfrak{F}(H_1) \cap \mathfrak{F}(H_2)$.

Perhaps the most useful consequence of the fundamental theorem is the following result, which will be used many times throughout this chapter.

COROLLARY 7: Let E/F be a Galois extension with Galois group G, and let $x \in E$. If $\sigma(x) = x$ for all $\sigma \in G$, then $x \in F$.

Proof: If $\sigma(x) = x$ for all $\sigma \in G$, then $x \in \mathfrak{F}(G)$. Now by the definition of G, $\mathcal{G}(F) = G$ and thus $\mathfrak{F}(\mathcal{G}(F)) = \mathfrak{F}(G)$. But by (10), $\mathfrak{F}(\mathcal{G}(F)) = F$. Therefore, $x \in F$.

THEOREM 8: Let E/F be a Galois extension and let $G = \mathrm{Gal}(E/F)$. Further, let D be an intermediate field and let $H = \mathcal{G}(D)$. Then

(1) D/F is a normal extension if and only if H is a normal subgroup of G.

(2) If H is a normal subgroup of G, then, D/F is a Galois extension and $\mathrm{Gal}(D/F) \approx G/H$.

Proof: Let $\sigma \in G$. Let us first show that

$$\mathcal{G}(\sigma D) = \sigma \mathcal{G}(D) \sigma^{-1}. \tag{11}$$

For if $\eta \in \mathcal{G}(D)$, then $\sigma \eta \sigma^{-1}$ is an F-automorphism of σD and therefore $\mathcal{G}(\sigma D) \supseteq \sigma \mathcal{G}(D) \sigma^{-1}$. If $\lambda \in \mathcal{G}(\sigma D)$, then $\sigma^{-1} \lambda \sigma$ is an F-automorphism of D, so that $\sigma^{-1}\lambda\sigma \in \mathcal{G}(D)$ and $\lambda \in \sigma\mathcal{G}(D)\sigma^{-1}$. Therefore, $\mathcal{G}(\sigma D) \subseteq \sigma\mathcal{G}(D)\sigma^{-1}$. Thus, (11) is proved.

(1) Note that

$$\begin{aligned} D/F \text{ is normal} &\Longleftrightarrow \sigma D = D \text{ for all } \sigma \in \mathrm{Gal}(E/F) \\ &\Longleftrightarrow \mathcal{G}(\sigma D) = \mathcal{G}(D) \text{ for all } \sigma \in \mathrm{Gal}(E/F) \quad \text{(by Theorem 6)} \\ &\Longleftrightarrow \sigma H \sigma^{-1} = H \text{ for all } \sigma \in \mathrm{Gal}(E/F) \quad \text{[by (11)]} \\ &\Longleftrightarrow H \text{ is a normal subgroup of } G. \end{aligned}$$

(2) By part (1), if H is a normal subgroup of G, then D/F is normal. Moreover, it is clear that D/F is finite. Thus, D/F is a Galois extension. Let us consider the mapping

$$\psi\colon \mathrm{Gal}(E/F) \longrightarrow \mathrm{Gal}(D/F)$$

which arises by mapping $\sigma \in \mathrm{Gal}(E/F)$ into its restriction to D. It is clear that this mapping is a group homomorphism. Moreover, since every F-automorphism of D can be extended to an F-automorphism of E, this homomorphism is surjective. Finally,

$$\begin{aligned} \ker \psi &= \{\sigma \in \mathrm{Gal}(E/F) \,|\, \sigma = \text{the identity on } D\} \\ &= H. \end{aligned}$$

Therefore, by the first isomorphism theorem,

$$G/H \approx \mathrm{Gal}(D/F).$$

9.4 *Exercises*

1. Show that the subfields of $\mathbf{Q}(\sqrt{5})$ are $\mathbf{Q}$ and $\mathbf{Q}(\sqrt{5})$.
2. Let $E = \mathbf{Q}(\sqrt{2}, \sqrt{3}, \sqrt{5})$.
   (a) Determine $\mathrm{Gal}(E/\mathbf{Q})$.
   (b) Show that E is normal over $\mathbf{Q}$.
   (c) Determine all subgroups of $\mathrm{Gal}(E/\mathbf{Q})$ and their corresponding fixed fields.
3. Let p be an odd prime, ζ a primitive pth root of 1. We have seen that $\mathbf{Q}(\zeta)$ is a Galois extension of $\mathbf{Q}$ with Galois group $\mathbf{Z}_p^\times$.
   (a) Show that $\mathbf{Z}_p^\times$ is cyclic. [*Hint:* $\mathbf{Z}_p^\times$ is a field and every element of $\mathbf{Z}_p^\times$ is a zero of $X^{p-1} - 1$. On the other hand, $\mathbf{Z}_p^\times$ is the direct product of cyclic groups of prime power order.]
   (b) Let r be a positive integer such that $r \mid p - 1$. Show that $\mathbf{Q}(\zeta)$ contains one and only one subfield E such that $\deg(E/\mathbf{Q}) = r$.
   (c) Show that $\mathbf{Q}(\zeta)$ contains one and only one subfield of the form $\mathbf{Q}(\sqrt{d})$ ($d \in \mathbf{Z}$).
   **(d) Prove that the subfield of c is given explicitly as $\mathbf{Q}(\sqrt{(-1)^{(p-1)/2}p})$. [*Hint:* Show that $(-1)^{(p-1)/2}p = [(1 - \zeta) \ldots (1 - \zeta^{(p-1/2)})]^2$.]
4. Let E/F be a Galois extension of prime degree. Show that if D is an intermediate subfield, then $D = E$ or $D = F$.

5. Let p be a prime, F a field containing the pth roots of 1, $a \in F$. Show that either a is a pth power in F or $X^p - a$ is irreducible in $F[X]$.
6. Is Exercise 5 valid if we drop the requirement that p is prime?
7. Let $f \in \mathbf{Q}[X]$ be a nonconstant irreducible polynomial, $\alpha_1, \ldots, \alpha_n$ its zeros in $\mathbf{C}$. Show that $\alpha_1 + \cdots + \alpha_n$ and $\alpha_1 \ldots \alpha_n$ belong to $\mathbf{Q}$. Show that $\alpha_1^3 + \cdots + \alpha_n^3$, $\alpha_1^4 + \cdots + \alpha_n^4$ belong to $\mathbf{Q}$.

9.5 The Galois Group of a Polynomial

In Section 1.1 we promised that we would associate to a polynomial a finite group which reflects the algebraic properties of the polynomial. We have now reached the point where this promise is very simple to fulfill.

Let F be a field and let

$$f = X^n + a_1X^{n-1} + \cdots + a_n \in F[X]$$

be a nonconstant polynomial. Let us fix an algebraic closure $\bar{F}$ of F. Then, in $\bar{F}[X]$, we can write

$$f = \prod_{i=1}^{n} (X - \alpha_i)$$

where $\alpha_1, \ldots, \alpha_n$ are the zeros of f in $\bar{F}$. By Theorem 1 of Section 9.1, $\alpha_1, \ldots, \alpha_n$ are all distinct. The splitting field E_f of f over F is given by

$$E_f = F(\alpha_1, \ldots, \alpha_n).$$

It is clear that E_f is a normal extension of F and is therefore a Galois extension of degree m, say.

DEFINITION 1: The *Galois group of f over F*, denoted $\mathrm{Gal}_F(f)$, is the finite group $\mathrm{Gal}(E_f/F)$.

By Theorem 4 of Section 9.3, we have

PROPOSITION 2: The order of $\mathrm{Gal}_F(f)$ is m.

A typical element σ of $\mathrm{Gal}_F(f)$ is an F-automorphism of E_f. But since E_f is generated over F by $\alpha_1, \ldots, \alpha_n$, σ is completely determined once $\sigma(\alpha_i)$ $(i = 1, \ldots, n)$ is specified. Let us try to pin down what $\sigma(\alpha_i)$ can be.

LEMMA 3: (1) The image of α_i is an α_j for some j depending on i. In other words $\sigma(\alpha_i) = \alpha_{j(i)}$ for some $j(i)$ $(1 \leq j(i) \leq n)$.

(2) If $\alpha_{j(i)} = \alpha_{j(i')}$, then $i = i'$.

Proof: (1) Clearly $\sigma(\alpha_i)$ equals some F-conjugate of α_i, so that $\sigma(\alpha_i)$ is a zero of $\mathrm{Irr}_F(\alpha_i, X)$ (by Proposition 2 of Section 9.2). But since $f \in F[X]$ has α_i as a zero, $\mathrm{Irr}_F(\alpha_i, X) \mid f$. Therefore, $\sigma(\alpha_i)$ is a zero of f and $\sigma(\alpha_i) = \alpha_{j(i)}$ for some $j(i)$ $(1 \leq j(i) \leq n)$.

(2) If $\alpha_{j(i)} = \alpha_{j(i')}$, then $\sigma(\alpha_i) = \sigma(\alpha_{i'})$. But since σ is an injection, this implies that $\alpha_i = \alpha_{i'}$. Therefore, $i = i'$, since all α_i are distinct.

Lemma 3 asserts that the set $\{\sigma(\alpha_1), \ldots, \sigma(\alpha_n)\}$ coincides with the set $\{\alpha_1, \ldots, \alpha_n\}$. For, indeed, $\{\sigma(\alpha_1), \ldots, \sigma(\alpha_n)\} \subseteq \{\alpha_1, \ldots, \alpha_n\}$ by (1); and $\{\sigma(\alpha_1), \ldots, \sigma(\alpha_n)\}$ contains n distinct elements by (2). In other words,

$$P_\sigma = \begin{pmatrix} \alpha_1 \ldots \alpha_n \\ \sigma(\alpha_n) \ldots \sigma(\alpha_n) \end{pmatrix}$$

is a permutation of the n distinct zeros $\{\alpha_1, \ldots, \alpha_n\}$. It is trivial to verify that the mapping

$$\begin{aligned} \psi: \mathrm{Gal}_F(f) &\longrightarrow S_n, \\ \psi(\sigma) &= P_\sigma \end{aligned} \tag{1}$$

is a homomorphism. Moreover, since σ is completely determined by $\sigma(\alpha_1)$, $\ldots, \sigma(\alpha_n)$, we see that ψ is an injection. Therefore, we have proved:

THEOREM 4: $\mathrm{Gal}_F(f)$ is isomorphic to a subgroup of the group of permutations of the zeros $\alpha_1, \ldots, \alpha_n$.

COROLLARY 5: $\mathrm{Gal}_F(f)$ is isomorphic to a subgroup of S_n.

COROLLARY 6: $\mathrm{Gal}_F(f)$ has order at most $n!$

Note that the isomorphism ψ of (2) will not generally be surjective, so that $\mathrm{Gal}_F(f)$ will not generally contain all the permutations on $\alpha_1, \ldots, \alpha_n$. The reader is referred to the examples below for instances of this phenomenon. In Examples 1–4, let $F = \mathbf{Q}$.

EXAMPLE 1: $f = X^2 - 2$. Then $E_f = \mathbf{Q}(\sqrt{2})$ and $\mathrm{Gal}_F(f)$ consists of the mappings

$$\psi_0: \begin{matrix} \sqrt{2} \longrightarrow \sqrt{2} \\ -\sqrt{2} \longrightarrow -\sqrt{2} \end{matrix}, \qquad \psi_1: \begin{matrix} \sqrt{2} \longrightarrow -\sqrt{2} \\ -\sqrt{2} \longrightarrow \sqrt{2} \end{matrix}.$$

Thus, $\mathrm{Gal}_F(\mathbf{Z}) \approx \mathbf{Z}_2$.

EXAMPLE 2: $f = (X^2 - 2)(X^2 - 3)$. Here $E_f = \mathbf{Q}(\sqrt{2}, \sqrt{3})$ and $\alpha_1 = \sqrt{2}, \alpha_2 = -\sqrt{2}, \alpha_3 = \sqrt{3}, \alpha_4 = -\sqrt{3}$. Then $\mathrm{Gal}_F(f)$ consists of the mappings

$$\psi_0: \begin{cases} \sqrt{2} \to \sqrt{2} \\ -\sqrt{2} \to -\sqrt{2} \\ \sqrt{3} \to \sqrt{3} \\ -\sqrt{3} \to -\sqrt{3} \end{cases}, \quad \psi_1: \begin{cases} \sqrt{2} \to -\sqrt{2} \\ -\sqrt{2} \to \sqrt{2} \\ \sqrt{3} \to \sqrt{3} \\ -\sqrt{3} \to -\sqrt{3} \end{cases},$$

$$\psi_2: \begin{cases} \sqrt{2} \to \sqrt{2} \\ -\sqrt{2} \to -\sqrt{2} \\ \sqrt{3} \to -\sqrt{3} \\ -\sqrt{3} \to \sqrt{3} \end{cases}, \quad \psi_3: \begin{cases} \sqrt{2} \to -\sqrt{2} \\ -\sqrt{2} \to \sqrt{2} \\ \sqrt{3} \to -\sqrt{3} \\ -\sqrt{3} \to \sqrt{3} \end{cases}.$$

Moreover, $G_1 = \{\psi_0, \psi_1\}$ and $G_2 = \{\psi_0, \psi_2\}$ are subgroups of $\mathrm{Gal}_F(f)$ isomorphic to $\mathbf{Z}_2$ and

$$\mathrm{Gal}_F(f) \approx G_1 \times G_2 \approx \mathbf{Z}_2 \times \mathbf{Z}_2.$$

EXAMPLE 3: $f = X^3 - 2$. Then $\alpha_1 = \sqrt[3]{2}, \alpha_2 = \sqrt[3]{2}\,\omega, \alpha_3 = \sqrt[3]{2}\,\omega^2$, where $\sqrt[3]{2}$ denotes a real cube root of 2 and ω is a primitive cube root of unity. Let us show that $\mathrm{Gal}_{\mathbf{Q}}(f) \approx S_3$, so that all permutations of the roots $\alpha_1, \alpha_2, \alpha_3$ correspond to elements of the Galois group. By theorem 4, $\mathrm{Gal}_{\mathbf{Q}}(f)$ is isomorphic to a subgroup of S_3. Moreover, $E_f/\mathbf{Q}$ is a Galois extension, so that by Theorem 4 of Section 9.3, the order of $\mathrm{Gal}_{\mathbf{Q}}(f)$ equals $\deg(E_f/\mathbf{Q})$. Therefore, it suffices to show that $\deg(E_f/\mathbf{Q}) \geq 6$. And this was proved in Example 3 of Section 7.6.

EXAMPLE 4: Let $f = X^5 - 6X + 3$. We will show that $\mathrm{Gal}_{\mathbf{Q}}(f) \approx S_5$. Let us begin by considering the function $f(x) = x^5 - 6x + 3$ for real x. From elementary calculus, we see that this function has a maximum for $x = -(6/5)^{1/4}$ and a minimum at $(6/5)^{1/4}$. Moreover, these are the only extrema of $f(x)$. Thus, we may sketch the graph of $f(x)$ as in Figure 9.1, and we see immediately that f has exactly three real zeros, α_3, α_4, and α_5. Let $\alpha_1 = a + b\sqrt{-1} \in \mathbf{C}$ be a nonreal zero of f.

If $\alpha = x + y\sqrt{-1}$ $(x, y \in \mathbf{R})$ is a complex number, let us define the *complex conjugate* of $\bar{\alpha}$ of α by

$$\bar{\alpha} = x - y\sqrt{-1}.$$

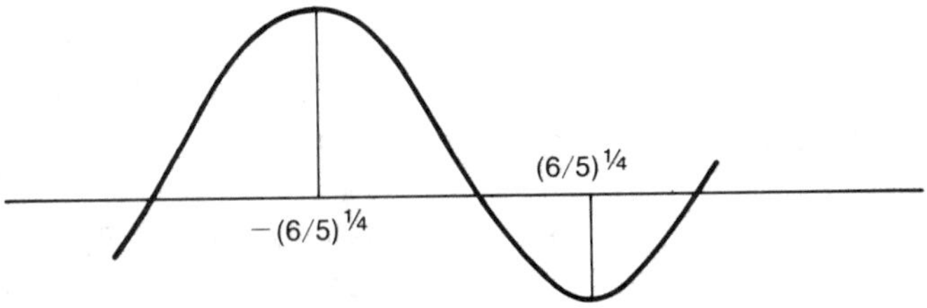

FIGURE 9-1: Graph of $f(x) = x^5 - 6x + 3$.

We leave it to the reader to check that the mapping $\alpha \longrightarrow \bar{\alpha}$ is an automorphism of $\mathbf{C}$. From this fact, it follows that

$$\begin{aligned} 0 = \bar{0} &= \overline{f(\alpha_1)} \\ &= \overline{\alpha_1^5 - 6\alpha_1 + 3} \\ &= \bar{\alpha}_1^5 - 6\bar{\alpha}_1 + 3 \\ &= f(\bar{\alpha}_1). \end{aligned}$$

Therefore, $\alpha_2 = \bar{\alpha}_1$ is a zero of f and $E_f = \mathbf{Q}(\alpha_1, \alpha_2, \alpha_3, \alpha_4, \alpha_5)$.

The mapping $\sigma: E_f \longrightarrow E_f$ defined by $\sigma(\alpha) = \bar{\alpha}$ is a $\mathbf{Q}$-automorphism of E_f and therefore $\sigma \in \mathrm{Gal}_{\mathbf{Q}}(f)$. Moreover,

$$\begin{aligned} \sigma(\alpha_1) &= \alpha_2, \\ \sigma(\alpha_2) &= \alpha_1, \\ \sigma(\alpha_i) &= \alpha_i \qquad (i = 3, 4, 5). \end{aligned}$$

Therefore, when σ is viewed as a permutation, σ is just the transposition (12), which interchanges α_1 and α_2 and leaves all other α_i fixed. Therefore, we have located one specific element of $\mathrm{Gal}_{\mathbf{Q}}(f)$.

By the Eisenstein irreducibility criterion, f is irreducible in $\mathbf{Q}[X]$, so that $\deg(\mathbf{Q}(\alpha_1)/\mathbf{Q}) = 5$ by Theorem 3 of Section 7.2. Therefore, since

$$\deg(E_f/\mathbf{Q}) = \deg(E_f/\mathbf{Q}(\alpha_1))\cdot\deg(\mathbf{Q}(\alpha_1)/\mathbf{Q}),$$

we see that $\deg(E_f/\mathbf{Q})$ is divisible by 5. Therefore, by Proposition 2 of Section 9.5 and Theorem 4 of Section 9.5, $\mathrm{Gal}_{\mathbf{Q}}(f)$ is a subgroup of the group of permutations of $\alpha_1, \ldots, \alpha_5$ and that the order of $\mathrm{Gal}_{\mathbf{Q}}(f)$ is divisible by 5. Moreover, by the first Sylow theorem, $\mathrm{Gal}_{\mathbf{Q}}(f)$ contains a subgroup H of order 5. In particular, $\mathrm{Gal}_{\mathbf{Q}}(f)$ contains an element η of order 5. Let

$$\eta = (i_1 \ldots i_r)(j_1 \ldots j_s)(k_1 \ldots k_t)\ldots$$

be the decomposition of η into a product of disjoint cycles. By Exercise 6 of 3.8, the order of η is the least common multiple of $r, s, t, \ldots$. Therefore, since η has order 5, η must be a 5-cycle, say

$$\eta = (1 i_2 i_3 i_4 i_5).$$

By replacing η by some power of η, we may assume that η has the form

$$\eta = (12 i_3 i_4 i_5).$$

However, by permuting and renumbering $\alpha_3, \alpha_4, \alpha_5$, we may assume that

$$\eta = (12345).$$

Therefore, we have shown that $\mathrm{Gal}_{\mathbf{Q}}(f)$ contains the permutations (12), (12345). However, it is easy to check that for $r = 0, 1, 2, 3$,

$$\begin{aligned} (12345)^{-r}(12)(12345)^r &= (r+1, r+2), \\ (12345)^{-4}(12)(12345)^4 &= (51). \end{aligned}$$

Therefore, $\mathrm{Gal}_{\mathbf{Q}}(f)$ contains the transpositions (12), (23), (34), (45), and (51). And these transpositions generate all transpositions. Therefore, since S_5 is generated by transpositions, $\mathrm{Gal}_{\mathbf{Q}}(f) = S_5$.

EXAMPLE 5: Let ζ be a primitive mth root of 1 and let F be a field containing ζ. Let $a \in F, f = X^m - a \in F[X]$, and let $E =$ the splitting field of f over F. If b is a zero of f in $\bar{F}$, then the zeros of f are given by

$$b, b\zeta, \ldots, b\zeta^{m-1},$$

so that $f = (X - b) \ldots (X - b\zeta^{m-1})$ in $\bar{F}[X]$, and $E = F(b)$ since $\zeta \in F$. If $\sigma \in \mathrm{Gal}_F(f)$, then there exists a unique integer $a(\sigma)$ such that

$$\sigma(b) = b\zeta^{a(\sigma)}, \qquad 0 \leq a(\sigma) \leq m - 1.$$

Let $\psi\colon \mathrm{Gal}_F(f) \longrightarrow \mathbf{Z}_m$ be defined by

$$\psi(\sigma) = a(\sigma) \pmod{m}.$$

Then ψ is an injection. Moreover,

$$\begin{aligned}(\eta\sigma)(b) &= \eta(b\zeta^{a(\sigma)})\\ &= \eta(b)\cdot\zeta^{a(\sigma)} \qquad \text{(since } \zeta \in F \text{ and } \eta \text{ is an } F\text{-automorphism of } E)\\ &= b\zeta^{a(\sigma)+b(\sigma)}.\end{aligned}$$

On the other hand, $(\eta\sigma)(b) = b\zeta^{a(\sigma)+b(\sigma)}$, so that $\zeta^{a(\sigma)+b(\sigma)-a(\eta\sigma)} = 1$ and $a(\sigma) + b(\sigma) - a(\eta\sigma) \equiv 0 \pmod{m}$. Thus,

$$\psi(\eta\sigma) = \psi(\eta) + \psi(\sigma),$$

and ψ is an isomorphism. We have proved that $\mathrm{Gal}_F(f)$ is isomorphic to a subgroup of $\mathbf{Z}_m$. In particular, since $\mathbf{Z}_m$ is a cyclic group, $\mathrm{Gal}_F(f)$ is cyclic. It is not usually true that $\mathrm{Gal}_F(f) \approx \mathbf{Z}_m$. We will give some examples in the exercises.

9.5 *Exercises*

1. What are the Galois groups over $\mathbf{Q}$ of the following polynomials?
   (a) $X^2 - 5$. (b) $X^2 - 4$.
   (c) $X^3 - 5$. (d) $X^3 - 1$.
   (e) $X^7 - 1$. (f) $X^5 - 2$.
   (g) $X^4 + X^3 + X^2 + X + 1$. (h) $X^2 - 2X - 4$.
   (i) $(X^2 - 2)(X^2 - 5)$. (j) $(X - 1)^3(X - 2)^5(X^2 - 2)$.
2. Suppose that F contains the pth roots of 1, where p is a prime, $a \in F$. Show that either $\mathrm{Gal}_F(X^p - a) \approx \mathbf{Z}_p$ or $\mathrm{Gal}_F(X^p - a) \approx \{1\}$, the last case occurring if and only if a is a pth power in F.
3. Suppose that $F = \mathbf{Q}(\zeta_m)$, where ζ_m is a primitive mth root of 1. Give an example of a polynomial $f = X^m - a \in F[X]$ such that $\mathrm{Gal}_F(f)$ is not isomorphic to $\mathbf{Z}_m$. [We know that $\mathrm{Gal}_F(f) \approx$ a subgroup of $\mathbf{Z}_m$.]

4. Let the notation be as in Exercise 3, and let $a_1, \ldots, a_r \in F$, $g = (X^m - a_1) \ldots (X^m - a_r)$.
   (a) Show that $\mathrm{Gal}_F(g)$ is an abelian group.
   (b) Show that if $\sigma \in \mathrm{Gal}_F(g)$, then $\sigma^m = 1$.
5. Let F be a field, E an extension of $F, f \in F[X]$. Show that $\mathrm{Gal}_E(f)$ is isomorphic to a subgroup of $\mathrm{Gal}_F(f)$.
6. Let F be a field, $f, g \in F[X]$. Show that $\mathrm{Gal}_F(f \cdot g) \approx$ a subgroup of $\mathrm{Gal}_F(f) \times \mathrm{Gal}_F(g)$.
7. Let E be a normal extension of F. Prove that there exists $\beta \in E$ such that $\mathrm{Gal}(E/F) \approx \mathrm{Gal}_F(\mathrm{Irr}_F(\beta, X))$.
8. (a) Exhibit polynomials $f_n \in \mathbf{Q}[X]$ $(n = 2, 3, 4, 5)$ such that $\mathrm{Gal}_{\mathbf{Q}}(f_n) \approx S_n$.
   (b) Exhibit a polynomial $f_7 \in \mathbf{Q}[X]$ such that $\mathrm{Gal}_{\mathbf{Q}}(f_7) \approx S_7$.
9. Let p be a prime.
   (a) If G is a subgroup of S_p containing a transposition and a p-cycle, then $G = S_p$.
   (b) Prove that there exists $f \in \mathbf{Q}[X]$ of degree p which is irreducible over $\mathbf{Q}$ and has precisely $p - 2$ real zeros.
   (c) Prove that S_p is the Galois group over $\mathbf{Q}$ of some polynomial in $\mathbf{Q}[X]$. (*Hint:* Look at Example 4.)

9.6 Solution of Equations in Radicals

In this section we return to a subject touched upon at the beginning of this book, the solution of equations in radicals.

Let $f \in F[X]$ be a nonconstant polynomial

$$f = X^n + a_1 X^{n-1} + \cdots + a_n,$$

First, let us clarify what we mean when we ask for a "solution in radicals" of the equation $f(x) = 0$. Roughly speaking, we mean a set of formulas which express the zeros of f in terms of the coefficients $a_1, \ldots, a_n$, using only addition, subtraction, multiplication, division, and extraction of roots. Let us reformulate this idea in terms of field extensions.

DEFINITION 1: Let E be a field extension of F. We say that E is a *radical extension* of F if $E = F(\alpha_1, \ldots, \alpha_t)$, where $\alpha_1^{m_1} \in F$, $\alpha_i^{m_i} \in F(\alpha_1, \ldots, \alpha_{i-1})$ $(i = 2, 3, \ldots, t)$ for some positive integers $m_1, m_2, \ldots, m_t$.

Remark: In the definition of a radical extension, let m denote the least common multiple of $m_1, m_2, \ldots, m_t$. Then for each $i (1 \leq i \leq t)$, $m = k_i m_i$ for some positive integer k_i. Therefore,

$$\alpha_1^{m_1} \in F \Longrightarrow \alpha_1^m = (\alpha_1^{m_1})^{k_1} \in F,$$

$$\alpha_i^{m_i} \in F(\alpha_1, \ldots, \alpha_{i-1}) \Longrightarrow \alpha_i^m = (\alpha_i^{m_i})^{k_i} \in F(\alpha_1, \ldots, \alpha_{i-1}) \qquad (i = 2, \ldots, t).$$

Thus, if we replace each m_i by m, we see that in the definition of a radical extension, we may assume that all m_i are equal.

Roughly speaking, a radical extension of F is obtained by successively adjoining a sequence of radicals to F.

Let $F = \mathbf{Q}$ in all the following examples:

EXAMPLE 1: $E = \mathbf{Q}(\sqrt{2})$ is a radical extension of $\mathbf{Q}$.

EXAMPLE 2: $E = \mathbf{Q}(\sqrt{2}, \sqrt[3]{\sqrt{2} - 3})$ is a radical extension of $\mathbf{Q}$.

EXAMPLE 3: $E = \mathbf{Q}(\zeta)$, where ζ is a primitive nth root of unity, is a radical extension of $\mathbf{Q}$, since $1 = \zeta^n \in \mathbf{Q}$.

DEFINITION 2: We say that f is *solvable in radicals* if the splitting field E_f of f over F is contained in some radical extension E of F.

The reader should have no difficulty convincing himself that this definition of solvability in radicals coincides with the roughly stated notion which we have previously introduced. For indeed, $E_f = F(\alpha_1, \ldots, \alpha_n)$, where $\alpha_1, \ldots, \alpha_n$ are the zeros of f. Therefore, to say that E_f is contained in some radical extension of F just means that the zeros α_i can be expressed in terms of radicals and addition, subtraction, multiplication, and division.

PROPOSITION 3: Let E/F be a radical extension. Then E is contained in a radical extension E' of F such that E'/F is a Galois extension.

Proof: Let $E = F(\alpha_1, \ldots, \alpha_t)$, where $\alpha_1^{m_1} \in F$, $\alpha_i^{m_i} \in F(\alpha_1, \ldots, \alpha_{i-1})$ $(i = 2, \ldots, t)$. Set $\alpha_i^{m_i} = a_i$ and let, for each $i (1 \leq i \leq t)$, $a_{ij} (1 \leq j \leq k_i)$ be the conjugates of a_i over F. Let E' be the smallest subfield of $\bar{F}$ containing F and all zeros of the polynomials $X^{m_i} - a_{ij} (1 \leq i \leq t, 1 \leq j \leq k_i)$. It is clear that E' is normal over F and hence E'/F is a Galois extension (since E' is a splitting field). Moreover, it is clear that $E' \supseteq E$, since for some j, $a_{ij} = a_i$ and hence $a_i \in E'$ $(1 \leq i \leq t)$. Finally, E' is a radical extension of F. We leave the verification of this last statement to the reader.

Let us now investigate the properties of the Galois group of a radical Galois extension E/F. Our main result is

THEOREM 4: Let E/F be a radical Galois extension with Galois group G. Then G is a solvable group.

Proof: Let us utilize the preceding Remark and write $E = F(\alpha_1, \ldots, \alpha_t)$ where $\alpha_1^n \in F$, $\alpha_i^n \in F(\alpha_1, \ldots, \alpha_{i-1})$ $(i = 2, \ldots, t)$. Let ζ be a primitive nth root of unity and let $E' = E(\zeta)$. Since E/F is a Galois extension, E is obtained from F by adjoining all roots of a finite collection of polynomials belonging to $F[X]$. In order to get E' from E, we adjoin all roots of the polynomial $X^n - 1 \in F[X]$. Therefore, E' is obtained by adjoining to F all the roots of a finite collection of polynomials belonging to $F[X]$, and hence E'/F is a Galois extension. Let $H = \text{Gal}(E'/F)$, and set

$$E_0 = F, \quad E_1 = F(\zeta), \quad E_i = F(\zeta, \alpha_1, \ldots, \alpha_{i-1}) \qquad (i = 2, \ldots, t+1).$$

Then

$$F = E_0 \subseteq \cdots \subseteq E_{t+1} = E'. \tag{1}$$

By the fundamental theorem of Galois theory, to the chain of intermediate fields (1), there corresponds a chain of subgroups of H:

$$H = H_0 \supseteq H_1 \supseteq \cdots \supseteq H_{t+1} = \{1\}, \tag{2}$$

where $H_i = \text{Gal}(E'/E_i)$. Let us show that the chain (2) of subgroups of H is a normal series with abelian factors. Indeed, E_1 is the splitting field over E_0 of the polynomial $X^n - 1$. Therefore, E_1/E_0 is a Galois extension with abelian Galois group. Therefore, by Theorem 8 of Section 9.4, $H_1 \lhd H_0$ and $\text{Gal}(E_1/E_0) = H_0/H_1$ is abelian. Next, note that $E_i (i \geq 1)$ contains the nth roots of unity and $E_{i+1} = E_i(\alpha_i)$, where α_i is a zero of $X^n - a_i$, where $a_i = \alpha_i^n \in E_i$. By Example 5 of Section 9.5, $E_{i+1}/E_i (i \geq 1)$ is a Galois extension with abelian Galois group. Therefore, by Theorem 8 of Section 9.4, $H_{i+1} \lhd H_i$ and $\text{Gal}(E_{i+1}/E_i) = H_i/H_{i+1}$ $(i = 1, 2, \ldots, t)$ is abelian. Thus, we have proved that (2) is a normal series for H having abelian factors. Therefore, H is a solvable group. It is now easy to show that G is solvable. We have

$$F \subseteq E \subseteq E', \qquad G = \text{Gal}(E/F), \qquad H = \text{Gal}(E'/F).$$

Let $J = \text{Gal}(E'/E)$. By the fundamental theorem of Galois theory, $J \leq H$. But E/F is a Galois extension, so that by Theorem 8 of Section 9.4, $J \lhd H$ and $G = \text{Gal}(E/F) = H/J$. But then G is a quotient of the solvable group H and hence is solvable by Corollary 9 of Section 8.6.

COROLLARY 5: Let $f \in F[X]$ be a nonconstant polynomial. If f is solvable by radicals, then $\text{Gal}_F(f)$ is solvable.

Proof: If f is solvable in radicals, then there exists a radical extension E of F such that

$$F \subseteq E_f \subseteq E.$$

Without loss of generality, by Proposition 3, we may assume that E/F is a Galois extension. Let $H = \text{Gal}(E/F)$. Since E_f/F is a normal extension, E_f/F is a Galois extension. Let $G = \text{Gal}(E/E_f)$. Then, by Theorem 8 of Section 9.4, we have $G \lhd H$ and

$$\text{Gal}_F(f) = \text{Gal}(E_f/F) = H/G.$$

But H is solvable by Theorem 4, so that H/G is solvable. Thus, $\text{Gal}_F(f)$ is solvable.

COROLLARY 6: There exist fifth-degree polynomials in $\mathbf{Q}[X]$ which are not solvable in radicals. Thus, the quintic (fifth-degree) equation has no general solution in radicals.

Proof: This is easy! We have seen in Section 9.5 that $f = X^5 - 6X + 3$ has the property that $\text{Gal}_{\mathbf{Q}}(f) = S_5$. If f is solvable in radicals, then, by Corollary 5, S_5 is solvable. But since A_5 is a simple group (Theorem 9 of Section 3.8) a composition series for S_5 is given by

$$S_5 \triangleright A_5 \triangleright \{1\},$$

and the factors are

$$S_5/A_5 \approx \mathbf{Z}_2, \qquad A_5/\{1\} \approx A_5.$$

But A_5 is nonabelian, so that S_5 is not solvable. Therefore, f is not solvable in radicals.

We have seen that if a polynomial f is solvable in radicals, then $\text{Gal}_F(f)$ is solvable. In the remaining part of this section we will prove the converse. Actually we will accomplish considerably more. If $\text{Gal}_F(f)$ is solvable, we will actually give a method for constructing a radical extension containing E_f. In particular, this will lead us to an expression of the zeros of f in terms of radicals. When the procedure of this section is applied to polynomials of degrees 2, 3, or 4, we will get the formulas which were heralded long ago in Chapter 1. In order to carry out our program, it will be necessary to study in some detail the structure of a Galois extension E/F of prime degree p, where F contains the pth roots of unity. Our main result will be

THEOREM 7: Let F be a field and let E/F be a Galois extension of prime degree p. Assume that F contains the pth roots of unity. Then there exists $a \in F$ such that $E = F(\alpha)$, where α is a zero of $X^p - a$.

The proof of Theorem 7 is accomplished via a trick going back to Lagrange. Let ζ be a primitive pth root of unity, η an arbitrary pth root of unity. Then η is of the form $\zeta^a (a = 0, 1, \ldots, p - 1)$. Let $\theta \in E$. Since E/F is a Galois extension of prime order p, $\text{Gal}(E/F)$ is cyclic of degree p. Let σ be a generator of $\text{Gal}(E/F)$. Then

$$\text{Gal}(E/F) = \{1, \sigma, \sigma^2, \ldots, \sigma^{p-1}\}.$$

Let us define the *Lagrange resolvent* $\langle \eta, \theta \rangle$ by

$$\langle \eta, \theta \rangle = \theta + \eta\sigma\theta + \cdots + \eta^{p-1}\sigma^{p-1}\theta.$$

It is clear that $\langle \eta, \theta \rangle \in E$. We may restate Theorem 7 as follows:

THEOREM 8: Let F be a field and let E/F be a Galois extension of prime degree p. Assume that F contains the pth roots of unity, and let $\theta \in E - F$. Then there exists a pth root of unity η such that

(1) $\langle \eta, \theta \rangle^p \in F$,
(2) $E = F(\langle \eta, \theta \rangle)$.

It is clear that Theorem 8 implies Theorem 7, upon setting $a = \langle \eta, \theta \rangle^p$.

Proof: First let us show that at least one of the Lagrange resolvents $\langle \zeta^a, \theta \rangle$ $(a = 1, \ldots, p - 1)$ is nonzero. Let us assume the contrary. Then

$$\begin{aligned} \langle 1, \theta \rangle &= \sum_{a=0}^{p-1} \zeta^{-a} \langle \zeta^a, \theta \rangle \\ &= \sum_{a=0}^{p-1} \zeta^{-a} \sum_{\nu=0}^{p-1} \zeta^{a\nu} \sigma^\nu \theta \\ &= \sum_{\nu=0}^{p-1} \sigma^\nu \theta \sum_{a=0}^{p-1} \zeta^{(\nu-1)a}. \end{aligned} \tag{3}$$

If η is a primitive pth root of unity, then

$$\begin{aligned} 0 = \eta^p - 1 &= (\eta - 1)(\eta^{p-1} + \eta^{p-2} + \cdots + 1) \\ &\Longrightarrow \eta^{p-1} + \eta^{p-2} + \cdots + 1 = 0. \end{aligned} \tag{4}$$

Let $0 \leq \nu \leq p - 1$, $\nu \neq 1$ and set $\eta = \zeta^{\nu-1}$. Then η is a primitive pth root of unity, so that by (4),

$$\sum_{a=0}^{p-1} \eta^a = \sum_{a=0}^{p-1} \zeta^{a(\nu-1)} = 0.$$

Therefore, by (3),

$$\begin{aligned} \langle 1, \theta \rangle &= p \cdot \sigma\theta, \\ \Longrightarrow \sigma\theta &= p^{-1} \langle 1, \theta \rangle. \end{aligned} \tag{5}$$

Note that

$$\langle 1, \theta \rangle = \theta + \sigma\theta + \sigma^2\theta + \cdots + \sigma^{p-1}\theta.$$

Therefore,

$$\begin{aligned} \sigma \langle 1, \theta \rangle &= \sigma\theta + \sigma^2\theta + \cdots + \theta \\ &= \langle 1, \theta \rangle, \end{aligned}$$

and thus $\sigma^a \langle 1, \theta \rangle = \langle 1, \theta \rangle$ $(a = 0, 1, \ldots, p - 1)$, so that $\langle 1, \theta \rangle$ is left fixed by every element of $\mathrm{Gal}(E/F)$. We therefore deduce, via the fundamental theorem of Galois theory, that $\langle 1, \theta \rangle \in F$. But, by (5), this implies that $\sigma\theta \in F$. And every element of F is invariant under σ^{-1}, so that $\theta = \sigma\theta \in F$. But this contradicts the hypothesis that $\theta \in E - F$. Finally, we can conclude that at least one of the Lagrange resolvents $\langle \zeta^a, \theta \rangle$ $(a = 1, \ldots, p - 1)$ is nonzero. Let $\eta = \zeta^a$, where $1 \leq a \leq p - 1$ is chosen so that $\langle \zeta^a, \theta \rangle \neq 0$.

Let us now calculate the effect of σ on $\langle\eta, \theta\rangle$. From the definition of $\langle\eta, \theta\rangle$, we have

$$\begin{aligned}\sigma\langle\eta, \theta\rangle &= \sigma(\theta + \eta\sigma\theta + \eta^2\sigma^2\theta + \cdots + \eta^{p-1}\sigma^{p-1}\theta)\\ &= \sigma\theta + \eta\sigma^2\theta + \eta^2\sigma^3\theta + \cdots + \eta^{p-1}\sigma^p\theta\\ &= \eta^{-1}(\eta\sigma\theta + \eta^2\sigma^2\theta + \eta^3\sigma^3\theta + \cdots + \eta^p\theta)\\ &= \eta^{-1}\langle\eta, \theta\rangle,\end{aligned}$$

where we have used the fact that $\eta \in F$ and therefore $\sigma(\eta) = \eta$. Therefore since $\eta^p = 1$,

$$\sigma(\langle\eta, \theta\rangle^p) = \langle\eta, \theta\rangle^p,$$

so that $\langle\eta, \theta\rangle^p$ is left fixed by all elements of $\text{Gal}(E/F) = \{1, \sigma, \ldots, \sigma^{p-1}\}$. But this implies that $\langle\eta, \theta\rangle^p \in F$, whence (1).

Since $\eta \neq 1$, $\sigma\langle\eta, \theta\rangle = \eta^{-1}\langle\eta, \theta\rangle \neq \langle\eta, \theta\rangle$. Therefore, $\langle\eta, \theta\rangle$ is not left fixed by σ, and $\langle\eta, \theta\rangle \notin F$ by the fundamental theorem of Galois theory. Thus, $F(\langle\eta, \theta\rangle) \neq F$. But, since $\deg(E/F) = p$, a prime, and $F \subseteq F(\langle\eta, \theta\rangle) \subseteq E$, we see that $\deg(F(\langle\eta, \theta\rangle)/F) = 1$ or p. In the former case, $F(\langle\eta, \theta\rangle) = F$, which is a contradiction. Therefore, $\deg(F(\langle\eta, \theta\rangle)/F) = p$, from which it follows that $\deg(E/F(\langle\eta, \theta\rangle)) = 1$, and thus $E = F(\langle\eta, \theta\rangle)$.

Let us immediately apply Theorem 7 to prove

THEOREM 9: Let $f \in F[X]$ be a nonconstant polynomial such that $\text{Gal}_F(f)$ is solvable. Then f is solvable in radicals.

Proof: Let $\deg(E_f/F) = n$ and let ζ be a primitive nth root of unity. Set $E_f(\zeta) = E'$, $F(\zeta) = F'$. The relationship between the fields E_f, F, E', and F' is illustrated in Figure 9.2, where a line connecting two fields denotes a containment relation. It suffices to show that E'/F is a radical extension. For then, since $F \subseteq E_f \subseteq E'$, we see that f is solvable in radicals. Note that E' is obtained from F by adjoining all zeros of f and $X^n - 1$, and thus E'/F is a Galois extension. Let $G = \text{Gal}(E'/F)$ and let H be the subgroup of G corresponding to E_f under the Galois correspondence. Then $\text{Gal}(E'/E_f) = H$. Moreover, since E_f/F is a Galois extension, $H \lhd G$ and $\text{Gal}(E_f/F) = G/H$. But $\text{Gal}(E_f/F) = \text{Gal}_F(f)$ is solvable. Therefore, G/H is solvable. Moreover, by Example 5 of Section 9.5, $H = \text{Gal}(E_f(\zeta)/E_f)$ is abelian and hence solvable. Thus, since H and G/H are solvable, G is solvable by Exercise 2 of Section 8.6. Let J be the subgroup of G corresponding to F' under the Galois correspondence. Then $J = \text{Gal}(E'/F')$ and J is a subgroup of a solvable group and hence is solvable. For the relationship between the various Galois groups we have defined, see Figure 9.2.

Since J is solvable, there exists a composition series

$$J = J_0 \rhd J_1 \rhd \cdots \rhd J_t = \{1\}$$

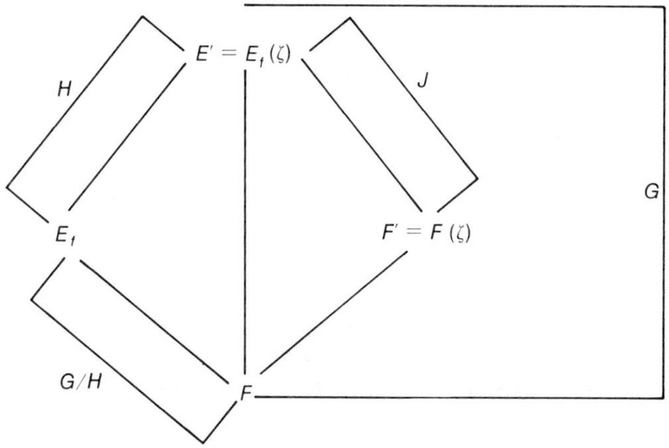

FIGURE 9-2: The Subfields of E' and their Galois Groups.

such that J_i/J_{i+1} is a cyclic group of prime order p_i $(0 \leq i \leq t-1)$. Let F_i be the fixed field of J_i. Then $J_i = \text{Gal}(E'/F_i)$ and

$$F' = F_0 \subseteq F_1 \subseteq \cdots \subseteq F_t = E'.$$

Moreover, since $J_i \rhd J_{i+1}$, we see that F_{i+1}/F_i is a Galois extension with Galois group J_i/J_{i+1}. Thus, F_{i+1}/F_i is Galois extension of prime degree p_i. Let us now show that F_i contains the p_ith roots of unity. This will allow us to apply Theorem 7 to the extension F_{i+1}/F_i.

If $\sigma \in \text{Gal}(E'/F')$, then the restriction of σ to E is an F-automorphism of E, that is, an element of $\text{Gal}(E/F)$ [because $E' = E(\zeta)$, $F' = F(\zeta)$]. Therefore, let us define the function

$$\psi: \text{Gal}(E'/F') \longrightarrow \text{Gal}(E/F),$$

by $\psi(\sigma) =$ the restriction of σ to E $(\sigma \in \text{Gal}(E'/F'))$. It is trivial to check that ψ is an isomorphism. Therefore, $J = \text{Gal}(E'/F')$ is isomorphic to a subgroup of $G/H = \text{Gal}(E_f/F)$. In particular, since p_i divides the order of J and since G/H has order n, we see that $p_i | n$ $(0 \leq i \leq t-1)$. If λ is a p_ith root of unity, then $\lambda^{p_i} = 1$, so that $\lambda^n = 1$ (since $p_i | n$). Therefore, every p_ith root of unity is an nth root of unity. But $F' = F(\zeta) \subseteq F_i$ $(i = 0, \ldots, t)$ and ζ is a primitive nth root of unity. Therefore, F_i $(i = 1, \ldots, t)$ contains the p_ith roots of unity, as asserted.

We may now apply Theorem 7 to each of the extensions F_{i+1}/F_i $(0 \leq i \leq t-1)$. We see that there exists $\alpha_{i+1} \in F_{i+1}$ such that (1) $F_{i+1} = F_i(\alpha_{i+1})$ and (2) α_{i+1} is a zero of a polynomial of the form $X^{p_i} - a_{i+1}$ $(a_{i+1} \in F_i)$. Thus, we derive that

$$E' = F_t = F(\zeta, \alpha_1, \ldots, \alpha_t)$$

and

$$\zeta^n = 1 \in F, \quad \alpha_1^{p_1} \in F(\zeta), \quad \alpha_{i+1}^{p_i} = a_{i+1} \in F_i = F(\zeta, \alpha_1, \ldots, \alpha_i) \qquad (1 \le i \le t-1).$$

Thus, E' is a radical extension of F.

THEOREM 10: Let $f \in F[X]$ be a nonconstant polynomial of degree at most 4. Then f is solvable in radicals.

Proof: By Corollary 5 of Section 9.5, $\mathrm{Gal}_F(f)$ is a subgroup of S_n ($n \le 4$). Since a subgroup of a solvable group is solvable, Theorem 9 implies that it suffices to show that S_1, S_2, S_3, and S_4 are solvable groups. This is obvious for S_1 and S_2, since these groups are abelian. A composition series for S_3 is

$$S_3 \rhd A_3 \rhd \{1\}$$

and the composition factors are $S_3/A_3 \approx \mathbf{Z}_2$, $A_3/\{1\} \approx \mathbf{Z}_3$. Thus, S_3 is solvable. A composition series for S_4 is given by

$$S_4 \rhd A_4 \rhd \{(1), (12)(34), (13)(24), (14)(23)\} \rhd \{(1), (12)(34)\} \rhd \{1\}.$$

(*Proof:* Exercise.) The composition factors are isomorphic to

$$\mathbf{Z}_2, \mathbf{Z}_3, \mathbf{Z}_2, \mathbf{Z}_2.$$

Thus, S_4 is solvable.

9.6 Exercises

1. Let $g, h \in \mathbf{Q}[X]$. Show that $g \cdot h$ is solvable in radicals if and only if g and h are solvable in radicals.
2. Let $f \in \mathbf{Q}[X]$ be a reducible polynomial of degree 5. Show that f is solvable in radicals.
3. Give an example of a polynomial $f \in \mathbf{Q}[X]$ of degree 6 and irreducible which is solvable in radicals.
4. Same as Exercise 3, except of degree 7.

9.7 The General Polynomial of Degree n

Let F be a field and let n be a positive integer. Let $X, X_1, X_2, \ldots, X_n$ be $n+1$ indeterminates over F, and let us consider the polynomial

$$\begin{aligned} f(X) &= \prod_{i=1}^{n} (X - X_i) \\ &= X^n - \sigma_1 X^{n-1} + \sigma_2 X^{n-2} - \cdots + (-1)^n \sigma_n, \end{aligned}$$

where

$$\sigma_1 = X_1 + X_2 + \cdots + X_n$$
$$\sigma_2 = X_1X_2 + X_1X_3 + \cdots + X_{n-1}X_n$$
$$\vdots$$
$$\sigma_n = X_1X_2 \cdots X_n.$$

The expressions $\sigma_1, \sigma_2, \ldots, \sigma_n$ are called the *elementary symmetric functions* of $X_1, \ldots, X_n$. They clearly have the property that they are unaltered when $X_1, \ldots, X_n$ are subjected to a permutation. Many other expressions in $X_1, \ldots, X_n$ also have this property. For example, consider

$$X_1^2 + X_2^2 + \cdots + X_n^2.$$

It is trivial to see, however, that

$$\begin{aligned} X_1^2 + X_2^2 + \cdots + X_n^2 &= (X_1 + \cdots + X_n)^2 \\ &\quad - 2(X_1X_2 + X_1X_3 + \cdots + X_{n-1}X_n) \\ &= \sigma_1^2 - 2\sigma_2. \end{aligned}$$

Thus, the function $X_1^2 + \cdots + X_n^2$ can be expressed in terms of the elementary symmetric functions. This is no accident, as we shall see below.

Let $F(X_1, \ldots, X_n)$ denote the smallest field containing F and $X_1, \ldots, X_n$. A typical element of $F(X_1, \ldots, X_s)$ is of the form

$$r(X_1, \ldots, X_n) = \frac{p(X_1, \ldots, X_n)}{q(X_1, \ldots, X_n)}, \qquad p, q \in F[X_1, \ldots, X_n].$$

If $\eta \in S_n$ and $r(X_1, \ldots, X_n) \in F(X_1, \ldots, X_n)$, we define the action of η on $r(X_1, \ldots, X_n)$ by

$$r^\eta(X_1, \ldots, X_n) = r(X_{\eta(1)}, \ldots, X_{\eta(n)}).$$

It is then trivial to verify that $\sigma_j^\eta = \sigma_j$ $(1 \leq j \leq n)$ for all $\eta \in S_n$. Further, it is trivial to show that the mapping

$$r \longrightarrow r^\eta \qquad (r \in F(X_1, \ldots, X_n)$$

is an automorphism of $F(X_1, \ldots, X_n)$ which is the identity on $F(\sigma_1, \ldots, \sigma_n)$. Let us denote this automorphism by $\tilde{\eta}$ and let us set $F(X_1, \ldots, X_n) = E_n$, $F(\sigma_1, \ldots, \sigma_n) = F_n$. Then by what we have said, $\tilde{\eta} \in \mathrm{Gal}(E_n/F_n)$.

LEMMA 1: E_n is a Galois extension of F_n with Galois group isomorphic to S_n. In fact, if

$$\psi: S_n \longrightarrow \mathrm{Gal}(E_n/F_n)$$

is defined by $\psi(\eta) = \tilde{\eta}$, then ψ is a surjective isomorphism.

Proof: E_n is the splitting field of the polynomial $f(X)$ over F_n. Therefore, E_n/F_n is a finite, normal extension. It is trivial to check that ψ is a homomorphism and it is clear that ψ is injective. But by Corollary 5 of Section 9.5, the order of $\text{Gal}(E_n/F_n)$ is at most $n!$ and we have already shown that the order is at least $n!$, so that the order is exactly $n!$ and ψ is surjective.

If $r(X_1, \ldots, X_n) \in F(X_1, \ldots, X_n)$, then we say that $r(X_1, \ldots, X_n)$ is a *symmetric function* if $r^\sigma(X_1, \ldots, X_n) = r(X_1, \ldots, X_n)$ for all $\sigma \in S_n$. Let us now prove

THEOREM 2 (FUNDAMENTAL THEOREM ON SYMMETRIC FUNCTIONS): Let $r \in F(X_1, \ldots, X_n)$ be a symmetric function. Then r is a rational function of the elementary symmetric functions $\sigma_1, \ldots, \sigma_n$. That is, there exists $s(X_1, \ldots, X_n) \in F(X_1, \ldots, X_n)$ such that

$$r(X_1, \ldots, X_n) = s(\sigma_1, \ldots, \sigma_n).$$

Proof: If r is symmetric, then $r^\eta = r$ for all $\eta \in S_n$. Therefore, $\tilde{\eta}(r) = r$ for all $\tilde{\eta} \in \text{Gal}(E_n/F_n)$ by Lemma 1. Thus, r is contained in the fixed field of $\text{Gal}(E_n/F_n)$, so that $r \in F_n$. But $F_n = F(\sigma_1, \ldots, \sigma_n)$, whence the theorem.

Let $X, a_1, \ldots, a_n$ be $n + 1$ distinct indeterminates over F. The polynomial

$$g(X) = X^n - a_1X^{n-1} + a_2X^{n-2} - \cdots + (-1)^n a_n$$

is called the *general polynomial of degree n over F*. The reason for the terminology "general polynomial" is as follows: The indeterminates $a_1, \ldots, a_n$ play the same role as "variables" in analysis. And as the "variables" $a_1, \ldots, a_n$ run independently over F, $g(X)$ runs over all monic polynomials of degree n over F. In the remainder of this section, we will study the properties of $g(X)$.

Let $F_0 = F(a_1, \ldots, a_n)$ and let us fix an algebraic closure $\bar{F}_0$ of F_0. If $y_1, \ldots, y_n$ are the zeros of g in $\bar{F}_0$, then

$$g(X) = \prod_{i=1}^{n} (X - y_i).$$

Let E_g denote the splitting field of g over F_0. Then

$$\begin{aligned} E_g &= F_0(y_1, \ldots, y_n) \\ &= F(a_1, \ldots, a_n, y_1, \ldots, y_n) \\ &= F(y_1, \ldots, y_n), \end{aligned}$$

since the a_i can be expressed in terms of $y_1, \ldots, y_n$. Since E_g/F_0 is a normal extension, it is also a Galois extension. Let $G = \text{Gal}(E_g/F_0)$. By the results of Section 9.5 we know that every $\sigma \in G$ can be identified with a permutation (also denoted σ) of $y_1, \ldots, y_n$, say $\sigma(y_i) = y_{\sigma(i)}$ $(1 \leq i \leq n)$. Our main result about the general equation of degree n is

THEOREM 3: $\text{Gal}_{F_0}(g) = S_n$.

The bulk of the proof is contained in the following lemma.

LEMMA 4: The mapping

$$\psi: F[X_1, \ldots, X_n] \longrightarrow F[y_1, \ldots, y_n]$$

defined by $\psi(p(X_1, \ldots, X_n)) = p(y_1, \ldots, y_n)$ is a surjective isomorphism.

Proof: It is immediate that ψ is a surjective homomorphism. Therefore, it suffices to show that $\ker(\psi) = \{0\}$. Let $f \in \ker(\psi), f \neq 0$. Then $f = f(X_1, \ldots, X_n) \in F[X_1, \ldots, X_n]$ and $f(y_1, \ldots, y_n) = 0$. For $\sigma \in S_n$, let us define

$$f^\sigma(X_1, \ldots, X_n) = f(X_{\sigma(1)}, \ldots, X_{\sigma(n)}).$$

Then it is easy to see that

$$f^\sigma(y_1, \ldots, y_n) = \sigma(f(y_1, \ldots, y_n)). \tag{1}$$

Set

$$g(X_1, \ldots, X_n) = \prod_{\sigma \in G} f^\sigma(X_1, \ldots, X_n).$$

If $\sigma \in S_n$, then

$$\begin{aligned} g^\sigma(X_1, \ldots, X_n) &= \left[\prod_{\eta \in S_n} f^\eta(X_1, \ldots, X_n)\right]^\sigma \\ &= \prod_{\eta \in S_n} f^{\sigma\eta}(X_1, \ldots, X_n) \\ &= \prod_{\lambda \in S_n} f^\lambda(X_1, \ldots, X_n) \qquad \text{(since } \eta\sigma \text{ runs over } S_n \text{ as } \eta \text{ does)} \\ &= g(X_1, \ldots, X_n), \end{aligned} \tag{2}$$

and $g(X_1, \ldots, X_n) \neq 0$ since $f(X_1, \ldots, X_n) \neq 0$. By (2) and Corollary 7 of Section 9.4, we have

$$g(X_1, \ldots, X_n) = s(\sigma_1, \ldots, \sigma_n),$$

where $\sigma_1, \ldots, \sigma_n$ are the elementary symmetric functions in $X_1, \ldots, X_n$, and $s \in F(X_1, \ldots, X_n)$. Since $g \neq 0$, we have $s \neq 0$. Note that

$$\psi(\sigma_i) = a_i \qquad (1 \leq i \leq n),$$

and therefore

$$\psi(g(X_1, \ldots, X_n)) = s(a_1, \ldots, a_n). \tag{3}$$

On the other hand,

$$\begin{aligned} \psi(g(X_1, \ldots, X_n) &= \prod_{\sigma \in S_n} \psi(f^\sigma(X_1, \ldots, X_n)) \\ &= f(y_1, \ldots, y_n) \cdot \prod_{\substack{\sigma \in S_n \\ \sigma \neq 1}} \psi(f^\sigma(X_1, \ldots, X_n)) \\ &= 0 \qquad [\text{since } f(y_1, \ldots, y_n) = 0]. \end{aligned} \tag{4}$$

Let $s = p/q$ $(p, q \in F[X_1, \ldots, X_n])$. Then $p \neq 0$ since $s \neq 0$. Thus, by (3) and (4), we see that $p(a_1, \ldots, a_n) = 0, p \neq 0$. But this contradicts the fact that $a_1, \ldots, a_n$ are distinct indeterminates. Therefore, $\ker(\psi) = \{0\}$.

Proof of Theorem 3: Let us define a mapping

$$\tilde{\psi}: F(X_1, \ldots, X_n) \longrightarrow F(y_1, \ldots, y_n)$$

by

$$\tilde{\psi}\left(\frac{p(X_1, \ldots, X_n)}{q(X_1, \ldots, X_n)}\right) = \frac{p(y_1, \ldots, y_n)}{q(y_1, \ldots, y_n)}, \qquad p, q \in F[X_1, \ldots, X_n], \quad q \neq 0.$$

This mapping is well defined, since if

$$\frac{p(X_1, \ldots, X_n)}{q(X_1, \ldots, X_n)} = \frac{p'(X_1, \ldots, X_n)}{q'(X_1, \ldots, X_n)}, \qquad p, q, p', q' \in F[X_1, \ldots, X_n], \quad q \neq 0, \quad q' \neq 0,$$

then

$$pq' - qp' = 0,$$
$$\Longrightarrow 0 = \psi(pq' - qp') = \psi(p)\psi(q') - \psi(q)\psi(p').$$

But by Lemma 4, $\psi(q) \neq 0, \psi(q') \neq 0$ since $q \neq 0, q' \neq 0$. Therefore,

$$\psi(p)/\psi(q) = \psi(p')/\psi(q'),$$
$$\Longrightarrow \tilde{\psi}(p/q) = \tilde{\psi}(p'/q').$$

Therefore, $\tilde{\psi}$ is well defined. It is trivial to verify that $\tilde{\psi}$ is a surjective isomorphism. With respect to the isomorphism $\tilde{\psi}$, $F(\sigma_1, \ldots, \sigma_n)$ corresponds to $F(a_1, \ldots, a_n)$. Therefore, the extension $F(X_1, \ldots, X_n)/F(\sigma_1, \ldots, \sigma_n)$ corresponds to the extension $F(y_1, \ldots, y_n)/F(a_1, \ldots, a_n)$. By Lemma 1, for every $\sigma \in S_n$, there exists an element of $\mathrm{Gal}(F(X_1, \ldots, X_n)/F(\sigma_1, \ldots, \sigma_n)$ which effects the permutation σ on $X_1, \ldots, X_n$. Therefore, there exists an element of $\mathrm{Gal}(F(y_1, \ldots, y_n)/F(a_1, \ldots, a_n))$ effecting the permutation σ on $y_1, \ldots, y_n$. In particular, $\mathrm{Gal}(F(y_1, \ldots, y_n)/F(a_1, \ldots, a_n))$ has order at least $n!$ However, $\mathrm{Gal}(F(y_1, \ldots, y_n)/F(a_1, \ldots, a_n))$ is isomorphic to a subgroup of S_n. Therefore, this group has exactly $n!$ elements and

$$\mathrm{Gal}_{F_0}(g) = \mathrm{Gal}(F(y_1, \ldots, y_n)/F(a_1, \ldots, a_n))$$
$$\approx S_n.$$

Let us now apply Theorem 3 of Section 9.7 to the question of which finite groups are Galois groups.

THEOREM 5: Let G be a finite group. Then there exists a Galois extension E/F such that $\mathrm{Gal}(E/F) \approx G$.

Proof: By Cayley's theorem (Theorem 1 of Section 3.7), there is an isomorphism $\psi: G \to S_n$ for some $n \geq 1$. Let us choose one such (ψ, n) and fix them. By Theorem 3, S_n is the Galois group of an extension E/F_0. Let F denote the fixed field of the subgroup $\psi(G) \subseteq S_n$. Then $F_0 \subseteq F \subseteq E$. Moreover, by the fundamental theorem of Galois theory,

$$\mathrm{Gal}(E/F) = \psi(G).$$

Thus, $\mathrm{Gal}(E/F) \approx G$.

Another easy application of Theorem 3 is

THEOREM 6: The general equation of degree n is solvable in radicals for $n \leq 4$. It is not solvable in radicals for $n \geq 5$.

Proof: In the proof of Theorem 10 of Section 9.6 we showed that S_n is solvable for $n \leq 4$. For $n \geq 5$, A_n is a simple, nonabelian group by Theorem 9 of Section 3.8. Therefore, a composition series for $\mathfrak{S}_n$ $(n \geq 5)$ is given by

$$S_n \rhd A_n \rhd \{1\};$$

the composition factors are $S_n/A_n \approx \mathbf{Z}_2$ and $A_n/\{1\} \approx A_n$. Therefore, since A_n is nonabelian, S_n is not solvable for $n \geq 5$. Therefore, the theorem follows by Theorem 5 of Section 9.6.

Later in this chapter we will find the solutions in radicals of the general equations of degrees 2, 3, and 4. For degree 2, we will get the usual quadratic formula. For degree 3, we will get the formulas of Tartaglia. For degree 4, we will get the formulas of Ferrari. As a preliminary to this discussion, let us now introduce the notion of the discriminant of a polynomial.

Let F be an arbitrary field and let $X_1, \ldots, X_n$ be any n indeterminates over F. Further, let $\sigma_1, \ldots, \sigma_n$ be the elementary symmetric functions of $X_1, \ldots, X_n$. The expression

$$\Delta(X_1, \ldots, X_n) = \prod_{1 \leq i < j \leq n} (X_i - X_j)^2$$

is a symmetric function of $X_1, \ldots, X_n$ and therefore can be expressed in terms of the elementary symmetric functions. Let us carry this computation out for a few values of n.

$$n = 1: \ \Delta(X_1) = X_1 = \sigma_1,$$

$$\begin{aligned} n = 2: \ \Delta(X_1, X_2) &= (X_1 - X_2)^2 \\ &= X_1^2 - 2X_1X_2 + X_2^2 \\ &= (X_1 + X_2)^2 - 4X_1X_2 \\ &= \sigma_1^2 - 4\sigma_2 \end{aligned}$$

$$n = 3: \ \Delta(X_1, X_2, X_3) = (X_1 - X_2)^2(X_2 - X_3)^2(X_3 - X_1)^2.$$

A somewhat tedious computation shows that

$$\Delta(X_1, X_2, X_3) = \sigma_1^2\sigma_2^2 - 4\sigma_2^3 - 4\sigma_1^3\sigma_3 - 27\sigma_3^2 + 18\sigma_1\sigma_2\sigma_3.$$

Let F be a field and let $f \in F[X]$,

$$f = X^n + a_1X^{n-1} + \cdots + a_n.$$

Further, let $\bar{F}$ be an algebraic closure of F and let $y_1, \ldots, y_n$ be the zeros of f in $\bar{F}$. The quantity

$$\begin{aligned} \Delta &= \Delta(y_1, \ldots, y_n) \\ &= \prod_{1 \leq i < j \leq n} (y_i - y_j)^2 \end{aligned}$$

is called the *discriminant of the polynomial* f. The discriminant has the following properties:

I. $\Delta \in F$. For Δ is unchanged when the y_i are permuted. Therefore, Δ is contained in the fixed field of $\text{Gal}(E_f/F)$, where E_f is the splitting field of f over F. Thus, $\Delta \in F$ by the fundamental theorem of Galois theory.

II. $\Delta = 0$ if and only if two of the zeros of f are equal.

III. Δ can be expressed rationally in terms of $a_1, \ldots, a_n$. For $\Delta(X_1, \ldots, X_n)$ can be expressed rationally in terms of $\sigma_1, \ldots, \sigma_n$. And on substituting y_1 for X_1, y_2 for $X_2, \ldots, y_n$ for X_n, σ_i becomes $(-1)^i a_i (1 \leq i \leq n)$ and $\Delta(y_1, \ldots, y_n) = \Delta$.

In particular, the discriminant of a quadratic polynomial $X^2 + a_1 X + a_2$ is $a_1^2 - 4a_2$, a quantity which we have become accustomed to thinking of as the discriminant in our use of the quadratic formula. Note also that II implies that once we know the formula for Δ in terms of $a_1, \ldots, a_n$, it is trivial to determine whether or not f has multiple roots.

Let us now consider another use of the discriminant. Consider the polynomial

$$\prod_{1 \leq i < j \leq n} (X_i - X_j) = \sqrt{\Delta(X_1, \ldots, X_n)}. \tag{5}$$

If we subject $X_1, \ldots, X_n$ to the transposition (ab), $a < b$, then in the product (5), the term $X_a - X_b$ becomes $X_b - X_a$, whereas all other terms are left unchanged. Therefore, under the transposition, (5) is transformed into its negative. Thus, with respect to a permutation which is a product of an odd number of transpositions, (5) goes over into its negative, whereas with respect to a permutation which is a product of an even number of transpositions, (5) remains unchanged. Therefore, we have proved

THEOREM 7: Let $\sigma \in S_n$. Then $\sigma \in A_n \Leftrightarrow \sigma$ transforms the product

$$\prod_{1 \leq i < j \leq n} (X_i - X_j)$$

into itself. If $\sigma \notin A_n$, then

$$\prod_{1 \leq i < j \leq n} (X_{\sigma(i)} - X_{\sigma(j)}) = -\prod_{1 \leq i < j \leq n} (X_i - X_j).$$

Let us now apply Theorem 7 to obtain some information about the splitting field of a polynomial f over a field F. Let $y_1, \ldots, y_n$ denote the zeros of f and let us consider

$$\sqrt{\Delta} = \prod_{1 \leq i < j \leq n} (y_i - y_j). \tag{6}$$

[It is clear that the right-hand side of (5) is a square root of Δ. Let us assume that the particular square root of Δ referred to on the left is defined by the expression on the right.]

By Theorem 7, if $\text{Gal}_F(f) \subseteq A_n$, then $\sigma(\sqrt{\Delta}) = \sqrt{\Delta}$ for all $\sigma \in \text{Gal}_F(f)$. Therefore, by the fundamental theorem of Galois theory, $\sqrt{\Delta} \in F$, so that Δ

is a square in F. Conversely, if Δ is a square in F, $\mathrm{Gal}_F(f) \subseteq A_n$, for otherwise if $\sigma \in \mathrm{Gal}_F(f) - A_n$, then we have $\sigma(\sqrt{\Delta}) = -\sqrt{\Delta} \Rightarrow \sqrt{\Delta} \notin F$.

Suppose that $\mathrm{Gal}_F(f) \nsubseteq A_n$. Then $H = \mathrm{Gal}_F(f) \cap A_n$ is a proper subgroup of $\mathrm{Gal}_F(f)$. Moreover, since $[S_n: A_n] = 2$, we see that $[\mathrm{Gal}_F(f): H] = 2$. (Exercise.) Therefore, the fixed field of H is a quadratic extension of F and this extension must clearly coincide with $F(\sqrt{\Delta})$. To summarize what we have proved:

THEOREM 8: (1) $\mathrm{Gal}_F(f) \subseteq A_n$ if and only if Δ is the square of some element of F.

(2) If $\mathrm{Gal}_F(f) \nsubseteq A_n$, then E_f contains the extension $F(\sqrt{\Delta})$, which is an extension of F of degree 2. In fact, $F(\sqrt{\Delta})$ is the fixed field of $\mathrm{Gal}_F(f) \cap A_n$.

This result has an interesting consequence.

COROLLARY 9: Let f be an irreducible cubic polynomial over F, Δ its discriminant.

(1) If Δ is the square of some element of F, then, $\mathrm{Gal}_F(f) \approx \mathbf{Z}_3$.

(2) If Δ is not the square of some element of F, then $\mathrm{Gal}_F(f) \approx S_3$.

Proof: Since f is irreducible over F, $\deg(E_f) \geq 3$. Therefore, $\mathrm{Gal}_F(f)$ is isomorphic to a subgroup of S_3 which is of order at least 3. If Δ is the square of some element of F, then $\mathrm{Gal}_F(f) \subseteq A_3$ by Theorem 8. Thus, since $A_3 \approx \mathbf{Z}_3$ and $\mathrm{Gal}_F(f)$ has order at least 3, we have (1). If Δ is not the square of some element of F, then $\mathrm{Gal}_F(f) \nsubseteq A_3$ by Theorem 8. Thus, $\mathrm{Gal}_F(f) \approx S_3$ since $\mathrm{Gal}_F(f)$ has order at least 3.

9.7 Exercises

1. In Exercise 9 of Section 9.5 we showed that if p is a prime, then there exists $f_p \in \mathbf{Q}[X]$ such that $\mathrm{Gal}_{\mathbf{Q}}(f_p) \approx S_p$. Use this result to show that every finite group is the Galois group of some extension of a finite extension of $\mathbf{Q}$. (*Hint:* Imitate the proof of Theorem 5.)
2. Express the following symmetric functions in terms of the elementary symmetric functions:
   (a) $X_1^2 + \cdots + X_n^2$.
   (b) $X_1^3 + \cdots + X_n^3$.
   (c) $\sum\limits_{\substack{1 \leq i,j < n \\ i \neq j}} X_i^2 X_j$.
   (d) $\sum\limits_{\substack{1 < i,j \leq n \\ i \neq j}} X_i^2 X_j^2$.
   (e) $\sum\limits_{\substack{1 \leq i,j,k \leq n \\ i,j,k \text{ distinct}}} X_i X_j X_k^2$.

3. (a) Prove that

$$1 + 2 + \cdots + n = \frac{n(n+1)}{2}.$$

(b) Using the formula of Exercise 2(a) and part (a) of this exercise, prove that

$$1^2 + 2^2 + \cdots + n^2 = \frac{n(n+1)(2n+1)}{6}.$$

(c) Find a formula for

$$1^3 + 2^3 + \cdots + n^3,$$

using the formula of Exercise 2(b) and parts (a) and (b) of this exercise.

4. Determine $\mathrm{Gal}_{\mathbf{Q}}(f)$ if $f = X^3 - 10X + 5$.
5. Prove the formula for $\Delta(X_1, X_2, X_3)$. Derive a similar formula for $\Delta(X_1, X_2, X_3, X_4)$.

9.8 The Solution in Radicals of Equations of Degree ≤ 4

Let

$$f = X^n + a_1X^{n-1} + \cdots + a_n \in \mathbf{C}(a_1, \ldots, a_n)[X]$$

be the general equation of degree n over $\mathbf{C}$, and let $y_1, \ldots, y_n$ be the zeros of f. In this section we will find formulas, in case $n = 2, 3, 4$, which express $y_1, \ldots, y_n$ in terms of radicals involving only the coefficients $a_1, \ldots, a_n$. For $n = 2$, we will get the well-known quadratic formula, while for $n = 3$ and 4, we will get the formulas of Tartaglia and Ferrari, respectively. Of course, we will make use of all the machinery which we have developed up to this point.

Before we begin the derivation of the formulas, let us make a few comments on our approach. First, we have limited our analysis to the general polynomial of degree n. This is because we already have accumulated much useful information about the general polynomial of degree n, and this information is the basis for our derivation of the desired formulas. It will turn out that the formulas which we derive are true formally. That is, they remain valid whatever values the a_i assume, and not only when the a_i are distinct indeterminates over $\mathbf{C}$. Thus, our formulas for the zeros of the general polynomial are really formulas valid for all polynomials of degree ≤ 4. Note that we have assumed that f $\mathbf{C}(a\ , \ldots, a_n)[X]$. The use of the field $\mathbf{C}$ is a convenience and is not absolutely necessary. However, it offers us the nice feature of not being required to adjoin roots of unity to the base field, since $\mathbf{C}$ contains all roots of unity.

The general plan of our attack is as follows: The Galois group of f over $\mathbf{C}(a_1, \ldots, a_n)$ is S_n, as we have seen in Section 9.7. If $n \leq 4$, S_n is a solvable group. Let

$$S_n = G_0 \rhd G_1 \rhd \cdots \rhd G_t = \{1\} \tag{1}$$

be a composition series for S_n. Then G_i/G_{i+1} is cyclic of prime order ($i = 0, 1, \ldots, t-1$). Let $\bar{F}$ be an algebraic closure of $F = \mathbf{C}(a_1, \ldots, a_n)$. Then in $\bar{F}[X]$ we have

$$f = \prod_{i=1}^{n} (X - y_i),$$

where the y_i are all distinct. The splitting field E_f is a Galois extension of F with Galois group S_n. To the chain of subgroups (1) corresponds a tower of intermediate fields:

$$F = E_0 \subseteq E_1 \subseteq \cdots \subseteq E_t = E_f, \tag{2}$$

where E_i is the fixed field of $G_i (0 \leq i \leq t)$. By (1) and the fundamental theorem of Galois theory, E_{i+1}/E_i is a Galois extension with Galois group G_i/G_{i+1}, which is a cyclic group of prime order. Therefore, as we showed in Section 9.6, $E_{i+1} = E_i(\alpha_i)$, where α_i is a Lagrange resolvent, a power of which belongs to E_i. Our plan is to express the Lagrange resolvents as radicals involving the elements of E_i. In principle, this allows us to express, every element of E_{i+1} in terms of radicals involving only elements of E_i. Thus, every element of E_1 can be written in terms of radicals involving $a_1, \ldots, a_n$. Then, every element of E_2 can be written in terms of radicals involving elements of E_1, and thus every element of E_2 can be written in terms of radicals involving $a_1, \ldots, a_n$. Proceeding in this way, we eventually can express every element of E_t in terms of radicals involving $a_1, \ldots, a_n$. But $y_1, \ldots, y_n$ belong to $E_t = E_f$. This gives us our solution in radicals.

Case 1: $n = 2$. Here we have

$$\begin{aligned} f &= X^2 + a_1 X + a_2 \\ &= (X - y_1)(X - y_2). \end{aligned}$$

Then

$$a_1 = -(y_1 + y_2), \tag{3}$$

$$a_2 = y_1 y_2. \tag{4}$$

A composition series for S_2 is given by

$$S_2 \rhd A_2 = \{1\}.$$

Therefore, $E_0 = \mathbf{C}(a_1, a_2)$, $E_1 = \mathbf{C}(a_1, a_2, y_1, y_2) = \mathbf{C}(y_1, y_2)$ [by (3) and (4)]. Now $E_1 = E_0(\alpha)$, where α is one of the Lagrange resolvents

$$\langle 1, y_1 \rangle = y_1 + y_2,$$

$$\langle -1, y_1 \rangle = y_1 - y_2.$$

Actually, it suffices to adjoin any nonzero Lagrange resolvent. And since $y_1 \neq y_2$, we have $\langle -1, y_1 \rangle \neq 0$. By our general theory, $\langle -1, y_1 \rangle^2 \in F = \mathbf{C}(a_1, a_2)$. In fact,

$$\begin{aligned}\langle -1, y_1 \rangle^2 &= y_1^2 + y_2^2 - 2y_1y_2 \\ &= (y_1 + y_2)^2 - 4y_1y_2 \\ &= a_1^2 - 4a_2.\end{aligned}$$

Therefore, $\langle -1, y_1 \rangle = \sqrt{a_1^2 - 4a_2}$ and

$$E_1 = E_0(\sqrt{a_1^2 - 4a_2}).$$

Also, since $\langle -1, y_1 \rangle = y_1 - y_2$ and $y_1 + y_2 = -a_1$, we have

$$\begin{aligned}2y_1 &= (y_1 + y_2) + (y_1 - y_2) \\ &= -a_1 + \sqrt{a_1^2 - 4a_2}.\end{aligned}$$

Therefore,

$$y_1 = \frac{-a_1 + \sqrt{a_1^2 - 4a_2}}{2}.$$

Similarly,

$$y_2 = \frac{-a_1 - \sqrt{a_1^2 - 4a_2}}{2}.$$

Thus, we get the familiar quadratic formula, derived in a seemingly very complicated way.

Case 2: $n = 3$. In this case,

$$\begin{aligned}f &= X^3 + a_1X^2 + a_2X + a_3 \\ &= (X - y_1)(X - y_2)(X - y_3).\end{aligned}$$

Moreover, by comparing coefficients in these two expressions for f, we have

$$a_1 = -(y_1 + y_2 + y_3), \tag{5}$$

$$a_2 = y_1y_2 + y_1y_3 + y_2y_3, \tag{6}$$

$$a_3 = -y_1y_2y_3. \tag{7}$$

A composition series for $\mathcal{S}_3$ is given by

$$S_3 \rhd A_3 \rhd \{1\}.$$

Therefore, we may set

$$G_0 = S_3, \qquad G_1 = A_3, \qquad G_2 = \{1\}.$$

By Theorem 8 of Section 9.7 we see that the fixed field of A_3 is $F(\sqrt{\Delta})$, where

$$\begin{aligned}\Delta &= (y_1 - y_2)^2(y_2 - y_3)^2(y_1 - y_3)^2 \\ &= a_1^2a_2^2 - 4a_2^3 - 4a_1^3a_3 - 27a_3^2 + 18a_1a_2a_3\end{aligned} \tag{8}$$

is the discriminant of f. Therefore, we may set $E_0 = F$, $E_1 = F(\sqrt{\Delta})$, $E_2 = E_f$. Let $\omega = (-1 + \sqrt{3}\,i)/2$. Then ω is a primitive cube root of 1. Note that E_f/E_1 is a Galois extension with cyclic Galois group $A_3 \approx \mathbf{Z}_3$. Let

$\mathrm{Gal}(E_f/E_1) = \{1, \sigma, \sigma^2\}$, where the zeros y_1, y_2, y_3 are numbered so that $\sigma(y_1) = y_2, \sigma(y_2) = y_3, \sigma(y_3) = y_1$. Our general theory asserts that E_f can be obtained by adjoining to E_1 the Lagrange resolvents

$$\langle 1, y_1 \rangle = y_1 + y_2 + y_3 = -a_1, \tag{9}$$

$$\langle \omega, y_1 \rangle = y_1 + \omega y_2 + \omega^2 y_3, \tag{10}$$

$$\langle \omega^2, y_1 \rangle = y_1 + \omega^2 y_2 + \omega y_1. \tag{11}$$

Again, according to our general theory, $\langle \omega, y_1 \rangle^3$ is an element of F_1 and thus can be expressed in terms of $\sqrt{\Delta}$. However, here we can carry out the computations quite explicitly:

$$\begin{aligned}\langle \omega, y_1 \rangle^3 &= y_1^3 + y_2^3 + y_3^3 + 3\omega[y_1^2 y_2 + y_2^2 y_3 + y_3^2 y_1] \\ &\quad + 3\omega^2[y_1 y_2^2 + y_2 y_3^2 + y_3 y_1^2] + 6y_1 y_2 y_3 \\ &= \sum_{i=1}^{3} y_i^3 - \tfrac{3}{2} \sum_{1 \leq i < j \leq 3} y_i^2 y_j + 6y_1 y_2 y_3 \\ &\quad + \tfrac{3}{2}\sqrt{3}\, i(y_1 - y_2)(y_2 - y_3)(y_1 - y_3) \\ &= (y_1 + y_2 + y_3)^3 - \tfrac{9}{2}(y_1 + y_2 + y_3)(y_1 y_2 + y_2 y_3 + y_1 y_3) \\ &\quad + \tfrac{27}{2} y_1 y_2 y_3 + \tfrac{3}{2}\sqrt{3}\, i(y_1 - y_2)(y_2 - y_3)(y_1 - y_3) \quad \text{(Check it!)} \\ &= -a_1^3 + \tfrac{9}{2} a_1 a_2 - \tfrac{27}{2} a_3 + \tfrac{3}{2}\sqrt{3}\, i\sqrt{\Delta}\end{aligned} \tag{12}$$

by (5), (6), and (7). Similarly,

$$\langle \omega^2, y_1 \rangle^3 = -a_1^3 + \tfrac{9}{2} a_1 a_2 - \tfrac{27}{2} a_3 - \tfrac{3}{2}\sqrt{3}\, i\sqrt{\Delta}). \tag{13}$$

Furthermore,

$$\begin{aligned}\langle \omega, y_1 \rangle \cdot \langle \omega^2, y_1 \rangle &= y_1^2 + y_2^2 + y_3^2 - y_1 y_2 - y_1 y_3 - y_2 y_3 \\ &= (y_1 + y_2 + y_3)^2 - 3(y_1 y_2 + y_1 y_3 + y_2 y_3) \\ &= a_1^2 - 3a_2.\end{aligned} \tag{14}$$

By (12),

$$\langle \omega, y_1 \rangle = \sqrt[3]{-a_1^3 + \tfrac{9}{2} a_1 a_2 - \tfrac{27}{2} a_3 + (3i\sqrt{3} \cdot \sqrt{\Delta})/2}, \tag{15}$$

for some value of the cube root. We will specify which value below. Similarly,

$$\langle \omega^2, y_1 \rangle = \sqrt[3]{-a_1^3 - \tfrac{9}{2} a_1 a_2 - \tfrac{27}{2} a_3 + (3i\sqrt{3} \cdot \sqrt{\Delta})/2}, \tag{16}$$

for some value of the cube root. The cube roots in (15) and (16) are not independent of one another because of (14). Therefore, let us choose the cube root of (15) in an arbitrary way and then let us choose the cube root of (16) so that (14) holds. Throughout what follows, let us assume that this special choice of cube roots has been made. Note that

$$1 + \omega + \omega^2 = 0. \tag{17}$$

Therefore,

$$\begin{aligned}\tfrac{1}{3}[\langle 1, y_1 \rangle + \langle \omega, y_1 \rangle + \langle \omega^2, y_1 \rangle] &= \tfrac{1}{3}(1 + 1 + 1)y_1 + \tfrac{1}{3}(1 + \omega + \omega^2)y_2 \\ &\quad + \tfrac{1}{3}(1 + \omega^2 + \omega)y_3 \\ &= y_1,\end{aligned} \tag{18}$$

by (9), (10), (11), and (17),. Similarly,

$$y_2 = \tfrac{1}{3}[\langle 1, y_1\rangle + \omega^2\langle \omega, y_1\rangle + \omega\langle \omega^2, y_1\rangle], \tag{19}$$

$$y_3 = \tfrac{1}{3}[\langle 1, y_1\rangle + \omega\langle \omega, y_1\rangle + \omega^2\langle \omega^2, y_1\rangle]. \tag{20}$$

Therefore, by (9), (15), (16), (18), (19), and (20), we have

$$\begin{aligned} y_1 = \tfrac{1}{3}\{-a_1 &+ \sqrt[3]{-a_1^3 + \tfrac{9}{2}a_1a_2 - \tfrac{27}{2}a_3 + \tfrac{3}{2}\sqrt{3}\,i\sqrt{\Delta}} \\ &+ \sqrt[3]{-a_1^3 + \tfrac{9}{2}a_1a_2 - \tfrac{27}{2}a_3 - \tfrac{3}{2}\sqrt{3}\,i\sqrt{\Delta}}\}, \\ y_2 = \tfrac{1}{3}\{-a_1 &+ \omega^2\sqrt[3]{-a_1^3 + \tfrac{9}{2}a_1a_2 - \tfrac{27}{2}a_3 + \tfrac{3}{2}\sqrt{3}\,i\sqrt{\Delta}} \\ &+ \omega\sqrt[3]{-a_1^3 + \tfrac{9}{2}a_1a_2 - \tfrac{27}{2}a_3 - \tfrac{3}{2}\sqrt{3}\,i\sqrt{\Delta}}\}, \\ y_3 = \tfrac{1}{3}\{-a_1 &+ \omega\sqrt[3]{-a_1^3 + \tfrac{9}{2}a_1a_2 - \tfrac{27}{2}a_3 + \tfrac{3}{2}\sqrt{3}\,i\sqrt{\Delta}} \\ &+ \omega^2\sqrt[3]{-a_1^3 + \tfrac{9}{2}a_1a_2 - \tfrac{27}{2}a_3 - \tfrac{3}{2}\sqrt{3}\,i\sqrt{\Delta}}\}. \end{aligned}$$

These last formulas do indeed provide a solution of the general cubic! And at last we have derived Tartaglia's sixteenth-century formulas.

Case 3: $n = 4$. In this case

$$\begin{aligned} f &= X^4 + a_1X^3 + a_2X^2 + a_3X + a_4 \\ &= (X - y_1)(X - y_2)(X - y_3)(X - y_4), \end{aligned}$$

and

$$a_1 = -(y_1 + y_2 + y_3 + y_4), \tag{21}$$

$$a_2 = y_1y_2 + y_1y_3 + y_1y_4 + y_2y_3 + y_2y_4 + y_3y_4, \tag{22}$$

$$a_3 = -(y_1y_2y_3 + y_1y_2y_4 + y_1y_3y_4 + y_2y_3y_4), \tag{23}$$

$$a_4 = y_1y_2y_3y_4. \tag{24}$$

In principle, it is possible to follow exactly the same procedure as used in case 1 and 2. A composition series for S_4 is given by

$$S_4 \rhd A_4 \rhd \{(1), (13)(24), (14)(23), (12)(34)\} \rhd \{(1), (12)\} \rhd \{1\}.$$

The subfield of E_f corresponding to A_4 is $F(\sqrt{\Delta})$, where $\Delta =$ the discriminant of f. Thus, if we were to follow the same technique as used in cases 1 and 2, it would be necessary to compute Δ. Fortunately, we can avoid this extremely messy computation by means of a simple trick. Consider the following three elements of E_f:

$$x_1 = (y_1 + y_2)(y_3 + y_4),$$

$$x_2 = (y_1 + y_3)(y_2 + y_4),$$

$$x_3 = (y_1 + y_4)(y_2 + y_3).$$

Let us construct the following cubic polynomial:

$$\begin{aligned} f^* &= (Y - x_1)(Y - x_2)(Y - x_3) \\ &= Y^3 + b_1Y^2 + b_2Y + b_3, \end{aligned}$$

where

$$b_1 = -(x_1 + x_2 + x_3),$$
$$b_2 = x_1x_2 + x_2x_3 + x_1x_3,$$
$$b_3 = -x_1x_2x_3.$$

Using (21), (22), (23), (24) and the definitions of x_1, x_2, x_3, it is not hard to show that

$$b_1 = -2a_2,$$
$$b_2 = a_2^2 + a_1a_3 - 4a_4,$$
$$b_3 = a_1a_2a_3 - a_1^2a_4 - a_3^2.$$

This computation is straightforward, but tedious, and therefore will be left to the reader. Thus, we see that f^* has coefficients belonging to F, and therefore by Tartaglia's formulas, the zeros of f^* can be expressed in terms of radicals in b_1, b_2, b_3. Thus, we can express x_1, x_2, x_3 in terms of radicals in a_1, a_2, a_3, a_4. Let us now express y_1, y_2, y_3, y_4 in terms of radicals in x_1, x_2, x_3. This will suffice to give us a solution of the general equation of the fourth degree. In order to make our formulas as elegant as possible, let us consider the special case of a fourth-degree polynomial f, for which $a_1 = 0$. In this case,

$$y_1 + y_2 + y_3 + y_4 = 0. \tag{25}$$

Using (25) and the definition of x_1, we see that

$$x_1 = -(y_1 + y_2)^2 \Longrightarrow y_1 + y_2 = \sqrt{-x_1}, \tag{26}$$

where $\sqrt{-x_1}$ is one of the square roots of $-x_1$, whose value will be pinned down later. By (25) and (26),

$$y_3 + y_4 = -(y_1 + y_2) \Longrightarrow y_3 + y_4 = -\sqrt{-x_1}. \tag{27}$$

Applying a similar argument to the definitions of x_2 and x_3, we derive

$$y_1 + y_3 = \sqrt{-x_2}, \qquad y_2 + y_4 = -\sqrt{-x_2}, \tag{28}$$
$$y_1 + y_4 = \sqrt{-x_2}, \qquad y_3 + y_2 = -\sqrt{-x_3}. \tag{29}$$

By solving Equations (26)–(29) for y_1, y_2, y_3, y_4, we have

$$y_1 = \tfrac{1}{2}\{\sqrt{-x_1} + \sqrt{-x_2} + \sqrt{-x_3}\},$$
$$y_2 = \tfrac{1}{2}\{\sqrt{-x_1} - \sqrt{-x_2} - \sqrt{-x_3}\},$$
$$y_3 = \tfrac{1}{2}\{-\sqrt{-x_1} + \sqrt{-x_2} - \sqrt{-x_3}\},$$
$$y_4 = \tfrac{1}{2}\{-\sqrt{-x_1} - \sqrt{-x_2} + \sqrt{-x_3}\}.$$

There are two points to be clarified concerning the above solution of the fourth-degree equation. The first concerns the choice of the square roots.

The choice cannot be arbitrary, since we must have

$$\begin{aligned}\sqrt{-x_1}\sqrt{-x_2}\sqrt{-x_3} &= (y_1 + y_2)(y_1 + y_3)(y_1 + y_4)\\ &= y_1^2(y_1 + y_2 + y_3 + y_4)\\ &\quad + (y_1y_2y_3 + y_1y_2y_4 + y_1y_3y_4 + y_2y_3y_4)\\ &= -a_3\end{aligned} \tag{30}$$

by (25). However, it is easy to verify that if the choice of the signs in the square roots is made so that (30) is satisfied, then the formulas for y_1, y_2, y_3, y_4 provide a solution to the fourth-degree equation:

$$X^4 + a_2X^2 + a_3X + a_4 = 0. \tag{31}$$

The second point involves the reduction of the general fourth-degree equation to the special equation (31). But this is easy. For if, in the general polynomial we substitute $Y = X + a_1/4$, we get a polynomial in Y with no term in Y^3. Therefore, $g(Y) = f(Y - a_1/4)$ is a polynomial of the special form considered above. If z_1, z_2, z_3, z_4 are the zeros of g, then the zeros of f are

$$z_1 - \frac{a_1}{4}, \qquad z_2 - \frac{a_1}{4}, \qquad z_3 - \frac{a_1}{4}, \qquad z_4 - \frac{a_1}{4}.$$

Thus, our solution in radicals of g leads to a solution in radicals of f.

9.8 Exercises

1. Solve the following equations:
   (a) $X^2 + 5X + 8 = 0$.
   (b) $X^3 + X^2 + X + 1 = 0$.
   (c) $X^3 + 3X + 2 = 0$.
   (d) $X^3 + 7X^2 + X + 1 = 0$.
   (e) $X^4 + X^3 + X^2 + X + 1 = 0$.
   (f) $X^4 + 7X^2 + 2 = 0$.
2. Let $f \in \mathbf{Q}[X]$ be an irreducible polynomial of degree 4. Determine a list of possible groups for $\mathrm{Gal}_{\mathbf{Q}}(f)$, and show that each possibility actually occurs by giving an example for f.
3. Let $f = \mathbf{Q}[X]$ be of degree 3. Describe the action of $\sigma \in \mathrm{Gal}_{\mathbf{Q}}(f)$ on the zeros of f by using the formulas for the zeros. [*Hint*: $\sigma(\sqrt{\Delta}) = \pm\sqrt{\Delta}$.]
4. Give an example of a cubic polynomial with three real zeros, for which the cubic formulas involve complex numbers. (This may seem wrong at first, but is in the nature of things.)
5. Let $f \in \mathbf{Q}[X]$ be an irreducible cubic. Prove that if $\mathrm{Gal}_{\mathbf{Q}}(f) \approx \mathbf{Z}_3$, then all zeros of f are real.

*6. Prove that all groups of order ≤ 8 are Galois groups of polynomials in $\mathbf{Q}[X]$.

7. Prove the formulas for b_1, b_2, b_3 in the solution of the fourth degree equation.

8. In the solution of the equation of the fourth degree, show that the discriminant of f equals the discriminant of f^*. Thus, derive a formula for the discriminant of f in terms of a_1, a_2, a_3, a_4.

10 CONCLUSION

In this book we have developed a number of algebraic ideas and have demonstrated how these ideas yield solutions to some classical mathematical questions. Let us conclude by applying the machinery developed in Chapter 9 to give a very simple proof of the fundamental theorem of algebra, and a proof of the assertion that every finite abelian group is the Galois group of some Galois extension of the field of rationals.

10.1 **C** *Is Algebraically Closed*

In Chapter 7 we stated Gauss's theorem:

THEOREM 1: Let $f \in \mathbf{C}[X]$ be a nonconstant polynomial of degree n. There exist $\beta, \alpha_1, \ldots, \alpha_n \in \mathbf{C}$ such that

$$f = \beta(X - \alpha_1) \cdots (X - \alpha_n).$$

Moreover, we saw that Gauss's theorem is equivalent to the statement that **C** is algebraically closed. To put this a little differently, we may state Gauss's theorem as follows:

THEOREM 2: If E is an algebraic extension of **C**, then $E = \mathbf{C}$.

Let us now give a simple proof of Theorem 2. Our proof is due to Emil Artin and uses, in addition to the tools we have developed, a simple result from calculus:

LEMMA 3: Let $a, b \in \mathbf{R}$, $a < b$, and let f: $\mathbf{R} \longrightarrow \mathbf{R}$ be a continuous function. If $f(a) < 0$, $f(b) > 0$, then there exists $c \in \mathbf{R}$, such that $a < c < b$ and $f(c) = 0$.

For a proof of Lemma 3 we refer the reader to a calculus book. We will need Lemma 3 in case f is the function defined by a polynomial in $\mathbf{R}[X]$.

LEMMA 4: Let $f \in \mathbf{R}[X]$ be a polynomial of odd degree n. Then there exists $c \in \mathbf{R}$ such that $f(c) = 0$.

Proof: Let $f = a_n X^n + \cdots + a_0$. By replacing f by $a_n^{-1}f$, we may assume that $a_n = 1$. Set $M = |a_0| + \cdots + |a_{n-1}| + 1$. Then

$$|f(x) - x^n| \leq |a_{n-1}||x|^{n-1} + \cdots + |a_0|,$$

so that

$$-(|a_{n-1}x^{n-1}| + \cdots + |a_0|) \leq f(x) - x^n \leq |a_{n-1}x^{n-1}| + \cdots + |a_0|.$$

In particular, since n is odd,

$$\begin{aligned} f(-M) &\leq -M^n + |a_{n-1}|M^{n-1} + |a_{n-2}|M^{n-2} + \cdots + |a_0| \\ &\leq -M^n + [|a_{n-1}| + \cdots + |a_0|]M^{n-1} \qquad \text{(since } M \geq 1) \\ &< -M^n + M \cdot M^{n-1} = 0. \end{aligned}$$

Therefore, $f(-M) < 0$. Similarly,

$$\begin{aligned} f(M) &\geq M^n - [|a_{n-1}|M^{n-1} + \cdots + |a_0|] \\ &\geq M^n - [|a_{n-1}| + \cdots + |a_0|]M^{n-1} \\ &> M^n - M \cdot M^{n-1} = 0. \end{aligned}$$

Thus, $f(M) > 0$. Thus, by Lemma 3, there exists $c \in \mathbf{R}$ such that $-M < c < M$ and $f(c) = 0$.

LEMMA 5: Let $\delta \in \mathbf{C}$. Then there exists $\eta \in \mathbf{C}$ such that $\eta^2 = \delta$.

Proof: If $\delta = 0$, then we may set $\eta = 0$, so assume that $\delta \neq 0$. Then δ can be written in polar form $\delta = r(\cos\theta + i\sin\theta)$, $r > 0$, $\theta \in \mathbf{R}$, and we may set $\eta = \sqrt{r}\,(\cos(\theta/2) + i\sin(\theta/2))$, where $\sqrt{r}$ denotes the positive square root of the real number r. Then $\eta^2 = \delta$ by de Moivre's theorem.

LEMMA 6: **C** has no algebraic extensions of degree 2.

Proof: Let E be an algebraic extension of $\mathbf{C}$ such that $\deg(E/\mathbf{C}) = 2$. Then $E \neq \mathbf{C}$. Let $\delta \in E - \mathbf{C}$. Then $E \supseteq \mathbf{C}(\delta) \supseteq \mathbf{C}$, $\mathbf{C}(\delta) \neq \mathbf{C}$, so that $\mathbf{C}(\delta) = E$ since $\deg(E/\mathbf{C}) = 2$. Let f be the irreducible polynomial of δ over $\mathbf{C}$. Then $\deg(f) = 2$. Let $f = X^2 + \alpha X + \beta$. Choose $\eta \in \mathbf{C}$ such that $\eta^2 =$

$\alpha^2 - 4\beta$. Such an η exists by Lemma 5. Then

$$f = \left(X - \frac{-\alpha + \eta}{2}\right)\left(X - \frac{-\alpha - \eta}{2}\right),$$

which contradicts the fact that f is irreducible in $\mathbf{C}[X]$.

LEMMA 7: Let G be a group of order p^n, p prime. Then there exists a subgroup H of G such that $H \lhd G$ and $[G:H] = p$.

Proof: By Exercise 3 of Section 8.6, G is solvable. Let

$$G = G_0 \rhd G_1 \rhd \cdots \rhd G_r = \{1\}$$

be a composition series such that G_i/G_{i+1} $(0 \leq i \leq r - 1)$ is a group of prime order. Then $|G_i/G_{i+1}| = p$ $(0 \leq i \leq r - 1)$. Thus, we may take $H = G_1$.

Proof of Theorem 2: Let E be an algebraic extension of $\mathbf{C}$, and let $\theta \in E$, $f = \mathrm{Irr}_{\mathbf{R}}(\theta, X)$ (since θ is algebraic over $\mathbf{C}$, θ is algebraic over $\mathbf{R}$). Let $E_0 =$ the splitting field of $(X^2 + 1)f$. Then E_0 is a Galois extension of $\mathbf{R}$ and $E_0 \supseteq \mathbf{C}$. Let $G = \mathrm{Gal}(E_0/\mathbf{R})$. Since $\mathbf{C}/\mathbf{R}$ is a normal extension, $E_0/\mathbf{C}$ corresponds to a subgroup H of G and $H \lhd G$. Let J be a 2-Sylow subgroup of G, and let E_1 be the subfield of E_0 corresponding to J. The various fields introduced are illustrated in Figure 10.1 Since J is a 2-Sylow subgroup of G, $[G{:}J]$

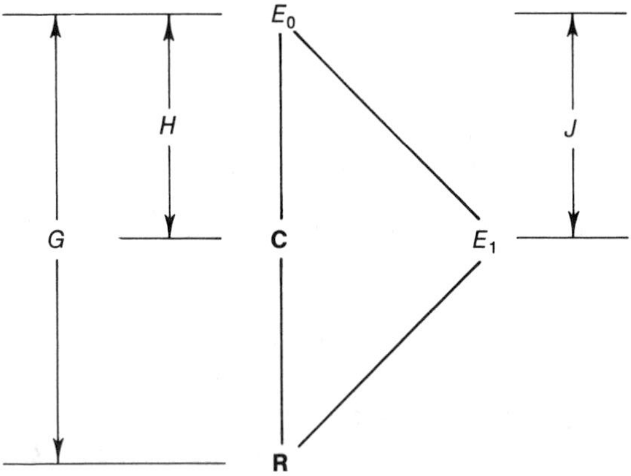

FIGURE 10-1: The Fields and their Galois Groups.

is odd. However, $\deg(E_1/\mathbf{R}) = [G{:}J]$ and thus $\deg(E_1/\mathbf{R})$ is odd. Let η be an arbitrary element of E_1 and let $g = \mathrm{Irr}_{\mathbf{R}}(\eta, X)$. Since $\mathbf{R} \subseteq \mathbf{R}(\eta) \subseteq E_1$, we see that $\deg(\mathbf{R}(\eta)/\mathbf{R}) \mid \deg(E_1/\mathbf{R})$, so that $\deg(\mathbf{R}(\eta)/\mathbf{R})$ is odd. But $\deg(\mathbf{R}(\eta)/\mathbf{R}) = \deg(g)$, so that we see that

$$\deg(g) \text{ is odd.}$$

By Lemma 4, g has a zero in $\mathbf{R}$. However, g is irreducible in $R[X]$, so that

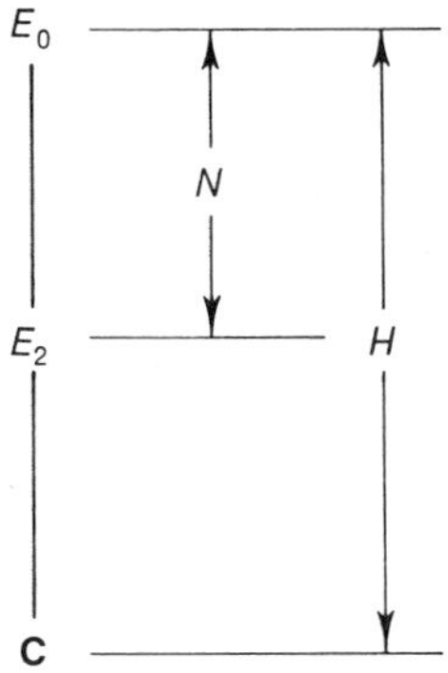

FIGURE 10-2

$\deg(g) = 1 = \deg(\mathbf{R}(\eta)/\mathbf{R})$. Thus, $\mathbf{R}(\eta) = \mathbf{R}$ and $\eta \in \mathbf{R}$. However, since η is an arbitrary element of E_1, we see that $E_1 = \mathbf{R}$, and consequently, $J = G$ and G is a group of order 2^k for some $k \geq 0$. Consequently H has order 2^l for some $l \geq 0$. Assume for the moment that $l \geq 1$. By Lemma 7, there exists N such that $N \lhd H$ and $[H:N] = 2$. Let E_2 be the subfield of E_0 coresponding to N. Since $N \lhd H$, E_2 is a Galois extension of **C** with Galois group H/N. In particular, $\deg(E_2/\mathbf{C}) = |H/N| = 2$, which contradicts Lemma 6. Thus, $l = 0$ and H is trivial, so that $E_0 = \mathbf{C}$. In particular, $\theta \in \mathbf{C}$. But θ was an arbitrary element of E, so that we may finally conclude that $E = \mathbf{C}$, which completes the proof.

10.2 *All Abelian Groups Are Galois Groups Over* **Q**

Let G be a finite group. A natural question to ask is: Does there exist a Galois extension E of **Q** whose Galois group is isomorphic to G? Note that if we drop our insistence that E be a Galois extension *of the rational numbers* then this question is answered in the affirmative by Theorem 5 of Section 9.7. However, the question as posed is exceedingly difficult and the answer is not known! The problem of determining which finite groups occur as Galois groups over the rationals was first posed by Hilbert in 1900 and has received wide attention from researchers in the last 70 years. The best result known as of this writing (1972) is due to the Russian mathematician Shafarevich, who proved that if G is solvable, then G is the Galois group of some Galois extension of **Q**. Shafarevich's theorem is very difficult to prove. However, the analogous assertion with "solvable" replaced by "abelian" is easy and will be the subject of this section. Thus, we will prove

THEOREM 1: Let G be a finite abelian group. Then G is the Galois group of a Galois extension E of **Q** the field of rational numbers.

A few words of motivation of our proof. A quite miraculous theorem of Kronecker and Weber asserts that if E is a Galois extension of $\mathbf{Q}$ having abelian Galois group, then E is contained in a cyclotomic field $\mathbf{Q}(\zeta_m)$, $\zeta_m =$ a primitive mth root of 1. Assuming that is the case, we see that if H is the subgroup of $\text{Gal}(\mathbf{Q}(\zeta_m)/\mathbf{Q})$ corresponding to E, then $\text{Gal}(E/\mathbf{Q}) \approx \text{Gal}(\mathbf{Q}(\zeta_m)/\mathbf{Q})/H)$. However, we saw in Chapter 9 that $\text{Gal}(\mathbf{Q}(\zeta_m)/\mathbf{Q} \approx \mathbf{Z}_m^\times$. Therefore, we see that $\text{Gal}(E/\mathbf{Q})$ is isomorphic to the quotient group of $\mathbf{Z}_m^\times$ by a certain subgroup. This fact suggests a plan of attack: Let us prove that every abelian group G is isomorphic to the quotient group of $\mathbf{Z}_m^\times$ by a certain subgroup A (for some value of m). Let E be the subfield of $\mathbf{Q}(\zeta_m)$ corresponding to A by Galois theory. Then since $\mathbf{Z}_m^\times$ is abelian, A is normal; and $E/\mathbf{Q}$ is a normal extension by the fundamental theorem of Galois theory. But then

$$\text{Gal}(E/\mathbf{Q}) \approx \mathbf{Z}_m^\times/A \approx G,$$

and we will have constructed an extension of $\mathbf{Q}$ having Galois group G. Thus, it suffices to prove the following result:

PROPOSITION 2: Let G be a finite abelian group. Then G is isomorphic to the quotient group of $\mathbf{Z}_m^\times$ by a certain subgroup A, for a suitable choice of m.

Before proving Proposition 2, we require a preliminary result.

PROPOSITION 3: Let p be a prime. Then $\mathbf{Z}_p^\times$ is a cyclic group.

Proof: Note that $\mathbf{Z}_p^\times$ is a finite abelian group having $p - 1$ elements. Therefore, by the fundamental theorem of finite abelian groups, there exist positive integers $n_1, \ldots, n_t$, greater than 1, such that

$$\mathbf{Z}_p^\times \approx \mathbf{Z}_{n_1} \times \mathbf{Z}_{n_2} \times \cdots \times \mathbf{Z}_{n_t},$$

where $n_1 \mid n_2 \mid n_3 \mid \cdots \mid n_t$. It is enough to prove that $t = 1$. For then $\mathbf{Z}_p^\times \approx \mathbf{Z}_{n_1}$ and $\mathbf{Z}_{n_1}$ is clearly cyclic. Assume, on the contrary, that $t > 1$. Then, since each n_i is greater than 1, and since $n_1 \ldots n_t = p - 1$, we see that $n_t < p - 1$. On the other hand, since $n_1 \mid n_2 \mid \cdots \mid n_t$, we see that every element of

$$\mathbf{Z}_{n_1} \times \cdots \times \mathbf{Z}_{n_t}$$

is of order dividing n_t. Therefore, every element of $\mathbf{Z}_p^\times$ has order dividing n_t. That is, if $x \in \mathbf{Z}_p^\times$, then $x^{n_t} = \bar{1}$. Thus, the polynomial $X^{n_t} - \bar{1} \in \mathbf{Z}_p[X]$ has at least $p - 1$ zeros in $\mathbf{Z}_p$. But this is impossible since the degree of $X^{n_t} - 1$ is n_t, which is less than $p - 1$. Thus, $t = 1$.

Proof of Proposition 2: By the fundamental theorem on finite abelian groups, there exist positive integers $n_1, \ldots, n_t$, such that

$$G \approx \mathbf{Z}_{n_1} \times \cdots \times \mathbf{Z}_{n_t}. \tag{1}$$

By Dirichlet's theorem on primes in arithmetic progressions, there exist an infinite number of primes p_i such that $p_i \equiv 1 \pmod{n_i}$. Thus, we may determine distinct primes $p_1, p_2, \ldots, p_t$ such that $p_i \equiv 1 \pmod{n_i}$ $(1 \leq i \leq t)$. Set $m = p_1 \cdots p_t$ and let $p_i - 1 = k_i n_i$, $k_i \in \mathbf{Z}$ $(1 \leq i \leq t)$. By Exercise 13 of Section 3.6 we have

$$\mathbf{Z}_m^\times \approx \mathbf{Z}_{p_1}^\times \times \cdots \times \mathbf{Z}_{p_t}^\times.$$

And since $\mathbf{Z}_{p_i}^\times \approx \mathbf{Z}_{p_i - 1}$ (Proposition 3), we see that

$$\mathbf{Z}_m^\times \approx \mathbf{Z}_{k_1 n_1} \times \cdots \times \mathbf{Z}_{k_t n_t} = H. \tag{2}$$

Let

$$C_i = \{\bar{0}, \bar{n_i}, \overline{2n_i}, \ldots, \overline{(k_i - 1)n_i}\} \subseteq \mathbf{Z}_{k_i n_i} \qquad (1 \leq i \leq t).$$

Then C_i is a subgroup of order k_i, so that $\mathbf{Z}_{k_i n_i}/C_i$ is a cyclic group of order n_i. Thus,

$$\mathbf{Z}_{k_i n_i}/C_i \approx \mathbf{Z}_{n_i}. \tag{3}$$

Set

$$A = C_1 \times \cdots \times C_t \subseteq \mathbf{Z}_{k_1 n_1} \times \cdots \times \mathbf{Z}_{k_t n_t}.$$

Then, by (1), (2), (3),

$$\begin{aligned} H/A &\approx \mathbf{Z}_{k_1 n_1}/C_1 \times \cdots \times \mathbf{Z}_{k_t n_t}/C_t \\ &\approx \mathbf{Z}_{n_1} \times \cdots \times \mathbf{Z}_{n_t} \\ &\approx G. \end{aligned}$$

Therefore, by (2), we have proved Theorem 1.

The Kronecker–Weber theorem cited above is one of the deep and important results in a branch of mathematics known as *class field theory*, and is far beyond the scope of this book. However, we have outlined in the exercises a proof of the special case where E is a quadratic extension of $\mathbf{Q}$.

10.2 *Exercises*

1. Let m and n be positive integers. Show that $\mathbf{Q}(\zeta_m, \zeta_n) \subseteq \mathbf{Q}(\zeta_{mn})$. Does equality necessarily hold?
2. Let E be a Galois extension of $\mathbf{Q}$ such that $\mathrm{Gal}(E/\mathbf{Q}) \approx \mathbf{Z}_2$. Prove that there exists an integer d, not divisible by any perfect square, such that $E = \mathbf{Q}(\sqrt{d})$.
3. Show that $\mathbf{Q}(\sqrt{-1}) = \mathbf{Q}(\zeta_4)$, $\mathbf{Q}(\sqrt{2}) \subseteq \mathbf{Q}(\zeta_8)$.
4. Let p be an odd prime. Prove that $\mathbf{Q}(\sqrt{(-1)^{(p-1)/2}p}) \subseteq \mathbf{Q}(\zeta_p)$. [*Hint*: Since $X^p - 1 = \prod_{a=0}^{p-1} (X - \zeta_p^a)$, differentiating both sides and setting $X = \zeta_p^b$ yields

$$p\zeta_p^{b(p-1)} = \prod_{\substack{a=0 \\ a \neq b}}^{p-1} (\zeta_p^b - \zeta_p^a).$$

Therefore, taking the product of these last expressions for $b = 0, 1, \ldots, p - 1$

yields

$$p^p \zeta_p^{(p-1)[0+1+\cdots+(p-1)]} = [\prod_{0 \leq a < b \leq p-1} (\zeta_p^b - \zeta_p^a)^2](-1)^{(p-1)/2}.$$

From this, deduce that $(-1)^{(p-1)/2}p$ is the square of an element of $\mathbf{Q}(\zeta_p)$.]

5. Use Exercises 2, 3 and 4 to show that every extension E of $\mathbf{Q}$ which is Galois and such that $Gal(E/\mathbf{Q}) \approx \mathbf{Z}_2$ is contained in a cyclotomic field. In fact, if d is as in Exercise 2, show that $\mathbf{Q}(\sqrt{d}) \subseteq \mathbf{Q}(\zeta_{4|d|})$.

INDEX